Undergraduate Lectu

Series Editors

Neil Ashby, University of Colorado, Boulder, CO, USA
William Brantley, Department of Physics, Furman University, Greenville, SC, USA
Matthew Deady, Physics Program, Bard College, Annandale-on-Hudson, NY, USA
Michael Fowler, Department of Physics, University of Virginia, Charlottesville, VA, USA
Morten Hjorth-Jensen, Department of Physics, University of Oslo, Oslo, Norway
Michael Inglis, Department of Physical Sciences, SUNY Suffolk County Community College, Selden, NY, USA

Undergraduate Lecture Notes in Physics (ULNP) publishes authoritative texts covering topics throughout pure and applied physics. Each title in the series is suitable as a basis for undergraduate instruction, typically containing practice problems, worked examples, chapter summaries, and suggestions for further reading.

ULNP titles must provide at least one of the following:

- An exceptionally clear and concise treatment of a standard undergraduate subject.
- A solid undergraduate-level introduction to a graduate, advanced, or non-standard subject.
- A novel perspective or an unusual approach to teaching a subject.

ULNP especially encourages new, original, and idiosyncratic approaches to physics teaching at the undergraduate level.

The purpose of ULNP is to provide intriguing, absorbing books that will continue to be the reader's preferred reference throughout their academic career.

More information about this series at http://www.springer.com/series/8917

Bryan H. Suits

Electronics for Physicists

An Introduction

Springer

Bryan H. Suits
Physics Department
Michigan Technological University
Houghton, MI, USA

ISSN 2192-4791   ISSN 2192-4805 (electronic)
Undergraduate Lecture Notes in Physics
ISBN 978-3-030-39087-7   ISBN 978-3-030-39088-4 (eBook)
https://doi.org/10.1007/978-3-030-39088-4

© Springer Nature Switzerland AG 2020
This work is subject to copyright. All rights are reserved by the Publisher, whether the whole or part of the material is concerned, specifically the rights of translation, reprinting, reuse of illustrations, recitation, broadcasting, reproduction on microfilms or in any other physical way, and transmission or information storage and retrieval, electronic adaptation, computer software, or by similar or dissimilar methodology now known or hereafter developed.
The use of general descriptive names, registered names, trademarks, service marks, etc. in this publication does not imply, even in the absence of a specific statement, that such names are exempt from the relevant protective laws and regulations and therefore free for general use.
The publisher, the authors and the editors are safe to assume that the advice and information in this book are believed to be true and accurate at the date of publication. Neither the publisher nor the authors or the editors give a warranty, expressed or implied, with respect to the material contained herein or for any errors or omissions that may have been made. The publisher remains neutral with regard to jurisdictional claims in published maps and institutional affiliations.

This Springer imprint is published by the registered company Springer Nature Switzerland AG
The registered company address is: Gewerbestrasse 11, 6330 Cham, Switzerland

Preface

Physicists rely heavily on electrical instrumentation to measure physical phenomena. There was a time, not long ago in history, when it was normal for a physicist to routinely design and construct their own electrical instrumentation. While there are exceptions, in general this is no longer the case; design and construction of most electronic instrumentation is left to specialists and engineers. However, a physicist, whether an experimentalist or theorist, will need some understanding of the basics of electronics and how the equipment works in order to fully understand and evaluate the results of measurements, and possibly in order to troubleshoot and make simple repairs. Much of this understanding must first come from a knowledge of the language of electronics—the lexicon as well as the symbolic representation of circuits using schematic diagrams. This book was written for use as an undergraduate physics text with such future use in mind. The details necessary for quality engineering design are generally excluded here in favor of a basic and practical understanding of what is going on.

A second utility of electronics in the physics curriculum is that many ideas and problem-solving strategies show up that are also used in other areas of science. Indeed, electrical analogies are often used as an explanation for situations that have little to do with electronics. One prime example of cross-use is the appearance of imaginary numbers. Imaginary numbers show up in all areas of science and engineering that include periodic signals. In addition to electronics, those areas include studies of vibration, including seismometry, acoustics, optics, radio and radar, and even "brain waves." It is important to understand how to use and interpret imaginary numbers since, after all, by definition no real measurement will ever give you an imaginary result. Other overlapping topics include resonance, solutions of linear equations, and the use of linearized models.

Each electronics instructor will have their own idea about what is most important. That cannot be helped. What is presented here represents my priorities. Based on my perspective and experience as a physicist and teacher of electronics, I chose introductory topics and problems that I find most interesting and potentially useful. There is, however, more in this book than can possibly be covered in a single semester course. Anyone using the book as a course textbook, in whole or in part, should feel free to skip those topics that do not match their interests.

Any electronics course would be expected to include a laboratory component. In fact, some of the material here originated as background material for such a laboratory experience. It is recognized that each instructor has their own laboratory priorities, and possibly a limited or specialized supply of laboratory equipment for such a purpose. It is hoped that the presentation here is written in a manner suitable for use in either the laboratory or classroom. The first obvious examples of laboratories that appear here are the sections related to the Wheatstone and Kelvin bridges, both of which originated primarily as laboratory exercises. Along with material in the last chapter, some practical experience using an embedded microcontroller—that is, some programming—is definitely useful. Programming varies from device to device and is not really "electronics," and so it is only included here in a very general way.

The book includes three broad categories of electronics. Chapters 1–5 cover passive linear electronics, Chaps. 6–11 look at nonlinear and active devices including diodes, transistors, and op-amps, and Chaps. 12–14 consider the basics of digital electronics and simplified computers.

This text originated as weekly handouts and laboratory write-ups for a course designed primarily for second-year university physics students. The level of the material here is appropriate for students who have successfully learned the material in introductory electricity and magnetism as well as mathematics up to the first course in integral calculus. As the extent of the material in the handouts expanded over the years, it got to the point that some students started referring to the handouts as "the book." That reference served as one motivating factor to formalize those course notes into a full volume. I thank those students for providing that inspiration, and I hope this proves to be a useful exercise for both of us.

Houghton, MI, USA   Bryan H. Suits

Contents

1	**The Basics**.	1
	Voltage and Current	2
	Simple Devices	3
	Kirchhoff's Laws	5
	Resistors in Series	5
	Resistors in Parallel	7
	Effective Resistance	9
	Resistors in Parallel–Notation	10
	Solving Circuits with Circuit Reduction	11
	Solving Circuits with Algebra	14
	Branch and Mesh Currents	14
	Example—Using Kirchhoff's Laws	15
	Nodal Analysis	17
	The Ideal Current Source	19
	The Ground and Common Connections	21
	Multiple Sources—The Superposition Theorem	22
	Electrical Power	23
	Additional Application—The Kelvin-Varley Divider	24
	Problems	26
2	**Additional Theorems**	31
	Thevenin and Norton Equivalents	31
	Determining the Thevenin and/or Norton Parameters	32
	How Is This Used for Circuit Reduction?	34
	Equivalent for an Infinite Array of Resistors	35
	The Wheatstone Bridge	36
	Wheatstone Bridge "Hieroglyphics"	38
	The Reciprocity Theorem	39
	Example—R-2R Ladder with Sources	40

	Delta-Y Conversion	42
	The Kelvin Bridge	44
	Additional Application—Resistivity of Lamellae	47
	Problems	49
	References	51
3	**Complex Impedances**	53
	What Is a Linear Device?	53
	Some Vocabulary	54
	Passive Linear Circuit Elements with Two Leads	55
	Idealized Sources	56
	RC and L/R Time Constants	57
	RC Time Constant Example	59
	Capacitors and Inductors with Sinusoidal Sources	60
	Superposition and Complex Impedances	62
	Series and Parallel Capacitors and Inductors	66
	Comments About Complex Arithmetic	67
	Solving Circuits Using Complex Impedances	68
	A.C. Power	71
	Condenser Microphones	73
	Problems	74
4	**More on Capacitors and Inductors**	79
	Real Capacitors and Inductors	79
	Measuring Capacitors and Inductors	80
	Capacitive Position Sensors	81
	A Simple Circuit for Measuring Inductors	82
	Switched Capacitor Methods	84
	Charging a Capacitor Efficiently	85
	Mutual Inductance and Transformers	86
	The Dot Convention for Transformers	89
	Inductive Position Sensors	89
	RLC Circuits	90
	Cable Models	93
	Cable Impedance	95
	Signal Speed in a Cable	96
	Impedance of Finite Cables	97
	Capacitor and Inductor Labels	99
	Duality	100
	Problems	101
	References	103

5 The Laplace Transform ... 105
The Transform ... 105
Laplace Transform Example 1 ... 108
   Method I ... 109
   Method II ... 110
Laplace Transform Example 2 ... 112
Laplace Transform Example 3 ... 114
Comment on Partial Fractions ... 116
Poles and Zeros ... 117
Problems ... 118

6 Diodes ... 121
Semiconductor Diodes ... 121
Diode Models ... 125
   Piece-Wise Linear Diode Models ... 125
   An Analytic Model for the Semiconductor Diode ... 126
Solving Circuits with Diodes ... 127
   The Ideal Diode ... 127
   Graphical Solutions ... 134
Diode Ratings ... 136
Diode Capacitance and Response Time ... 136
Specialty and Other Diodes ... 136
Problems ... 139

7 FETs ... 143
Junction Field Effect Transistors ... 143
Circuit Analysis with a JFET ... 146
   Example 1—Determine Circuit Components ... 146
   Example 2—Determine Operating Point ... 147
The FET A.C. Model ... 150
FET Amplifier Configurations ... 153
The Ohmic Region ... 156
MOSFETs ... 156
Additional Application—Dynamic Memory ... 158
Problems ... 160

8 Bipolar Junction Transistors ... 163
BJT D.C. Model ... 165
BJT A.C. Model ... 168
BJT Large Signal Example ... 169
   Graphical Solutions ... 169
   Single Supply Operation ... 171
   Solutions from Parameters ... 172

BJT Amplifiers.. 174
  Common Emitter Amplifier Example........................ 177
  Common Collector Amplifier Example...................... 179
Using the Saturation Region................................... 181
Problems... 184

9 More on Amplifiers.. 187
Miller's Theorem... 187
Two-Transistor Configurations.................................. 189
  The Cascode Configuration................................. 189
  The Darlington Pair....................................... 190
  Complementary Symmetry Amplifier ("Push-Pull")............ 190
  Differential Amplifier.................................... 192
  Current Mirror.. 193
  Silicon Controlled Rectifiers (SCR) and Triacs............ 193
Connecting Amplifiers.. 195
  Impedance Matching.. 196
Problems... 197

10 The Ideal Op-Amp... 199
Ideal Op-Amp Properties.. 199
Linear Op-Amp Circuits... 201
  Example 1—Buffer.. 201
  Example 2—Inverting Amplifier............................. 202
  Example 3—Non-inverting Amplifier......................... 203
  Example 4—Difference Amplifier............................ 203
  Example 5—Summing Amplifier............................... 204
  Example 6—Integrator...................................... 205
  Example 7—Low-Pass Filter................................. 207
  Example 8—Instrumentation Amplifier....................... 209
  Example 9—A Capacitive Sensor for Smaller Values of
    Capacitance... 210
  Example 10—Negative Resistor.............................. 212
  Example 11—Constant Current Source........................ 213
Other Op-Amp Circuits.. 214
  Example 12—Non-linear Element in Feedback................. 214
  Example 13—Ideal Diode.................................... 215
  Example 14—Peak Follower.................................. 216
  Example 15—Log Amplifier.................................. 217
  Example 16—Absolute Value Circuit......................... 219

Contents   xi

    More Power 220
    Less Than Ideal Difference Amplifiers 220
        Finite Input Resistance and Gain 220
        Finite Frequency Range. 223
        Small Signals and Drift 223
    Oscillations 224
    The Transconductance Amplifier 225
    Problems 226
    References 230

11 Non-linear Uses of Op-Amps 231
    Limited Output Range 231
    The Op-Amp Comparator 232
        Example 1—Low-Level Warning. 234
        Example 2—Pulse Generator 235
        Example 3—Simple Oscillator. 236
        Example 4—A Voting Circuit 236
        Example 5—Sine to Pulse Train Converter. 238
        Example 6—Zero Crossing Detector 238
        Example 7—Pulse Conditioner/Lengthener 239
    Using the Comparator for Feedback. 240
        Automatic Gain Control Amplifier 240
    Putting Pieces Together. 242
        Simple Phase Sensitive Detector 242
    Problems 243

12 Digital I 247
    Boolean Algebra. 247
    Useful Rules and Theorems for Boolean Algebra 248
    Digital Logic Circuits 250
    Combinations of Digital Logic Gates 252
        Example 1—Solving with Boolean Algebra 252
        Example 2—Solving with a Truth Table 253
        Example 3—Solving Both Ways 254
    Equivalent Circuits 256
    Gates Versus Logic Functions 257
    Decoders and Encoders. 258
    Multiplexing. 261
    Flip-Flop Circuits 261
    Edge-Triggered Flip-Flops. 265
    A Directional Electric Eye 266
    Combinations of Flip-Flops. 267
        Shift Register 267
        Binary Counter 268

	Other Non-logical Applications	271
	Very Short Pulse Generator	271
	Oscillators	273
	Problems	273
13	**Digital II**	**277**
	Binary and BCD Numbers	277
	Binary Numbers	277
	BCD Numbers	278
	Hexadecimal and Octal Notation	279
	Other Weighted Binary Codes	279
	The 4221 Code	280
	2 of 5 Codes	281
	Non-weighted Codes	282
	Gray Code	282
	The ASCII Code	283
	Bar Codes	283
	Interleaved 2 of 5	284
	UPC Codes	284
	Some Numeric Code Conversions	286
	Binary to Gray Code	286
	Gray Code to Base-2 Binary	287
	Decimal to Gray Code	287
	BCD to Binary Conversion	288
	Binary to BCD Conversion	289
	Digital to Analog Conversion	290
	The 1-Bit D/A	291
	A Summing D/A	292
	Analog to Digital Conversion	293
	Voltage to Frequency Conversion	293
	Timing Schemes	294
	Search Schemes	294
	Analog to Gray Code Conversion	295
	Quantization Noise	296
	Problems	297
	Reference	297
14	**Calculators and Computers**	**299**
	Adding Base-2 Numbers	299
	Two's Complement Arithmetic	299
	A Simple Arithmetic Logic Unit (ALU)	302
	Base-2 Multiplication	305

Some Recursive Computations	307
Compute $1/x$	308
Compute $(1/x)^{1/2}$	308
Compute $x^{1/2}$	309
Compute $\tan(x)$	310
Compute $K(k)$	310
Communications	311
Tri-state Outputs	311
Simplified CPU	312
Other Uses for Tri-state Devices	312
Measuring a Small Capacitance	312
Charlieplexing	315
Problems	316
References	317
Appendix	319
Index	327

Chapter 1
The Basics

In this chapter, the basic concepts, definitions and results from electricity and magnetism that are applicable to electronics are reproduced, along with some additional definitions more specific to electronics. Simple circuits involving resistors, wires, and power sources are represented by schematic diagrams and then solved using Ohm's law and Kirchhoff's laws. Simple theorems are developed for series and parallel resistors which can be used to simplify a circuit, leading to a simpler solution.

Historically, physicists have often been required to design and create their own specialized electronic circuits as part of their vocation. During the last few decades that has become less and less of a requirement. One may ask, then, why it is a standard practice to have modern physics students study electronics at all. Certainly, there is no expectation that most will be designing state of the art electronic circuitry. There are electrical engineers who are usually better equipped to do that. The goal here, however, is not to teach the level of detail necessary for engineering work, but to give an appreciation for what is involved, some understanding of how circuits work, and at the same time, focus on some basic skills and techniques that show up in other areas of science. Such an understanding is an essential part of figuring out what goes into and what comes out of both experimental and theoretical scientific studies.

Hence, while venturing through this material try to keep an eye out for the "big picture" that goes beyond the specifics of the electronics. Of course, the first key to understanding is to have some knowledge of what you are talking about. That often arises from learning basic definitions, the lexicon, and the basic rules that apply. The second key is to be able to apply logic and mathematics to deduce an end result. With that in mind, the starting point here becomes the simplest definitions and rules, in this case for electronics. It is expected that at least some of this should be review. Following that, new results will be gradually introduced that can be used to solve and understand more advanced electronic circuits.

Voltage and Current

In electronics "voltages" and "currents" are of primary concern.

Voltage—what is measured by a voltmeter. Voltage arises from an electromotive force (emf). If the force involved is conservative, then the voltage is directly related to an electrostatic potential difference. In the case of some time-dependent emf's, such as arise due to Faraday's law of induction, the emf is not associated with an electrostatic potential but its effect is similar.[1] In electronics, a distinction between the two is rare. They are both simply referred to as "voltages."

Voltages are measured in "volts," abbreviated V.[2]

Current—a net movement of electric charge past a particular point measured over time. Current is measured with an Ammeter.[3]

Currents are measured in "amperes," or "amps" for short, abbreviated A, where 1 ampere = 1 coulomb of charge per second.

An applied voltage can be thought of as being somewhat like a force (or a pressure) and the current as being a motion—a flow—in response to that force.

In mechanics, one is given a force and tries to find the motion (i.e., the response) and/or is given a motion and attempts to deduce the force. In electronics, the same is done with voltages and currents. To "solve" a circuit is to find the various voltages and currents—that is, to predict what you would measure with a voltmeter and/or an ammeter.

Volts and amps are Standard International (SI) units. An appropriate standardized prefix can be added to indicate powers of ten. A list of some of the more common of these can be found in the Appendix. For example, 1000 V = 1 kV, 10^{-3} A = 1 mA, etc. On some older electronics devices one may see "mm" (literally milli-milli-) instead of "µ" (micro-) to indicate 10^{-6}. In addition, some older electronic devices may be labeled using "m" or "M" instead of "µ" to mean micro- and hence "mm" (or "MM") becomes pico- (10^{-12}). These peculiarities will have to be resolved from the context or by direct measurement. This rather confusing font issue resulted from the fact that Greek fonts were not readily available to (U.S.) manufacturers until the latter part of the 20th century. Fortunately, this prefix confusion does not occur very often for more recently manufactured electronic components.[4] To avoid confusion, the use of standard SI prefixes is encouraged.

---

[1] In other contexts, EMF may be used to stand for electromagnetic fields, which, incidentally, may be the cause of an electromotive force.

[2] Named after Alessandro Volta (1745–1827), an Italian scientist credited with many discoveries leading to the electric battery.

[3] Named after French mathematician and scientist, André Marie Ampère (1775–1836), who, among many important discoveries in physics, did many measurements to quantify currents.

[4] In some fields of study, it is still the practice to use the ambiguous prefix "mc" for "micro," though that use is not usually seen in electronics.

Since electronics deals with the relationship between voltage and current, the ratio of 1 volt/1 amp = 1 ohm,[5] the unit of electrical resistance, will also show up quite often. The ohm is abbreviated with an upper-case Greek omega (Ω) and is also an SI unit.

As a practical matter, the "Ω" symbol is often omitted in electronics when its presence is clearly implied by circumstance. That practice will often be used in this text. Also, in some circumstances decimal points can may be hard to see. If that is a concern, prefixes may be used to substitute for a decimal point. For example, a value of 4.7 kΩ (4700 Ω) may appear as "4.7k" and sometimes as "4k7." When there is no prefix, the letter R (or r) is sometimes used for a decimal point. Hence, a value of "4R7" would be 4.7 Ω. Less often the units may also be missing for units other than ohms when their presence is clearly implied by circumstance.

Simple Devices

Any electronic device will have one or more connection points, called "leads" (pronounced lēds, as in "the captain leads his troops into battle"). A device with one lead will not be useful as part of an electronic circuit, hence only devices with two or more need be considered.

The electrical characteristics of a device are specified by the relationship between the currents into (or out of) each lead corresponding to a given voltage between the leads—there is a "device rule" that specifies that relationship. Two generic devices are illustrated in Fig. 1.1. By convention, currents for each lead are usually considered positive going into the device. As will be seen in more detail later, "net current in" = "net current out" so, for the two-lead device shown, $I_1 = -I_2 \equiv I$ and so only the current I going *through* the device need be considered. For the three-lead device shown, it must be the case that $I_3 + I_4 + I_5 = 0$ so one of the currents can always be expressed using the sum of the other two. Three-lead devices will be seen later on.

Device leads are connected together to form circuits. A circuit will include one or more paths through connected devices that allow a return to any starting point without retracing steps. If there is no such path, then one has an "open circuit." There can be no current in an open circuit. An open circuit corresponds to a dead-end street. There is no net flow (over time) of cars into a dead-end street.

A very important consideration for circuit analysis is that any two (or more) devices that have the same device rule, that is, the same relationship between their currents and voltages, will behave exactly the same when placed in any circuit. Such devices are electronic equivalents. Being able to replace a device with another (possibly hypothetical) device will often prove convenient for solving and understanding circuits.

---

[5]Named after Georg Simon Ohm (1789–1854), a German mathematician and scientist.

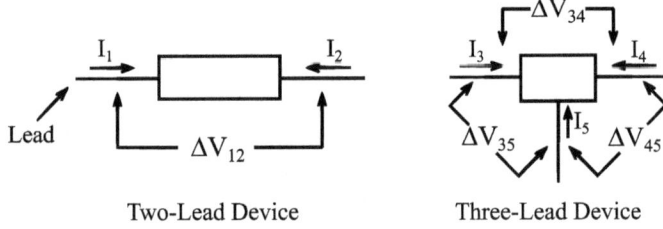

Fig. 1.1 An illustration of two- and three-lead devices, and how the currents into and voltage differences between the leads might be formally defined

Table 1.1 The simplest devices and their ideal behavior

Device	Symbol	Current–Voltage relation
Wire, junction of wires		$\Delta V = 0$ between all leads, no matter what the current
Switch		$\Delta V = 0$ if switch is closed, $I = 0$ if the switch is open
Resistor		$\Delta V = IR$ ("Ohm's Law"), where R is a constant and is called the "resistance." Voltage decreases in the direction of current flow
Battery (ideal voltage source)		$\Delta V = V_0$, where V_0 is a constant called the "battery voltage." Voltage increases from the "small" side to the "large" side

A circuit is described on paper using a "schematic diagram" where each device is represented by a symbol, referred to as a "schematic symbol." Many of these symbols are standardized, though some variation will be seen. Examples of symbols used for the simplest "ideal" devices, previously introduced in introductory electricity and magnetism, are shown in Table 1.1. A "node" is a connection point, a junction, between two or more devices. In practice, nodes where three or more devices are connected together will be most important.

Voltages are always measured as a voltage difference between two points in a circuit. These voltage differences might be better described using a prefix delta (Δ) to indicate one is speaking of a difference, such as "ΔV." However, in electronics the "Δ" will routinely be omitted to simplify the notation. The current-voltage relationship for a device will, of course, always involve such a difference. For example, Ohm's law for resistors, which describes the current-voltage relationship for resistors, is usually written simply as $V = IR$, rather than $\Delta V = IR$, even though the latter might be clearer and/or more appropriate. When the voltage decreases across a device in the direction of current flow, the change is referred to as a voltage drop.

Simple Devices

The devices in Table 1.1 are all linear, meaning that if you change the voltage (current) by any factor, the current (voltage) changes by the same factor. Circuits that are constructed solely using linear devices are referred to as linear circuits. Linear circuits can always be solved using linear mathematics. Additional linear devices will be added in due course.

Kirchhoff's Laws

Kirchhoff's laws are obeyed by all circuits[6] and provide the fundamental relations used to solve for currents and voltages in a circuit.

- Kirchhoff's Current Law (KCL): The sum of the currents entering any point in a circuit equals the sum of the currents leaving that point. (That is, current is not created or destroyed within the circuit—current is not "used up".)
- Kirchhoff's Voltage Law (KVL): The sum of the changes in the voltage, ΔV_i, around any closed path is zero. (A closed path is one that ends at the same position where it starts.)

Example Application of Kirchhoff's Laws in a simple case.

Apply Kirchhoff's laws to the circuit illustrated by the schematic in Fig. 1.2 to get:

- KCL: The current through the 3k resistor must be 2 mA (5 mA = 2 mA + 3 mA).
- KVL: $V = 11$ V because one must have $V - (5 \text{ mA})(1k) - (3 \text{ mA})(2k) = 0$.

(Here Ohm's law, $V = IR$, was used to find the voltage change across the resistors.)

Resistors in Series

Two resistors are connected in series if they are connected so that the same current *must* flow through both of them—there is no alternative current path.

---

[6]As a fine point, Kirchhoff's laws assume that the devices interact with each other only through connecting wires. All of the basic physics inside the device is hidden. On the other hand, if two devices interact to a significant degree via an electric or magnetic field, the rules may not apply *as written*. In those cases, however, additional (hypothetical) devices can often be used to model the effect of the interaction(s). Once those additional devices are included in the analysis, the rule is again satisfied.

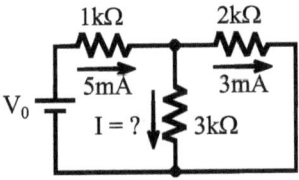

Fig. 1.2 A simple circuit to illustrate Kirchhoff's laws

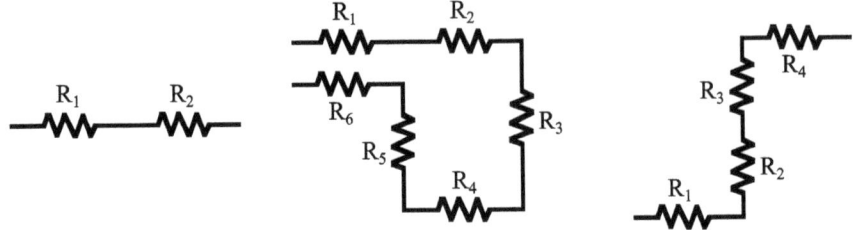

Fig. 1.3 Examples of resistors in series. In each case there is only one path for the current through all the resistors

Figure 1.3 shows several examples of resistors in series. Resistors in series form a voltage divider. If the voltage across all of them is V_0, then the voltage across the resistors that are between two points A and B, designated V_{AB}, will be given by the "voltage divider equation[7]:"

$$V_{AB} = V_0 \frac{\sum R \text{ for all resistors between points A and B}}{\sum R \text{ for all the series resistors}}. \quad (1.1)$$

This is a direct consequence of Ohm's law and Kirchhoff's voltage law.

Examples Use of the voltage divider equation.

- The voltage divider equation can be applied to find V_{AB} for the circuit of Fig. 1.4a, to get

$$V_{AB} = V_0 \frac{R_3 + R_4}{R_1 + R_2 + R_3 + R_4 + R_5 + R_6}. \quad (1.2)$$

- The voltage divider equation can be used to solve for V_1 for the circuit of Fig. 1.4b to get

---

[7]In this context, the upper-case Greek sigma, Σ, indicates a summation and is read as "the sum of," and the symbol R is a shorthand to represent "resistance values."

Resistors in Series

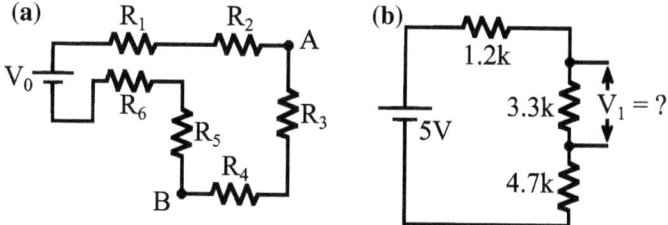

Fig. 1.4 Two examples of the use of the voltage divider equation for series resistors

$$V_1 = 5\,\text{V}\frac{3.3}{1.2+3.3+4.7} = 1.8\,\text{V}. \tag{1.3}$$

The voltage divider equation (e.g., in the examples above) is only valid if *all* the resistors are in series. In practice, to measure the resulting voltage, a voltmeter must be connected in parallel with a portion of the circuit. For example, to measure V_1 in the second example (Fig. 1.4b), the voltmeter would be connected across the 3.3 kΩ resistor—one lead from the voltmeter goes to each side of the resistor. If the resistance of the voltmeter is too small, the current through the voltmeter will be significant and that will affect the result—the voltmeter provides an alternate path for some of the current so the resistors are no longer in series. The result will be reasonably valid, however, if the current through the voltmeter is much less than the current through the resistor. In this example, the voltmeter must have a resistance very large compared to 3.3 kΩ for the reading to be accurate.

An "ideal voltmeter" has an infinite resistance (i.e., there will be no current through the ideal meter). Most modern digital voltmeters will have an input resistance of 1 MΩ or even larger, which is usually, but not always, large enough to be of little consequence. Some older (analog) voltmeters may have a resistance small enough so that the current through the meter cannot be routinely neglected.

Resistors in Parallel

Resistors are connected in parallel if the voltage across all of them *must* be the same because they all are connected by (ideal) wires on both sides. Remember that there is no voltage change across an ideal wire.

Figure 1.5 shows some examples of resistors in parallel. Parallel resistors form a current divider. If the total current into the parallel combination is I_0, then the current through any given resistor, I_k, can be computed using the current divider equation:

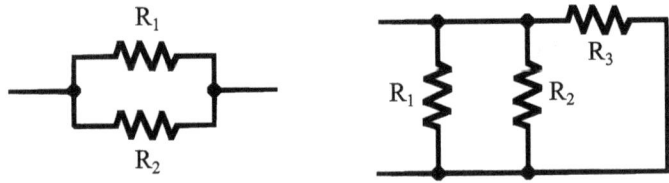

Fig. 1.5 Examples of resistors in parallel. In each case the voltage across all the resistors must be the same

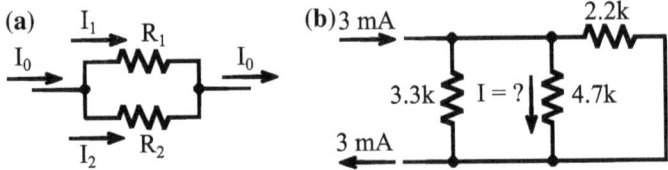

Fig. 1.6 Two examples of the use of the current divider equation for parallel resistors

$$I_k = I_0 \frac{1/R_k}{\sum 1/R \text{ for all the parallel resistors}}. \tag{1.4}$$

This is a direct consequence of Ohm's law and Kirchhoff's current law.

Examples Use of the current divider equation.

- Figure 1.6a illustrates the simplest case of two resistors in parallel. For that case, the current divider equation yields

$$I_1 = I_0 \frac{1/R_1}{1/R_1 + 1/R_2} = I_0 \frac{R_2}{R_1 + R_2}. \tag{1.5}$$

Note the simple form of the current divider equation on the right when only two resistors are involved—the current through one resistor is the total current multiplied by the value of the *other* resistor, divided by the sum of the two. That simple form is convenient to remember and use, but keep in mind that it only works for *pairs* of resistors and does not generalize to three or more parallel resistors.
- For the portion of the circuit shown in Fig. 1.6b, the current I is given by

$$I = 3 \text{ mA} \frac{1/4.7}{1/3.3 + 1/4.7 + 1/2.2} = 0.66 \text{ mA}. \tag{1.6}$$

The current divider equation is only valid when *all* the resistors involved are in parallel. To measure a current, you must insert a current measuring device in series with some portion of the circuit—the circuit is broken and the ammeter is connected

Resistors in Parallel

across the break. In the second example above, the ammeter would be inserted in series with the 4.7 kΩ resistor. If the resistance of the measuring device is significant compared to the resistance in the circuit, the meter's presence will affect the results. In the second example, the resistance of the ammeter must be very small compared to 4.7k in order to get an accurate reading.

An "ideal ammeter" has no resistance (0 Ω). The internal resistance for many ammeters will depend on the scale setting of the meter—that is, whether it is set to measure larger or smaller currents.

Effective Resistance

If a circuit contains a number of resistors and some are in parallel and/or in series, those combinations can be replaced with an "effective" or "equivalent" resistance, often designated R_{eff} or R_{eq}, without changing the behavior of the rest of the circuit. It will be shown later that all combinations of resistors that ultimately connect to two leads will have an equivalent, though it might not be simple to compute. The equivalents for series and parallel resistors are easily derived using Kirchhoff's laws:

- For series resistors: $R_{eff} = \sum R$ for all the resistors
- For parallel resistors: $1/R_{eff} = \sum 1/R$ for all the resistors.

Example Equivalent resistance—parallel resistors.

The portion of a circuit in Fig. 1.7 has the three resistors are in parallel, so they can be replaced with effective resistance:

$$1/R_{eff} = 1/3.3k + 1/4.7k + 1/2.2k = 0.97(1/k)$$
$$R_{eff} = \frac{1}{0.97}k = 1.03k.$$
(1.7)

Example Equivalent resistance—combinations of parallel and series resistors.

Consider the portion of a circuit shown on the left in Fig. 1.8. As a first step, note that all *except* the 15k resistor are in series. We replace only those resistors with an effective resistance:

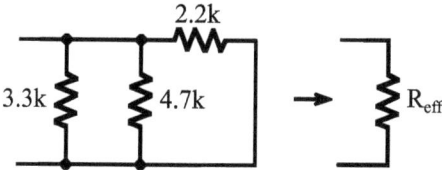

Fig. 1.7 An example showing a group of parallel resistors that can be replaced with a single resistor without affecting the rest of the circuit

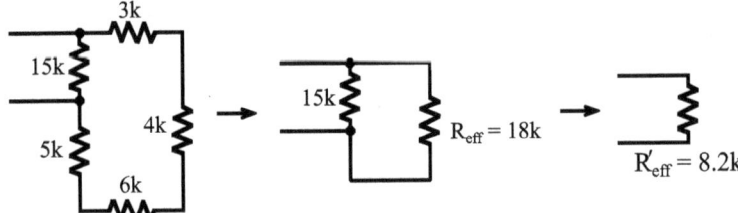

Fig. 1.8 An example showing a portion of a circuit that can be reduced in two steps to a single equivalent resistor

$$R_{eff} = 3k + 4k + 6k + 5k = 18k. \tag{1.8}$$

Once that is done, the 15k resistor and the new effective resistance are in parallel. They can be replaced, if desired, with a second effective resistance,

$$R'_{eff} = \left(\frac{1}{15k} + \frac{1}{18k}\right)^{-1} = \frac{15 \cdot 18}{15 + 18} k = 8.2k. \tag{1.9}$$

That is, when placed in a circuit, the original combination of resistors will behave the same as a single 8.2k resistor, as shown on the right of Fig. 1.8.

Being able to replace series and parallel components with equivalents is one of the most powerful ways to simplify circuits as part of an analysis. The process of simplifying a circuit using equivalents is sometimes referred to as "circuit reduction" and is discussed in more detail below. Additional circuit reduction techniques will arise later.

Resistors in Parallel–Notation

Two resistors in parallel, R_1 and R_2, can be replaced by an effective resistance, R_{eff}, that is calculated using

$$1/R_{eff} = 1/R_1 + 1/R_2 \rightarrow R_{eff} = \frac{R_1 R_2}{R_1 + R_2}. \tag{1.10}$$

For convenience, define an operator, ∥, that acts on two variables and which produces this result. That is,

$$R_1 \| R_2 \equiv \frac{R_1 R_2}{R_1 + R_2}. \tag{1.11}$$

Effective Resistance

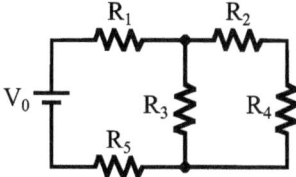

Fig. 1.9 A circuit used to illustrate the description of the equivalent resistance using the "||" operator

This notation is used as a shorthand only, and be warned that the operator may not behave like other mathematical operators, such as for multiplication and addition. For example, the distributive law does not work, that is,

$$R_1 || (R_2 + R_3) \neq R_1 || R_2 + R_1 || R_3. \tag{1.12}$$

Example Describing combinations using the "||" operator.

Writing down the steps that would be used to compute an equivalent resistance, but without solving at each step, is a way of describing the calculation. For example, the total effective resistance as seen by the battery for the circuit of Fig. 1.9 can be described symbolically as

$$R_{\text{eff}} = R_1 + R_3 || (R_2 + R_4) + R_5, \tag{1.13}$$

that you might read as

> R-one in series with the combination of R-three in parallel with the series combination of R-two and R-four, all of which is in series with R-five.

Can you provide a convincing argument to prove[8] that for any three positive values, A, B, and C,

$$A || (B || C) = (A || B) || C = (A || C) || B? \tag{1.14}$$

That result means that the parenthesis can be left out without changing the result, however remember that the || operator acts on only two values at a time.

Solving Circuits with Circuit Reduction

As mentioned above, many circuits can be solved using a process known as circuit reduction. That is, a more complicated circuit is reduced to something simpler using an electronic equivalent. The effective resistance of series and parallel resistors can be most useful for this purpose.

---

[8]Hint: this can be done rigorously without any algebra! Use some simple physics instead.

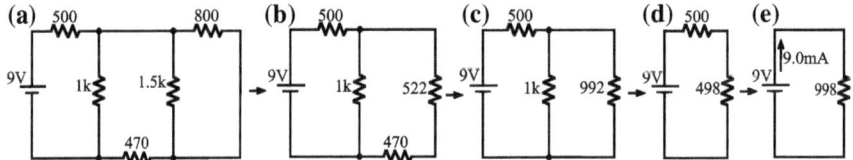

Fig. 1.10 Using equivalents for series and parallel resistors, the circuit shown in (**a**) is reduced in several steps to the simple circuit shown in (**e**)

Example Circuit reduction using parallel and series resistors.

Determine the currents through all of the resistors for the circuit in Fig. 1.10a.

Recall:

- Resistors are in series if the current through them MUST be the same (because there is no other path). In the circuit of Fig. 1.10a there are NO series resistors.
- Resistors are in parallel if the voltage across them MUST be the same (i.e., both ends are connected by wires.)

In the circuit of Fig. 1.10a, the 800 Ω and 1.5 kΩ resistors are in parallel (*and no others*). As far as the rest of the circuit is concerned, these two can be replaced with a single resistor with a value

$$R_1 = 800\,\Omega \| 1.5\,\text{k}\Omega = \frac{800 \cdot 1500}{800 + 1500}\,\Omega = 522\,\Omega, \tag{1.15}$$

and so now the circuit is as shown in Fig. 1.10b.

Examining this new circuit, there are no resistors in parallel, but the 470 and 522 Ω resistors are in series. Hence, as far as the rest of the circuit is concerned, they can be replaced with a single resistor

$$R_2 = 470\,\Omega + 522\,\Omega = 992\,\Omega, \tag{1.16}$$

and now the circuit is as shown in Fig. 1.10c.

Examining Fig. 1.10c, there are no resistors in series, but the 1k and 992 Ω resistors are in parallel. Hence, as far of the rest of the circuit is concerned, they can be replaced with a single resistor with a value

$$R_3 = 1\,\text{k}\Omega \| 992\,\Omega = \frac{1000 \cdot 992}{1000 + 992}\,\Omega = 498\,\Omega, \tag{1.17}$$

and the circuit now is that of Fig. 1.10d.

Examining Fig. 1.10d, there are no parallel resistors, but the 500 and 498 Ω resistors are in series. Hence, as far as the rest of the circuit is concerned (which now is only the battery and wires) they can be replaced with a single resistor

Solving Circuits with Circuit Reduction

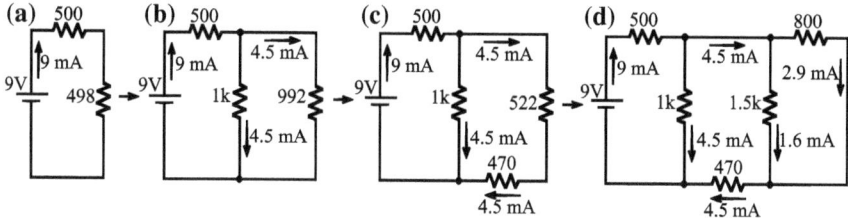

Fig. 1.11 Starting with the result from Fig. 1.10e, the currents through each resistor are determined by working in reverse

$$R_4 = 500\,\Omega + 498\,\Omega = 998\,\Omega. \tag{1.18}$$

Now the circuit is the simple circuit of Fig. 1.10e. The current delivered by the 9 V battery to the rest of the circuit (which, to the battery, looks like a single 998 Ω resistor) is easily computed, using Ohm's law, to be

$$I_1 = \frac{9\,\text{V}}{998\,\Omega} = 9.0\,\text{mA}. \tag{1.19}$$

At this point all of the original circuit has been lost except for the battery. Working backwards, as needed, the currents through (and hence the voltages across) all of the original resistors can be determined. Refer to Fig. 1.11.

Going back one step, whatever current goes through the 998 Ω equivalent resistor is really going through the 500 and 498 Ω resistors which are in series (and hence they must have the same current).

Now the 498 Ω resistor came from two resistors in parallel. Those two must have the same voltage across them and that voltage is the same as the voltage across the 498 Ω resistor. That is, $V = 9.0$ mA $\cdot$ 498 Ω = 4.48 V, so find the currents that give this voltage for the two parallel resistors, or use the current divider equation, which is equivalent, to get the currents shown in Fig. 1.11b.

Now the 992 Ω resistor came from two resistors in series, each of which must have the same current (4.52 mA, Fig. 1.11c). The 522 Ω resistor is really two resistors in parallel, so the voltage across both of them is $V = 4.52$ mA $\cdot$ 522 Ω = 2.36 V. Once again, find the appropriate currents that give you this voltage for each of the two resistors to get the currents shown in Fig. 1.11d. Now "everything" is known and the circuit has been solved.

Depending upon what question is to be answered, it may not be necessary to proceed this far. For example, if only the current delivered by the battery was required, the process could have stopped several steps ago. If the voltage across the 800 Ω resistor was desired, it was available at the previous step (it's 2.36 V).

Solving Circuits with Algebra

Not all circuits contain components that can be easily replaced with an equivalent. In those cases, Kirchhoff's laws and algebra can be used. Several different approaches are useful in order to generate the equations to be solved.

Branch and Mesh Currents

There are many ways to solve circuits algebraically using the equations that result from Kirchhoff's laws. Two common procedures to set up those equations are the branch and mesh methods. A third method, known as nodal analysis, will be shown separately. All of these methods start with Kirchhoff's laws to generate a number of equations. For linear circuits, those equations will be mathematically linear and can always be solved, at least in principle.

- Branch Method:
  A current is defined for each portion of the circuit that may have a different current. That is, between all adjacent pairs of nodes where three or more devices connect.[9]
- Mesh (or Loop) Method:
  Each possible loop in the circuit is given a label, up to the number of independent loops necessary to include each component at least once. The advantage is that Kirchhoff's current law (KCL) will be automatically satisfied and there are fewer equations to solve. The disadvantage is that the current in any particular portion of the circuit may be the sum of several such mesh currents, which may be an inconvenience.

A simple circuit is illustrated in Fig. 1.12a. The description of the circuit using branch currents is shown in Fig. 1.12b, and using mesh currents in Fig. 1.12c. Of course, what actually happens in the circuit must not depend on the description. Hence, it is always the case that one description can be converted to the other by equating the currents through components. In this example,

$$I_a = I_p, \; I_c = I_q, \; I_b = I_p - I_q. \tag{1.20}$$

Note that the variable names you use are up to you. It is very common to use the same letter (e.g. "I") with different subscripts (whether you use number, letters, or

---

[9] Defining the current means giving it a name and defining which direction is to be considered positive. If a negative value for the current results, that simply means the arrow was drawn opposite to what really occurs. This process is analogous to defining the positive the x- and y-axes for a mechanics problem.

Solving Circuits with Algebra

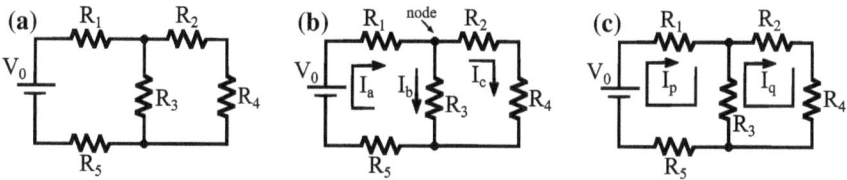

Fig. 1.12 The circuit shown in (a) is to be solved. For the branch and mesh methods, the first step is to define the currents such as is shown in parts (b) and (c) respectively

signs of the Zodiac as subscripts, is up to you).[10] As a further note, in electronics the symbol "i" (lower case) is sometimes used for currents. In this text the lower case "i" will be used for another purpose and so its use for describing currents is avoided.

Example—Using Kirchhoff's Laws

As an example, the currents in a circuit will be found with Kirchhoff's laws using both branch currents and mesh currents. Of course, the actual current found in any part of the circuit *must be the same* for the two solutions. For many circuits, the mesh method is more efficient and therefore is preferred. If the correctness of the answer is particularly important, solve using both methods and compare the results.

Goal: Find all the currents for the circuit of Fig. 1.12.

Using branch currents:

(1) Define the currents in each branch (i.e. give them a name and direction). This is illustrated in Fig. 1.12b.
(2) Write down KCL equations at nodes (junctions). Remove any duplicate or redundant equations. KCL is trivially satisfied elsewhere. In this case there is one unique KCL equation,

$$I_a = I_b + I_c. \tag{1.21}$$

(3) Write down KVL for enough loops to include all components at least once, adding the voltage changes along the way. The loops can be in either direction. There are three (simple) loops here, generating the following equations:

---

[10]What is now simply called "current" used to be called "current intensity," or in French, the language of Ampère, *intensité du courant*. Hence the use of "I" to represent that "intensity."

$$V_0 - I_a R_1 - I_b R_3 - I_a R_5 = 0$$
$$V_0 - I_a R_1 - I_c R_2 - I_c R_4 - I_a R_5 = 0 \qquad (1.22)$$
$$I_b R_3 - I_c R_2 - I_c R_4 = 0.$$

There are three unknown currents so three independent equations will be necessary. The KCL equations are (almost) always required. Here there is one unique KCL equation. Hence only two of these KVL equations are needed. Note, for example, in this case the third KVL equation can be generated from the first two (by subtracting the first from the second), hence, given any two of these KVL equations, the third will provide no additional information.

(4) Now solve the equations for the desired quantity(ies).

As an example, use the first and third KVL equations. The third equation can be rearranged to get

$$I_c = \frac{R_3}{R_2 + R_4} I_b, \qquad (1.23)$$

and putting this into the KCL equation gives

$$I_a = I_b \left(1 + \frac{R_3}{R_2 + R_4} \right) = I_b \left(\frac{R_2 + R_4 + R_3}{R_2 + R_4} \right)$$
$$I_b = \left(\frac{R_2 + R_4}{R_2 + R_3 + R_4} \right) I_a, \qquad (1.24)$$

that can be substituted into the first KVL equation to get

$$I_a = V_0 \left(R_1 + \frac{R_3 (R_2 + R_4)}{R_2 + R_3 + R_4} + R_5 \right)^{-1}, \qquad (1.25)$$

and this value can be substituted back into the previous equation relating I_a and I_b to find I_b, and then that value is used to find I_c.

Using mesh currents:

(1) Define mesh currents (each current goes around a complete loop, and the total current at any location may be the sum of several mesh currents). Each component must have at least one mesh current through it. There are three simple loops here: smaller loops on the left and right, as in Fig. 1.12c, and a larger loop around the outside (not shown). Very complicated loops, such as figure eights or involving multiple passes through a device, are also possible, but not recommended. Keep it simple.
(2) Since mesh currents automatically satisfy KCL, there are no KCL equations to write down.

(3) Write down KVL equations—at a minimum, there should be enough to include all components at least once. The loops used for these equations do not need to be the same as those used to define the currents. Note that for this example, the current through R_3 involves the sum of two mesh currents with opposite signs. Three equations generated from the three simple loops are

$$V_0 - I_p R_1 - (I_p - I_q) R_3 - I_p R_5 = 0$$
$$V_0 - I_p R_1 - I_q R_2 - I_q R_4 - I_p R_5 = 0 \quad (1.26)$$
$$(I_p - I_q) R_3 - I_q R_2 - I_q R_4 = 0.$$

There are two unknown currents and so only two of these equations will be necessary. Rewriting the second and third,

$$I_p(R_1 + R_5) + I_q(R_2 + R_4) = V_0$$
$$I_q = \frac{R_3}{R_2 + R_3 + R_4} I_p, \quad (1.27)$$

and putting the second of these two results into the first yields

$$I_p \left(R_1 + R_5 + \frac{R_3(R_2 + R_4)}{R_2 + R_3 + R_4} \right) = V_0$$
$$\rightarrow I_p = V_0 \left(R_1 + R_5 + \frac{R_3(R_2 + R_4)}{R_2 + R_3 + R_4} \right)^{-1}, \quad (1.28)$$

which can then be substituted back into the equation just above get I_q.

Note the connection between this solution and the solution using branch circuits. Here $I_a = I_p$, $I_c = I_q$, and $I_b = I_p - I_q$, as they should.

For both of these algebraic methods it is easy to generate more equations than necessary. Such a situation can be easily resolved. It is also possible to generate too few *independent* equations. If the equations are not solvable or reduce to trivial results (such as 0 = 0) then go back and try a different set of equations.

Nodal Analysis

An additional way to analyze a circuit is known as nodal analysis. Nodal analysis may be somewhat inefficient when applied to an entire circuit, but it may be very useful for smaller circuits or for the analysis of a small portion of a circuit. Nodal analysis also works well for automated (i.e., computerized) solutions.

To apply nodal analysis, the voltage (relative to some convenient common reference) at each unique node is given a label. Kirchhoff's voltage law then allows you to find the current between each pair of nodes in terms of those voltages and the device rules (e.g., Ohm's law). Kirchhoff's current law is then used at each node,

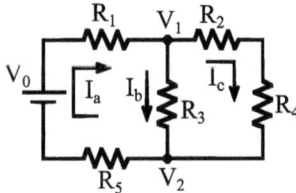

Fig. 1.13 The first step when solving the circuit of Fig. 1.12a using nodal analysis is to define a variable associated with each of the branch currents and the node voltages

writing the currents in terms of the node voltages. If you have N nodes, there will be $N - 1$ unique equations that can then be solved. Again, in the general case all the devices must be included at least once to get a complete solution.

Example A solution using nodal analysis.

Start by labeling the nodes and defining the currents, such as is illustrated in Fig. 1.13. In this example there are two nodes, one of which can be taken to be the voltage reference. Kirchhoff's current law gives, as previously seen above for the analysis using branch currents,

$$I_a - I_b - I_c = 0, \tag{1.29}$$

where

$$I_a = \frac{V_2 - V_1 + V_0}{R_1 + R_5}; \quad I_b = \frac{V_1 - V_2}{R_3}; \quad I_c = \frac{V_1 - V_2}{R_2 + R_4}. \tag{1.30}$$

If the lower node is taken as the reference, equivalent to using $V_2 = 0$ V, then KCL gives

$$V_0 \left(\frac{1}{R_1 + R_5} \right) + V_1 \left(\frac{-1}{R_1 + R_5} + \frac{1}{R_3} + \frac{1}{R_2 + R_4} \right) = 0, \tag{1.31}$$

where the only unknown is V_1. It is straightforward to solve for V_1, and that value is used to solve for the currents. Of course, the resulting currents through the devices must agree with those found above using the other methods.

With nodal analysis, there may be a larger number of equations to solve, but many of them will often be, in a practical sense, simpler than what is encountered when using branch or mesh currents.

The Ideal Current Source

Another basic linear device seen in electronics is the "ideal current source." The ideal current source is a device that provides a constant current output regardless of the voltage across it. That is $I = I_0$, and V is whatever is necessary to maintain that current. Schematically the current source appears as shown in Fig. 1.14, where the arrow indicates the direction for positive current (the "I" may or may not be present).

Example Simple example involving a current source.

In the circuit of Fig. 1.15a, what is the voltage across the current source?

Using circuit reduction, the resistors can be replaced by an equivalent with a value $1k + (2k\|3k) = (1 + 6/5)k = 2.2k$. The equivalent circuit as seen by the current source is shown in Fig. 1.15b.

The current I is 30 mA and so, using Ohm's law, the voltage across the equivalent resistor is $2.2k \cdot 30$ mA $= 66$ V. In this case KVL requires that that must also be the magnitude of the voltage across the current source.

Most commercial power supplies can be modeled as constant voltage sources, though there are exceptions and constant current supplies are certainly available. Constant current sources will be most prominent later, principally as part of a simplified model for a transistor.

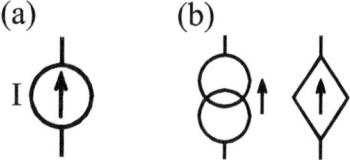

Fig. 1.14 The schematic symbol for a constant current source is shown in (**a**). Some alternate symbols are sometimes used for current sources, such as those in (**b**). The diamond shaped symbol is often used to indicate a "dependent source," where the current depends on what is happening somewhere else in the circuit

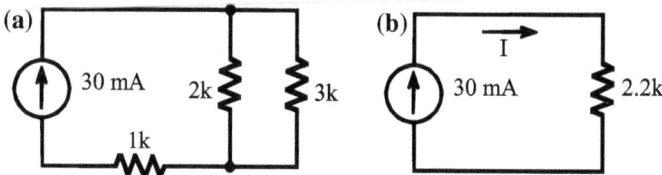

Fig. 1.15 One method to find the voltage across the current source for the circuit shown in (**a**), is to use circuit reduction to find an equivalent resistance, as shown in (**b**)

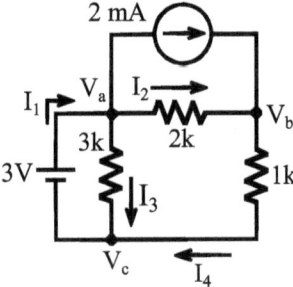

Fig. 1.16 A circuit involving both a current source and a voltage source that is to be solved using nodal analysis

Example Nodal analysis with both a voltage and current source.

The goal is to solve for the currents in the circuit of Fig. 1.16. There are three node voltages labeled, V_a, V_b, and V_c, and five branch currents. Four of the branch currents are labeled with variables I_1 to I_4. The fifth branch current is that of the current source, which has a known value (2 mA).

Kirchhoff's current law (KCL) will need to be applied at each of the nodes, generating the following equations:

$$\begin{aligned} I_1 &= I_2 + I_3 + 2\,\text{mA} \\ I_3 + I_4 &= I_1 \\ I_4 &= I_2 + 2\,\text{mA}. \end{aligned} \quad (1.32)$$

Now look at the voltage changes along the paths between adjacent nodes to compute the currents:

$$\begin{aligned} I_3 &= (V_a - V_c)/3\text{k} = 3\,\text{V}/3\text{k} = 1\,\text{mA} \\ I_2 &= (V_a - V_b)/2\text{k} \\ I_4 &= (V_b - V_c)/1\text{k}. \end{aligned} \quad (1.33)$$

There are three node voltages, any one of which could be taken to be the reference—remember that only voltage *differences* will matter when determining the currents. For the sake of this example, use V_c as the reference. That is equivalent to taking $V_c = 0$ V (since the difference between V_c and itself is 0 V, why not just call it 0 V?).

Note that there are 6 unknowns: four currents and two voltages. Thus, six independent equations are required. (If the mesh current method above was used, there would have been only 3 unknowns.)

Solving this circuit yields:

$V_a = 3$ V

$I_3 = 1$ mA

$3\text{V} = (2\text{k})I_3 + (1\text{k})I_4 = (2\text{k})I_3 + (1\text{k})(I_2 + 2\,\text{mA}) \rightarrow I_2 = \dfrac{1\,\text{V}}{3\text{k}} = 0.33\,\text{mA}$

$I_4 = 2.33$ mA

$V_b = 2.33$ V

$I_1 = 2.33\,\text{mA} + 1\,\text{mA} = 3.33\,\text{mA}$

and the voltage across the current source is 0.666 V.

The Ground and Common Connections

The schematic representation of the "ground" and "common" connections are shown in Fig. 1.17. When these appear in circuit diagrams, all of the "common" connection points are to be connected together with a wire (or equivalent). The "ground" is a special type of common connection that includes a wire that is actually connected to the Earth. Sometimes such a connection is called "earth" instead of the "ground" connection. Inside a building, this connection to the Earth might not be obvious and there may be a long route to get there.

The Earth is very large, is able to conduct electricity, and is presumed to be a good source and sink for excess charge. When you ground (or "earth") a circuit you are connecting it to the Earth. Such a connection is presumed to provide a constant electric potential, usually taken to be 0 V. The person operating an electronic device is typically in contact with the Earth, either directly or indirectly, so there will be no voltage difference between the operator and such a ground connection. Hence such a location will not pose a shock hazard.

Sometimes the ground connection will be used in a schematic diagram when there is no actual connection to the Earth. What is usually meant is a common connection and/or a connection to a chassis or other larger metal object that acts like an (idealized) earth connection, at least for the intended use of the circuit. There is some sloppiness in the use of the ground symbol.

Fig. 1.17 Schematic symbols for ground and common connections

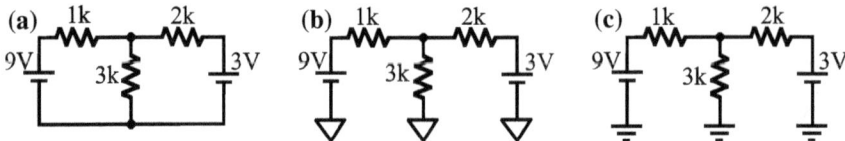

Fig. 1.18 Since all common connections, shown in (**b**), are connected to each other and all ground connections, shown in (**c**) are connected to each other, the analysis for all three of these circuits will be identical

For the sake of circuit analysis, the three schematics of Fig. 1.18 would be solved in exactly the same way.

Some complicated circuits may have more than one common connection. In such cases a label will appear inside or next to the symbol. All common connections with the same label are connected together, but those with different labels are not connected to each other.

Multiple Sources—The Superposition Theorem

The superposition theorem can be used to solve circuits containing multiple sources using any of the methods described above. Superposition will also be very important for the treatment of time-dependent voltages and currents.

The superposition theorem says that for any *linear circuit* containing more than one independent source, the circuit can be solved by considering one source at a time, with all the other source(s) "turned off," and then adding those results together. That is, the solution for the sum is the sum of the solutions. This is a special case of a more general result from basic electricity and magnetism (E&M) and, even more generally, linear mathematics.

In this context:

- A voltage source that is "turned off" is a voltage source fixed at 0 V—such a source is equivalent to a wire.
- A current source that is "turned off" is a current source fixed at 0 A—such a source is equivalent to an open circuit (no connection).

Example Use of the superposition principle.

Determine the current, *I*, delivered by the 9 V battery in the circuit of Fig. 1.19a. The current delivered by the 9 V battery is the same as the current through the 1k resistor. First, look at that current due to the 9 V battery when the 3 V battery is "off" (i.e., replaced with a wire, Fig. 1.19b). Then $I_1 = 9\ V/R_{eff}$ where $R_{eff} = 1k + 2k \| 3k = 2.2k$, so $I_1 = 4.09$ mA.

A common mistake is to claim that I_1 is the current delivered by the 9 V battery. This is not true. The analysis is not yet complete.

Multiple Sources—The Superposition Theorem

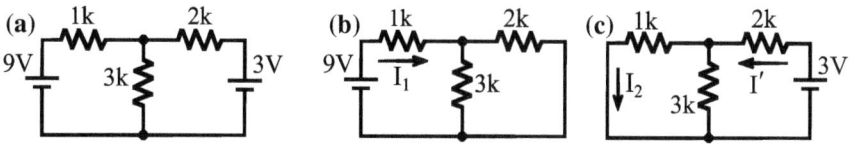

Fig. 1.19 The circuit shown in (**a**), that contains two sources, is to be solved using superposition by considering each source separately with the other "off," as shown in (**b**) and (**c**).

Now look at the current due to the 3 V battery, with the 9 V battery "off" (Fig. 1.19c). Then $I_2 = I'\ 3k/(1k + 3k) = 3I'/4$ (using the current divider), and $I' = 3\ V/R_{\mathit{eff}}$ where $R_{\mathit{eff}} = 2k + 3k\|1k = 2.75k$. Hence, $I' = 1.09$ mA and so $I_2 = 0.82$ mA.

To get the result for the original circuit, add these together *taking direction into account*. Since the directions used for I_1 and I_2 were opposite, and I_1 is in the direction of the current desired (I), we subtract I_2 from I_1.

Hence, $I = (4.09 - 0.82)$ mA $= 3.3$ mA is the current delivered by the 9 V battery.

As an exercise, compute the current through the 3k resistor and the current delivered by the 3 V source for the same circuit.[11]

Electrical Power

Power is the rate of change of energy and is measured in watts (W),[12] where 1 W = 1 J/s. In mechanics, power, P, is the dot product of force and velocity. Power is supplied to an object if the force on the object and the velocity of the object are in the same direction. In electronics, the power supplied to a device, or supplied by a device, is given by the product of the voltage across the device and the current through the device, $P = VI$. The power is being supplied by the device if the current through the device goes from a lower voltage to a higher voltage. Power is being absorbed by the device (e.g., turned into another form of energy, such as heat, chemical, or mechanical energy) if the current flows through the device from a higher voltage to a lower voltage.

For the devices shown in Fig. 1.20 (and assuming I is positive), power is being supplied *to* the device on the left, while power is being supplied *by* the device on the right. For a resistor, Ohm's law can be used to see that power, P, is always supplied *to* a resistor, and never *by* the resistor, and $P = VI = I^2 R = V^2/R$.

The watt (W) is an SI unit. Equivalent units are: $1\ W = 1\ V \cdot A = 1\ A^2 \cdot \Omega = 1\ V^2/\Omega = 1\ J/s = 1\ kg \cdot m^2/s^2$.

---

[11] Answers: 1.9 mA down through the 3k resistor, and -1.4 mA delivered by the 3 V battery—that is 1.4 mA is going into the 3 V battery.

[12] Named for the Scottish inventor and engineer, James Watt (1736–1819).

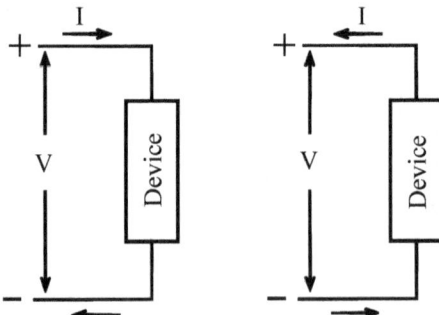

Fig. 1.20 If the current goes from higher to lower voltage, shown on the left, power is being supplied to the device. If the current goes from lower to higher, shown on the right, power is being supplied by the device

Note that since power involves the product of voltage and current, superposition does not apply directly to power. That is, the power supplied when there is more than one power source is not the sum of the power from each source considered separately.

Additional Application—The Kelvin-Varley Divider

A potentiometer is a variable resistor with three leads. Hence, the potentiometer can be used as a variable voltage divider. In the schematic (Fig. 1.21), the connection with the arrow can move and is called the "wiper." The total resistance between the outer two leads is fixed and is cited as the value, R, of the potentiometer. The resistance from the wiper to the two outer contacts (R_1 and R_2) varies from 0 to the full value R, and no matter what the position, it is always the case that $R_1 + R_2 = R$. In simple potentiometers the wiper makes a spring-loaded contact with a resistive material and a continuous range of values is obtained. The potentiometer can also be used as a variable resistor (a "rheostat") if the center lead and only one of the outer leads is used.

The wiper is often moved using a dial or a slider. The two common types of such devices will use a "linear" or "audio" taper. That is, the relationship between R_1 and

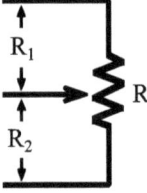

Fig. 1.21 The schematic symbol for a potentiometer shows three leads and is an adjustable voltage divider. The resistance between the outer two leads is fixed. The resistance between the center lead and each outer lead is adjusted, typically by turning a knob

Additional Application—The Kelvin-Varley Divider

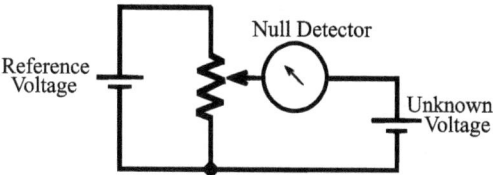

Fig. 1.22 A potentiometer can be used with a known reference voltage to determine an unknown voltage to high accuracy. A calibrated potentiometer is adjusted until the null detector (e.g., an ammeter) reads zero

the dial or slider position is either linear or non-linear (in a manner which is approximately exponential, and is called an "audio" or "log taper"). One reason the latter is convenient is because our ears perceive volume approximately linearly with the logarithm of output amplitude, and the log is the inverse of the exponential—our hearing straightens out the exponential changes in amplitude.

While potentiometers are often used as simple variable voltage dividers, a precision voltage divider can be used to compare two voltages (or "potentials") very precisely. Connected as shown in Fig. 1.22, the divider is adjusted for a null reading giving the ratio of the known reference to an unknown being measured. When adjusted for a null, no current, and hence no power, will be drawn from the unknown source, and hence the reference and potentiometer act as if they have infinite resistance to the unknown.

The Kelvin-Varley voltage divider is a potentiometer design that allows for great precision and a discrete readout. The potentiometer is made from discrete, matched, precision resistors with a schematic such as is shown in Fig. 1.23. In each of the "digital" sections there are eleven resistors. Two of those eleven will be in parallel with a resistance equal to those two in series. As you turn each dial, two contacts move together to adjacent positions along the voltage divider. The last stage is usually a continuous reading dial.

For the switch positions shown in the schematic, $V_{out} = 0.164xx\ V_{in}$, where the last digits ('xx') depend on the position of the wiper on the 80 Ω potentiometer. Here the effective resistance seen at the input (if nothing is connected on the output) is 10k no matter what the dial settings.

Note that when the null is achieved for the comparison circuit above, there will be no current into or out of the final output of the divider and so the output voltage (V_{out}) can be determined as if nothing were connected and accuracy is maintained. For accurate results in other cases, the output of the divider must be attached to a device or circuit that draws only a negligible current.

To analyze this Kelvin-Varley potentiometer in the case where there is no output current, start from the left and note that between the two points connected by any pair of switches, half of the current proceeds down and half goes through the switch connections to the right. For example, for the circuit of Fig. 1.23, if I_{200} is the current entering the string of 200 Ω resistors, then the current going to the 40 Ω resistors is $I_{200}/2$.

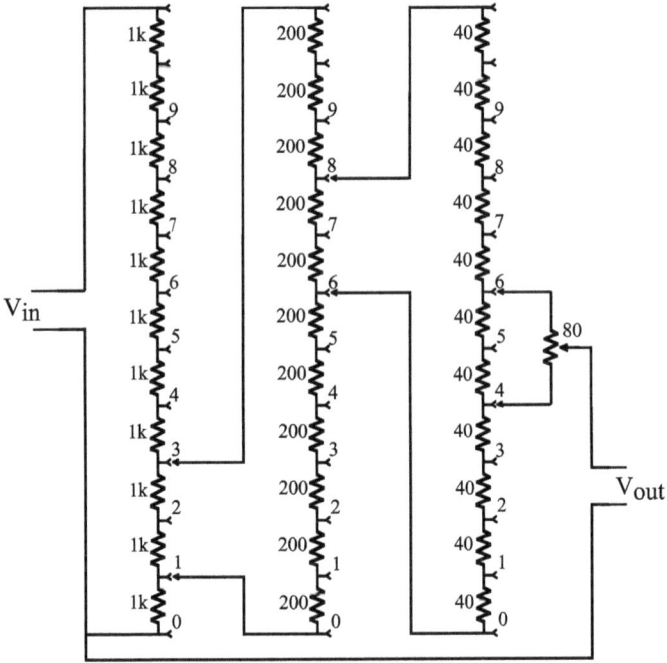

Fig. 1.23 A schematic showing how a 10k Kelvin-Varley potentiometer is constructed

Problems

1. (a) For the circuits shown in Fig. 1.P1, find the current through each of the resistors using Kirchhoff's current and voltage laws to set up equations, then solve the equations (do not use parallel/series resistor equivalents, circuit reduction, etc.). (b) Using your results from (a), find the voltage across the 3k resistor for each circuit.

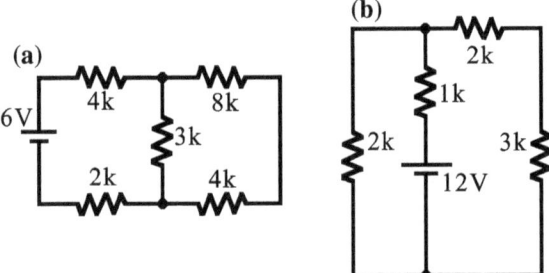

Fig. 1.P1 Problem 1

Problems   27

2. For the circuits of Fig. 1.P2, use circuit reduction and superposition to determine the current through each of the resistors. What is the power delivered by each of the sources?

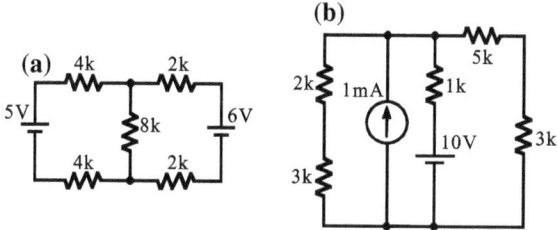

Fig. 1.P2 Problem 2

3. For the circuit of Fig. 1.P3, the current through the 1k resistor is 1.0 mA. What is the battery voltage, V?

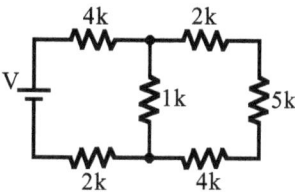

Fig. 1.P3 Problem 3

4. Use Kirchhoff's and Ohm's laws to derive the (a) voltage divider, (b) current divider, and effective resistance results for (c) series and (d) parallel resistors.
5. A real voltmeter that acts as a 1 MΩ resistance is used to measure the voltage across a resistor, as shown in the circuits of Fig. 1.P5. For each circuit, estimate the difference between the measured voltage and the voltage that would be measured with an ideal voltmeter.

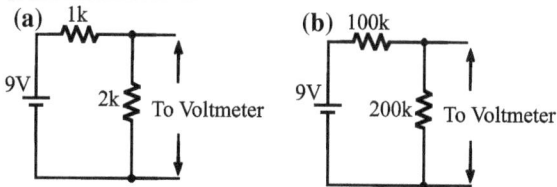

Fig. 1.P5 Problem 5

6. For the circuit shown in Fig. 1.P6, (a) what is the current through R_0 and (b) show that when R_0 gets large, in particular when $R_0 \gg R_1 + R_2$, that the voltage across the current source is close to $V - I_0(R_1 + R_2)$, independent of the value of R_0, and that if it is also the case that $R_0 \gg V/I_0$, then $I_1 \approx I_0$. That is, if these two conditions are satisfied, the circuit acts the same as if R_0 where not present.

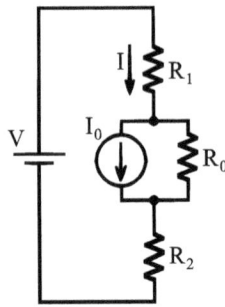

Fig. 1.P6 Problem 6

7. What is the voltage across the 3k resistor in the circuit of Fig. 1.P7?

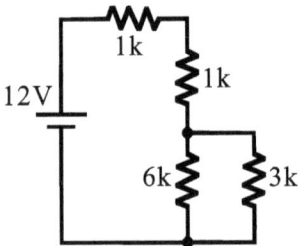

Fig. 1.P7 Problem 7

8. The circuit of Fig. 1.P8 consists of a resistor "tree" having four levels connected in series and powered by a battery with voltage, V_0. The uppermost level

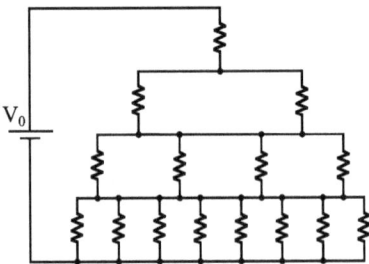

Fig. 1.P8 Problem 8

has one resistor and each lower level has twice as many resistors as the level above it. If all of the resistors have the same value, R, what is the current from the battery for this four-level tree? What is the current in the limit that there are an infinite number of such levels?

9. Using "=" and ">", rank order the magnitudes of the five branch currents for the circuit of Fig. 1.P9 from largest to smallest, without first solving for the currents.

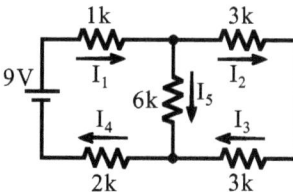

Fig. 1.P9 Problem 9

10. Resistors are sometimes used to model the electrical behavior of incandescent light bulbs. What resistance value should be used for a 50 W bulb designed to work with (a) a 12 V supply, (b) a 120 V supply, and (c) a 240 V supply.

11. Show that (a) the effective resistance of a series combination of resistors is always larger than the largest individual resistance, and (b) that the effective resistance of a parallel combination of resistors is always smaller than the smallest individual resistance.

Chapter 2
Additional Theorems

Kirchhoff's laws, parallel and series resistor equivalents, along with the superposition principle, will go a long way towards determining the voltage-current relationships in many simple circuits involving resistors and ideal sources. Later, these staples of circuit analysis will be applied more generally to include other linear circuit elements. However, before adding additional circuit elements, a few additional theorems, and examples of their use, are presented.

Thevenin and Norton Equivalents

Consider a box with two external leads. Inside the box is any number of resistors and linear sources (voltage sources and/or current sources) connected in any way (for example, see Fig. 2.1a). When this box is connected into a circuit, and assuming appropriate values are chosen for V_{th}, R_{th}, I_n, and R_n, the circuit will behave the same if either the Thevenin or Norton equivalent is used instead (Fig. 2.1b, c).

The Thevenin and Norton equivalents are used, for example, to analyze or better understand a circuit by reduction. That is, even a very complicated piece of a circuit can be replaced by a very simple equivalent, simplifying the rest of the analysis. The problem is to find appropriate values for the parameters V_{th}, R_{th}, I_n, and R_n.

It is important to realize that once a complicated circuit has been replaced with one of these equivalent circuits, all details of what might be going on inside the "box" that was replaced are hidden from view.

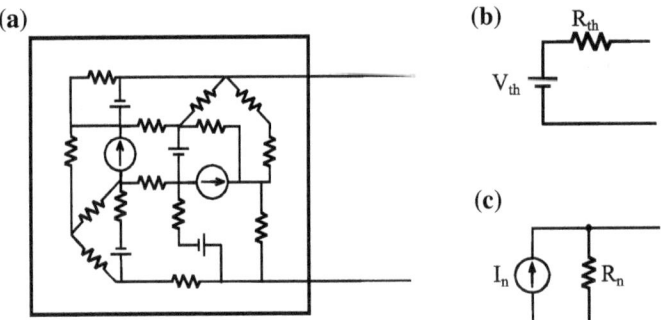

Fig. 2.1 Any configuration of resistors and ideal sources, such as illustrated in (**a**), can be simplified to the (**b**) Thevenin equivalent or (**c**) Norton equivalent circuit if appropriate component values are used

Determining the Thevenin and/or Norton Parameters

The appropriate values needed for the equivalents can often be determined by considering limiting cases. The voltage-current relationship for both the Thevenin and Norton equivalents can be described by a single straight-line plot of current as a function of voltage, such as shown in Fig. 2.2. Thus, the behavior of the entire complicated "box" of components is also described graphically by the same straight line. That line can be determined by finding any two distinct points on that line. On paper, at least, the intercepts with the axes form a convenient pair of points to use.

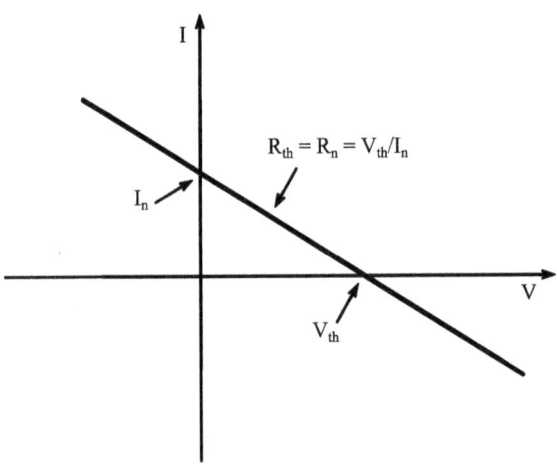

Fig. 2.2 The Thevenin and Norton equivalents, and thus the more complicated circuit that they replaced, will have a straight-line relationship between output current and output voltage. The intercepts between that line and the axes are convenient values to use to determine V_{th} and I_N

Thevenin and Norton Equivalents

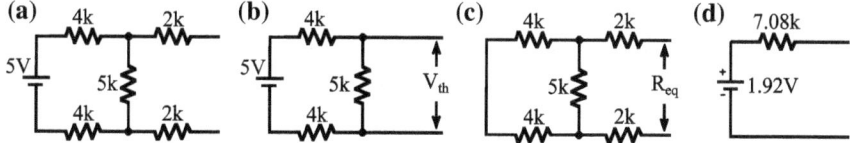

Fig. 2.3 The circuit shown in (**a**) is replaced by its Thevenin equivalent. Intermediate steps are shown in (**b**) and (**c**) and the result in (**d**)

Hence, the values for the equivalent circuits can be found by noting the following:

- The intercept with the *V*-axis is the voltage between the two wires from the box when the box is not connected to anything. If it is not connected, the net current out of the box is zero. This intercept is the "open circuit voltage." Comparing the original circuit to the equivalent circuits, in each case with nothing connected, that open circuit voltage is the same as the Thevenin equivalent voltage, V_{th}.
- The intercept with the *I*-axis is the current through the two leads of the box when the output is short-circuited—that is, an ideal wire is attached between them forcing the voltage to zero.[1] That current is the "short circuit current" and is equal to the Norton equivalent current, I_n.
- The equivalent resistance is given by the negative inverse slope of the line, equal to V_{th}/I_n. Note that it is always the case that for a given circuit, $R_{th} = R_n$.

In general, any two points on the line can be used—they do not need to be the intercepts. The intercepts are often algebraically convenient to find, and from them the slope. The Norton equivalent can always be found from the Thevenin equivalent, and vice versa, using the relations $R_{th} = R_n$ and $I_n R_n = V_{th}$. Sometimes it is easier to find one or the other equivalent and then convert as needed.

An alternate (and often easier) way to determine $R_{th} = R_n$ is to find the equivalent resistance of the components in the box when all the sources within it are adjusted so their outputs are zero. That is, they are "turned off." This method is directly related to the superposition theorem seen in Chap. 1. Remember that in this context a voltage source that is "off" (0 V) behaves the same as a wire, and a current source that is "off" (0 A) behaves the same as an open circuit. While it might not be possible to turn off the sources for a real box, it can always be accomplished on paper as part of circuit analysis.

Example Finding Thevenin (and Norton) Equivalents.

As an example, find the Thevenin equivalent for the portion of a circuit shown in Fig. 2.3a.

---

[1] Note that you can think about these limiting cases on paper, but may not be able to realize them in practice without destroying some circuit components.

The Thevenin equivalent voltage equals the voltage between the two free wires when nothing is attached to them. With nothing attached, there will be no current through the 2k resistors so the change in voltage across those resistors is zero (from Ohm's law). Hence, start by looking at the simpler circuit of Fig. 2.3b, which will have the same V_{th}. As long as nothing is connected to the wires on the right, the two 4k resistors and the 5k resistor are in series—whatever current goes through one must go through the others—and so the voltage across the 5k resistor, which is equal to V_{th}, can be found using the voltage divider equation,

$$V_{th} = \frac{5k}{4k+5k+4k} 5\,\mathrm{V} = \frac{25}{13}\,\mathrm{V} = 1.92\,\mathrm{V}. \tag{2.1}$$

To find R_{th}, return to the original circuit but turn the voltage source off (i.e., replace it with a wire) and find the equivalent resistance as seen by anything connected to this circuit, that is, *as seen from the right* as shown in Fig. 2.3c. Note that the simplification made above to find V_{th} (i.e., ignoring the 2k resistors) obviously does not apply when finding R_{th}. Hence,

$$R_{th} = 2k + 5k \| (4k+4k) + 2k = 5k \| 8k + 4k = (5 \cdot 8)k/(5+8) + 4k = 7.08k. \tag{2.2}$$

Hence, the original piece of the circuit in Fig. 2.3a can be replaced with the simpler equivalent piece shown in Fig. 2.3d. Such a replacement will not change the behavior of the rest of the circuit. The Norton equivalent current will be the current when the outputs of the Thevenin equivalent are short-circuited. In this case, I_n = 1.92 V/7.08 kΩ = 0.271 mA.

How Is This Used for Circuit Reduction?

Suppose the circuit in Fig. 2.4a is to be analyzed to determine the current through the 1 kΩ resistor. Then, using the analysis from the previous example, everything *except* the 1k resistor can be replaced by its Thevenin equivalent, resulting in the circuit of Fig. 2.4b. That circuit is easily solved to give a current I = 1.92 V/(8.08 kΩ) = 0.24 mA.

A common mistake at this point is to say that the current from the original 5 V battery is also 0.24 mA. This, of course, is not correct. The 5 V battery also sends current through the 5k resistor. After the Thevenin equivalent is created, the current through the 5k resistor is no longer visible in the diagram—that current is hidden, as is the original 5 V battery.[2]

---

[2] For this particular circuit, the actual current delivered by the 5 V battery is 0.48 mA, exactly twice what flows through the 1k resistor. Can you see why it is exactly twice for this example?

Thevenin and Norton Equivalents 35

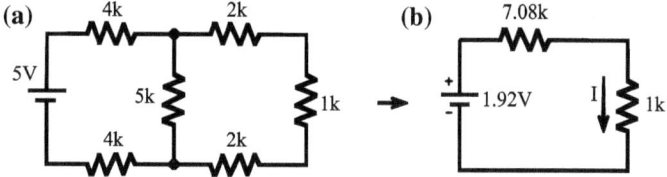

Fig. 2.4 To determine the current through the 1k resistor, the rest of the circuit can be replaced with its Thevenin equivalent

Equivalent for an Infinite Array of Resistors

Thevenin's (or Norton's) equivalent for any number of resistors connected in any way, and in the absence of voltage or current sources, will be a single equivalent resistor, R_{eq}. In addition, there are times when an array of components, considered in the limit that their number becomes infinite, can be a useful model for a real system. Examples include models for certain transmission lines (long cables). With that in mind, consider the infinite array of resistors of Fig. 2.5a.

The trick for solving many problems is to exploit a symmetry of the system. For this infinite array the basic symmetry is translational. If the first pair of resistors is removed (or the first n pairs), the remaining array is still infinite and looks exactly the same as the original. Therefore, the remaining array can be replaced with R_{eq}, which is not yet known, but is presumed to exist. Then the circuit looks like Fig. 2.5b.

Using circuit reduction,

$$R_{eq} = R_1 + R_2 \| R_{eq} = \frac{R_1(R_2 + R_{eq})}{R_2 + R_{eq}} \frac{R_2 R_{eq}}{R_2 + R_{eq}}, \qquad (2.3)$$

and then multiplying to get rid of the denominator on the right, and rearranging results in the quadratic

$$R_{eq}^2 - R_1 R_{eq} - R_1 R_2 = 0. \qquad (2.4)$$

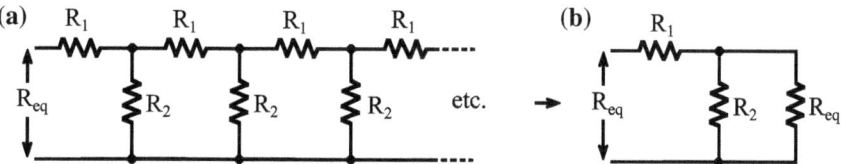

Fig. 2.5 The equivalent resistance of the infinite array of resistors shown in (**a**) can be determined using the translational symmetry (**b**), which is then solved using parallel and series resistor substitutions

Using the quadratic equation, the solutions are

$$R_{eq} = \frac{R_1 \pm \sqrt{R_1^2 + 4R_1R_2}}{2}, \qquad (2.5)$$

where clearly one must choose the plus sign (or else R_{eq} is negative, which means it supplies energy rather than dissipates energy).

One special case is when $R_1 = R_2$ in which case

$$R_{eq} = R_1 \left(\frac{1+\sqrt{5}}{2}\right). \qquad (2.6)$$

The ratio on the right is known as the "Golden Ratio" that is claimed to have some aesthetic appeal in the Arts. Apparently, it can show up in electronics as well.

Solutions for some two-dimensional infinite resistor arrays are also possible. Some such solutions are presented by Atkinson and Van Steenwijk (1999).

The Wheatstone Bridge

The Wheatstone bridge circuit provides a means to match resistance values to very high precision.[3] The bridge circuit when used to make such a measurement is often drawn as shown in Fig. 2.6a. The circuit element "M" is a meter. Using Thevenin's theorem, the meter can be replaced with a resistor (this assumes the meter does not also act as a source). The goal is to predict the reading on the meter. To solve the circuit using Kirchhoff's laws and equations, the setup as shown in Fig. 2.6b could be used. With loop currents, three loops currents are required and three equations will result. It is possible, though a bit cumbersome, to produce values using those three equations. Such a solution can be found in many electronics texts. It is much simpler to understand and produce a result using the theorems that have been presented above.

To solve for the voltage across, and current through, the detector, R_5, it is straightforward to show that the rest of the circuit can be reduced using a Thevenin equivalent (Fig. 2.7) where

$$V_{th} = V_0 \left(\frac{R_3}{R_1+R_3} - \frac{R_4}{R_2+R_4}\right) \text{ and } R_{th} = R_1 \| R_3 + R_2 \| R_4. \qquad (2.7)$$

---

[3]The circuit was originally described by Samuel H. Christie. Charles Wheatstone, a British Scientist and inventor, popularized it, giving proper citation to Christie, however Wheatstone's name became attached to the circuit.

The Wheatstone Bridge

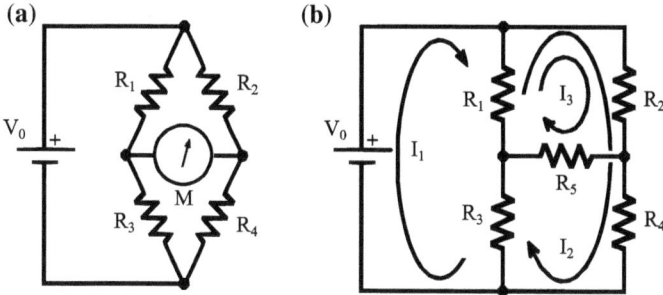

Fig. 2.6 The Wheatstone bridge circuit as it is often drawn is shown in (**a**), where the device "M" is a meter. The circuit as it might be drawn as the first step in an algebraic solution is shown in (**b**). The meter is replaced with an equivalent resistance, R_5

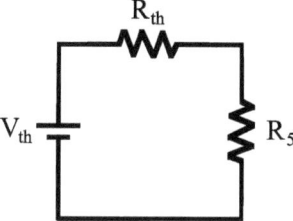

Fig. 2.7 The bridge circuit when all but R_5 (the meter) is replaced with its Thevenin equivalent

Using the voltage divider equation, the voltage across the meter, the output voltage, is

$$V_{out} = V_0 \left(\frac{R_5}{R_{th} + R_5} \right). \tag{2.8}$$

With some effort, (Eq. 2.8) can be written in terms of the original resistors in the form

$$V_{out} = V_0 \frac{R_2 R_3 R_5 - R_1 R_4 R_5}{R_1 R_2 R_3 + R_1 R_2 R_4 + R_1 R_2 R_5 + R_1 R_3 R_4 + R_1 R_4 R_5 + R_2 R_3 R_4 + R_2 R_3 R_5 + R_3 R_4 R_5}, \tag{2.9}$$

where the denominator includes all the possible combinations of three resistors except $R_1 R_3 R_5$ and $R_2 R_4 R_5$.

The Wheatstone bridge is often used for a null measurement where the bridge is balanced so that $V_{out} = 0$. To do that, at least one of the resistors should be variable in a known way. The balance occurs when $R_2 R_3 - R_1 R_4 = 0$, or equivalently,

$$\frac{R_1}{R_3} = \frac{R_2}{R_4}. \tag{2.10}$$

If three of the resistance values are known, then it is clear that the fourth can then be determined. Experimentally two of the values do not need to be known. Suppose R_1 and R_2 are fixed values, R_3 is adjustable and R_4 is the unknown to be determined. The bridge is first balanced by adjusting R_3 until no current flows through the meter. Then $R_4 = R_2 R_3 / R_1$. Now the fixed resistors R_1 and R_2 are swapped and R_3 is once again adjusted to rebalance the bridge. Call the new value of R_3, R'_3. Then $R_4 = R_1 R'_3 / R_2$. Taking the product of these two results gives $R_4^2 = R_3 R'_3$. Values for R_1 and R_2 are not required to find R_4 and any uncertainty in those values will not contribute to the uncertainty of the result—quite the experimental convenience.

By taking appropriate derivatives (e.g., such as dV_{out}/dR_3) evaluated at the balance point, it can be shown that near the balance condition, V_{out} is most sensitive to a *change* in the value of any of the bridge resistors when $R_1 \approx R_2 \approx R_3 \approx R_4$.

Null measurements can be made to be very sensitive since it is comparatively easier to sense the difference between zero and non-zero values than it is to sense the difference between two non-zero, but similar, values.

Wheatstone Bridge "Hieroglyphics"

Sometimes scientists come up with alternate descriptions and/or alternate ways of describing problems in order to help understand them better.[4] Here we show how one *might* do this for the Wheatstone bridge. What follows is by no means a standard notation. Here the bridge solution is shown in a different way to see what is special about the terms that are *not* included.

Representing the original bridge circuit with the stick figure shown in Fig. 2.8, each product term is described by drawing the corresponding line segments. Hence, the term $R_1 R_3 R_4$ would be represented by the symbol ⩘, $R_2 R_3 R_5$ by the symbol ⩗, and so on. Then the output voltage (2.9), is given by the picture equation

$$V = V_0 \frac{⩗ - ⩘}{⩗ + ⩖ + ⩕ + ⩔ + ⩓ + ⩒ + ⩑ + ⩐} , \qquad (2.11)$$

where the denominator, as before, includes of all the combinations of three resistors except two. The two missing combinations correspond to the symbols ← and →. Can you see what is different about these missing "terms" compared to the others?[5] This

---

[4] For example, the so-called "Feynman diagrams" are often used in advanced quantum mechanics. They are a well-established way of representing complex and abstract mathematical expressions using pictures.

[5] If you want to learn more about this, a course on graph theory might be of interest. In fact, a method to solve arbitrary arrays of equal valued resistors using graph theory is presented by Shapiro (1987).

The Wheatstone Bridge

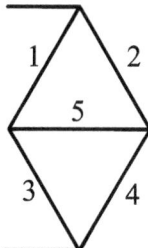

Fig. 2.8 A stick figure as a simplified representation of the original Wheatstone bridge circuit of Fig. 2.6a

pictorial way of representing the equation also makes it easy to remember, should that ever be a concern—there is the difference between the "lightning bolts" in the numerator and all combinations of three except the "satellite dishes" in the denominator.

The Reciprocity Theorem

The reciprocity theorem is a general result from electricity and magnetism. Only a limited statement of that result, which can be applied to circuit analysis, is presented here.

In a single source circuit, a voltage source can be considered to be providing an "excitation." Somewhere else in the circuit, there is a current, a "response," to that excitation. With a linear circuit, the size of the response is always proportional to the size of the excitation.

The reciprocity theorem says that:

> In a (linear) single voltage source circuit, the ratio of the source voltage to the current, measured somewhere else in the circuit, is the same when the positions are interchanged.

Consider a (possibly hypothetical) box with any number of passive linear circuit elements (passive here meaning there are no sources) inside. The elements are connected in any way and there are *four* wires sticking out of the box. The reciprocity theorem says that if an ideal voltage source is connected to one pair of wires, and the current measured with an ideal ammeter connected across the other pair, you will get the same reading on the ammeter if you switch the two devices. This is illustrated in Fig. 2.9.

As a concrete example, the two circuits in Fig. 2.10 have the same 6 V "excitation," and so will have the same reading on the ammeter (recall that an "ideal ammeter" measures the current but otherwise looks like a wire). That is, reciprocity gives $I_2 = I'_1$.

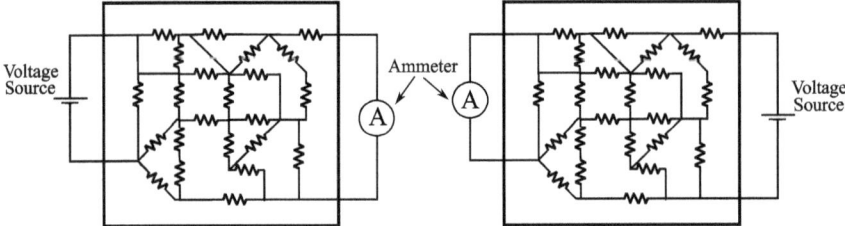

Fig. 2.9 The reciprocity theorem says that the current through the ammeter will be the same for these two circuits

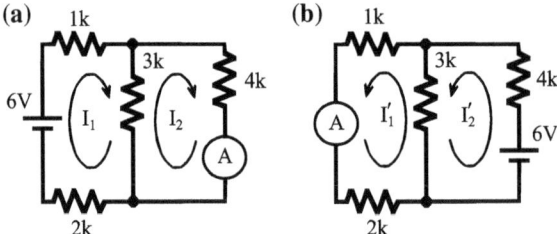

Fig. 2.10 A simple example of the reciprocity theorem. The current through the ammeter will be the same for these two circuits. Note that the current supplied by the battery is not the same

Hence, I_2 can be found by finding I'_1 instead. The only trick here is to keep track of minus signs. Note, however, that the theorem does _not_ say that $I_1 = I'_2$.[6]

The reciprocity theorem may allow the solution of one problem by the substitution of another (hopefully simpler) problem. The reciprocity theorem also works with current sources and ideal voltage measurements swapped in the same way.

Example—R-2R Ladder with Sources

Consider the "ladder" circuit with four voltages sources shown in Fig. 2.11. The goal here is to determine V_{out}. Note that if the current I (shown) can be determined, then $V_{out} = V_4 - (2R)I$.

With considerable effort, this circuit can be solved using circuit reduction or by solving equations generated using Kirchhoff's laws. The solution below is not only simpler, but also easily generalizes to the case of a much larger ladder circuit with N sources. The solution relies on superposition, reciprocity, and parallel and series resistor equivalents.

---

[6] In the example circuits shown, the reading on the ammeter is 0.55 mA. The current out of the battery is 1.27 mA for the circuit on the left, and 1.09 mA for the circuit on the right.

The Reciprocity Theorem

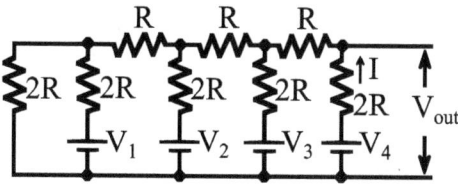

Fig. 2.11 A type of ladder circuit with multiple sources

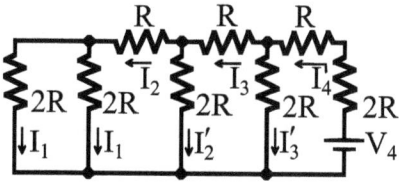

Fig. 2.12 The ladder circuit with only the source V_4, along with reciprocity and superposition, can be used to determine V_{out} for the original circuit of Fig. 2.11

First consider a calculation with the source V_4 only, with all of the other sources "turned off." Since they are all voltage sources that means they are replaced with wires (Fig. 2.12). From the symmetry, it is clear that the current I_2 splits into two paths with $I_1 = I_2/2$.

The two resistors on the left of Fig. 2.12 are in parallel and hence can be replaced with an equivalent, $2R \| 2R = R$, shown in Fig. 2.13a. Once that is done, the two resistors now on the left are in series giving an equivalent resistance of $2R$. Hence, $I'_2 = I_2 = I_3/2$. This process is repeated until there is only one loop left as shown in Fig. 2.13b. So $I_4 = V_4/(4R)$, and going backwards, the currents everywhere from the *single* source V_4 are $I_3 = I'_3 = I_4/2 = V_4/(8R)$, $I_2 = I_3/2 = V_4/(16R)$, and $I_1 = I_2/2 = V_4/(32R)$.

Reciprocity says that the location of the voltage source and the location where the current is measured can be switched and the resulting value (*at that location only*) will be the same. Hence, suppose the source V_4 is moved to another location, such as the position for source 2, as shown in Fig. 2.14. Then the current on the right will be the corresponding current computed above. For the position shown it will equal $I_2 = V_4/(16R)$. From this result it can be concluded that with the original source V_2 alone, there will be a current on the right equal to $V_2/(16R)$ in the

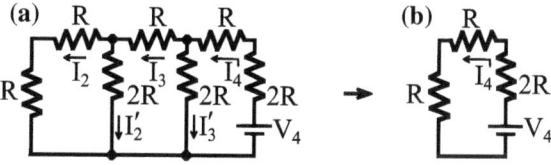

Fig. 2.13 Circuit reduction is used to find the current from the one source V_4 of Fig. 2.12

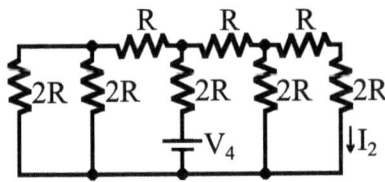

Fig. 2.14 Reciprocity says that if the source is moved, the corresponding current also moves, as shown

direction shown. By solving for the source V_4 alone, the solution for V_2 alone is also obtained. To see how powerful the theorems are, try computing that result using one of the other techniques!

Using the same process for each of the original sources, the total current on the right of the original circuit can be computed using superposition and the solution above for the source V_4 alone. Note that there is a sign difference between the current contributed to I from V_4 and those arising from the other source locations. Adding the currents from each source considered alone, the resulting total current is

$$I = (V_4/4 - V_3/8 - V_2/16 - V_1/32)/R, \qquad (2.12)$$

and hence

$$V_{out} = V_4 - 2IR = V_4/2 + V_3/4 + V_2/8 + V_1/16. \qquad (2.13)$$

Using induction, one can conclude that a similar circuit with N sources will have an output

$$V_{out} = \frac{1}{2}\sum_{k=1}^{N}\frac{V_k}{2^{N-k}} = \frac{1}{2^{N+1}}\sum_{k=1}^{N}V_k 2^k. \qquad (2.14)$$

Circuits similar to this example can be useful for digital to analog (D to A) conversions, considered later. In that case, each of the voltage sources takes on one of two values: for example, 0 V representing a binary digit 0 and a fixed non-zero value, such as 3.3 V or 5 V, representing a binary digit 1.

Delta-Y Conversion

The Delta-Y conversion (sometimes called Delta-Wye, Delta-T, or pi-T) provides another tool that can be useful for circuit reduction. It can also be useful if, for some application, the necessary component values for one configuration become impractical (which can often happen for certain filter and attenuator circuits). The values for the other configuration may be more reasonable and will produce the same result—they are equivalent as far as the rest of the circuit is concerned.

Delta-Y Conversion

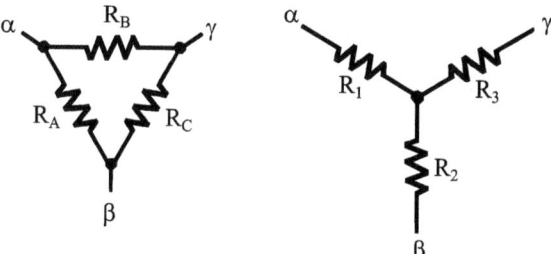

Fig. 2.15 The delta (left) and "Y" (right) configurations for three resistors. The delta-Y transformation can be used to replace one with the other

Referring to Fig. 2.15, the combination on the left is the "delta configuration" (as often drawn, it is an upside-down Δ). The combination on the right is the "Y-configuration". One can be replaced with the other using the Delta-Y conversion. That is,

$$R_A = \frac{R_1 R_2 + R_1 R_3 + R_2 R_3}{R_3} \qquad R_1 = \frac{R_A R_B}{R_A + R_B + R_C}$$
$$R_B = \frac{R_1 R_2 + R_1 R_3 + R_2 R_3}{R_2} \qquad R_2 = \frac{R_A R_C}{R_A + R_B + R_C} \qquad (2.15)$$
$$R_C = \frac{R_1 R_2 + R_1 R_3 + R_2 R_3}{R_1} \qquad R_3 = \frac{R_B R_C}{R_A + R_B + R_C}$$

This transformation should not be memorized, but be aware that it exists and refer to it when needed. The biggest challenge when using this transform is bookkeeping —carefully use labels on the resistors and the connection points, as shown, to be sure everything is correct.

Example Equivalent of Bridge Resistors

As an example, find the current I for the bridge circuit of Fig. 2.16a. Note that there are no parallel or series resistors. Nevertheless, the Delta-Y conversion can be used for circuit reduction.

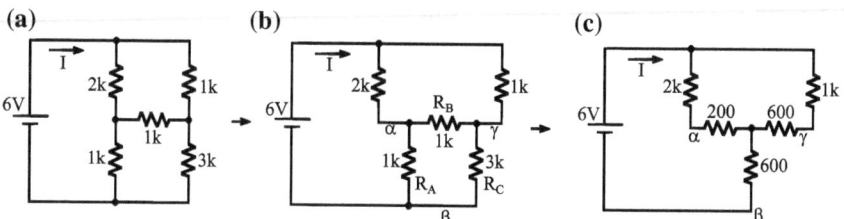

Fig. 2.16 As an example, the delta-Y transformation is used to reduce the bridge circuit in (**a**) to a circuit easily solved using parallel and series resistor equivalents. The steps for the transform are shown in (**b**) and (**c**)

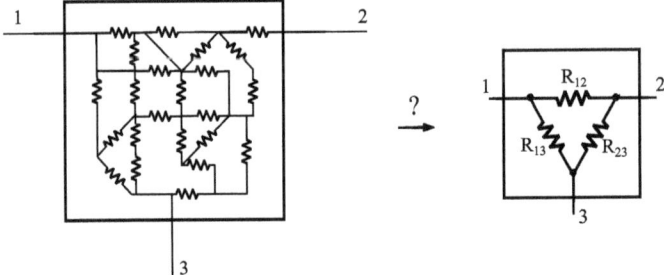

Fig. 2.17 Is it always possible to convert an arbitrary box with three external leads (left) to a simple delta (right)?

First, redraw and carefully label the circuit. There are several ways to do this. One possibility is shown in Fig. 2.16b. Now transform the Delta (R_A, R_B, and R_C) to a Y, as shown in Fig. 2.16c. From there, it is straightforward to use series and parallel resistor equivalents to find

$$R_{eq} = (2k + 200) \| (1k + 600) + 600 = 1526 \, \Omega. \tag{2.16}$$

The current is then $I = 6 \text{ V}/1526 \, \Omega = 3.93$ mA. Try this by replacing the top resistors instead of the bottom resistors. The result should be the same.

Can you now convince yourself that a box containing any number of resistors, connected in any way, and with three leads coming out of it, such as illustrated in Fig. 2.17, can always be replaced by a delta (or Y) circuit if appropriate values are found? If so, how might you go about finding those values? If not, can you find an example where it cannot be done?

The Kelvin Bridge

The Kelvin bridge is used to measure a very small resistance, where the resistance of the connections and wires might be large compared to the resistance to be measured.[7] This occurs, for example, when the resistance of a short wire or a piece of metal is to be measured. Of importance for these determinations is the use of a four-wire resistance measurement. The circuit analysis strategy here is to use circuit

---

[7]The Kelvin bridge is named for the physicist known as Lord Kelvin, also known as William Thomson (1824–1907). Sometimes you will see this same bridge referred to as the Thomson bridge. Thomson was knighted by Queen Victoria for his various contributions; most notably work on the transatlantic telegraph. Sir William Thomson adopted the title Baron Kelvin (of Largs) partly for political reasons, using the name of the Kelvin river which flowed near his laboratory. He was elevated to the House of Lords where people then began referring to him as Lord Kelvin. At least that's the way the story goes.

The Kelvin Bridge

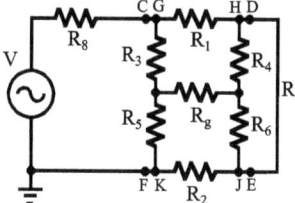

Fig. 2.18 The Kelvin bridge circuit, which can be used to determine smaller resistance values precisely

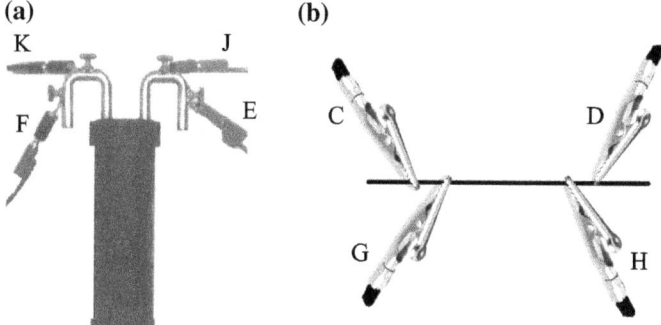

Fig. 2.19 An illustration showing how the four separate contacts might be made (**a**) to a standard resistor and (**b**) a wire to be measured. The letters correspond to those in Fig. 2.18

reduction, in particular the Δ-Y conversion, to transform the Kelvin bridge to a Wheatstone bridge where the solution is already known (see above).

A schematic for the Kelvin bridge is shown in Fig. 2.18, where

- R_1 = the sample to be measured
- R_2 = a small valued standard resistor
- R_7 = resistance of a wire in series with contact resistances at D and E (all unknown, but possibly of comparable size or even much larger than R_1)
- R_g = meter or detector
- R_3, R_4, R_5, R_6, and R_8 = larger valued resistors.

Four *separate* contacts are made to R_1 and R_2, such as is illustrated in Fig. 2.19, and there will be some small, but unknown resistance at each of those contacts. Aside from those included in R_7, the contact resistance at the connections are small compared to the series fixed resistance attached to it (e.g., the resistance at contacts C and F are small compared to R_8, at G it is small compared to R_3, etc.).

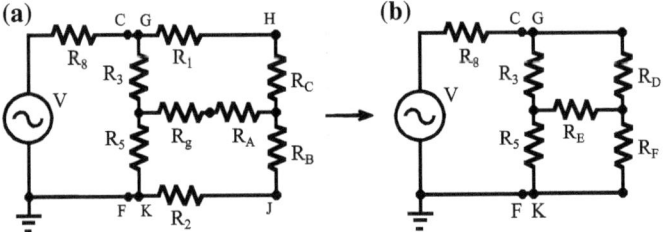

Fig. 2.20 To solve the Kelvin bridge, the delta-Y transformation is used to convert it to a Wheatstone bridge, where the solution is already known

Now use the Δ-Y transformation to convert the "Δ" formed by R_4, R_6 and R_7 to a "Y". Defining a value D, $D = R_4 + R_6 + R_7$, for convenience, the transformed values are

$$R_A = R_4 R_6 / D$$
$$R_B = R_6 R_7 / D \qquad (2.17)$$
$$R_C = R_4 R_7 / D.$$

The circuit is then converted as shown in Fig. 2.20, where $R_D = R_1 + R_C$, $R_E = R_g + R_A$, $R_F = R_2 + R_B$, and the current through R_E is the same as the current through R_g.

The null condition can then be found as for the Wheatstone bridge (2.10). Rewriting the equation for the null condition for the equivalent Wheatstone bridge in terms of the original resisters of the Kelvin bridge, the null condition is

$$(R_4 + R_6 + R_7)(R_2 R_3 - R_1 R_5) + R_7 (R_3 R_6 - R_4 R_5) = 0. \qquad (2.18)$$

The null can be found without knowledge of the unknown small value R_7 by simultaneously satisfying both $(R_3 R_6 - R_4 R_5) = 0$ and $(R_2 R_3 - R_1 R_5) = 0$. These two conditions are simultaneously satisfied when

$$\frac{R_1}{R_2} = \frac{R_3}{R_5} = \frac{R_4}{R_6}. \qquad (2.19)$$

Note that such a four-lead resistance measurement technique is used for all measurements of low value resistances, whether or not the Kelvin bridge is used. Two of the connections supply the current through the sample and the other two to measure a voltage difference. When using comparative words such as "low," it is always important to ask, "low compared to what?" Here "low value" means comparable to or smaller than the resistance of the wires and electrical contacts. For some measurements, such as to determine soil conductivity, a low value may not be particularly small in absolute terms—it is not as easy to connect wires to soil as it is to connect wires to other wires.

Additional Application—Resistivity of Lamellae

A thin sheet, foil, or layer of material with uniform thickness, a "lamella," can be connected to the Kelvin bridge using four contacts such as illustrated in Fig. 2.21. The current introduced through two of the contacts will create a voltage that can be measured between the other two. The ratio of the measured voltage to the supplied current will yield a value with units of ohms, and can be treated as a resistance. Of course, the value for that resistance will depend on where the contacts are made. L. J. van der Pauw showed that two such measurements could be combined to find the resistivity of the material.[8] The resistivity is a material property, like density, which is independent of size and shape.

In order to use the van der Pauw technique, the lamella should have uniform electrical properties, be of uniform thickness, and the sample should be "simply connected," meaning there are no holes. A resistance measurement is made with the current connections (C and D) and the voltage connections (G and H) on the edge of the sample and adjacent to each other (see the labels in Fig. 2.21). Then all connections to the Kelvin bridge are rotated by one place and another measurement is made (see the labels in parentheses in Fig. 2.21). Call the result of the first measurement R_A, and the second R_B. Then the resistivity is given by

$$\rho = \frac{\pi d}{\ln 2} \frac{R_A + R_B}{2} f(r), \qquad (2.20)$$

where $f(r)$ is the solution to the wonderfully complicated equation

$$\frac{r-1}{r+1} = \frac{f(r)}{\ln 2} \cosh^{-1}\left[\frac{\exp(\ln 2/f(r))}{2}\right], \qquad (2.21)$$

for which no analytic solution is known. Using $r = R_A/R_B$ or R_B/R_A, whichever is larger, only values for $r \geq 1$ need to be known. It is not difficult to make a plot of $f(r)$, such as Fig. 2.22, or to create a table of values. A number of approximate relations have also been developed that can be used to find $f(r)$ to good accuracy.

Notice that the actual shape of the sample is not important for the van der Pauw measurement as long as the connections are made at the edge and the sample is simply connected. A symmetric sample with symmetric connections, such as a square foil with contacts at the corners, will have $R_A = R_B$ and hence $r = 1$. Noting that $f(1) = 1$, the measured resistance will be

$$R_A = R_B = \frac{2\rho \ln 2}{\pi d}, \qquad (2.22)$$

---

[8]This was done using conformal mappings in the complex plane—a topic from Complex Analysis. See van der Pauw (1958/59).

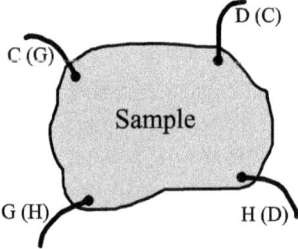

Fig. 2.21 To determine the resistivity of lamella, four contacts are used around the edges. The letters correspond to the connection points in the Kelvin bridge (Fig. 2.18). A second measurement is made using the connection points identified in parentheses

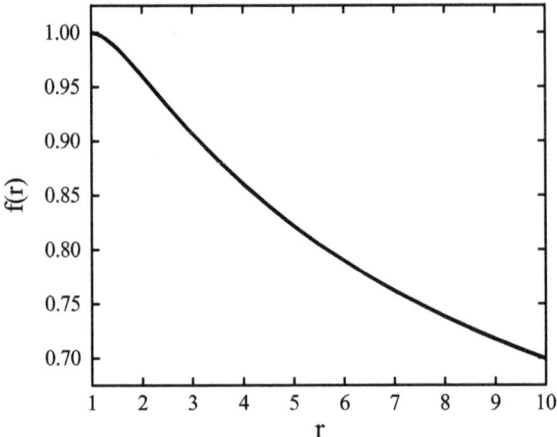

Fig. 2.22 A graphical representation of van der Pauw's function, $f(r)$

that depends on the thickness and resistivity of the foil, but is independent of the size of the square. Such a result can thus be expressed using units such as "ohms per square."[9] For layered materials of known quality, a measurement of ohms per square can be used to determine the thickness of the layer. For a material of known thickness, such a measurement can be used to characterize the quality of the material. The latter measurement can be used, for example, to characterize the quality of semiconductor wafers used during the production of integrated circuits.

---

[9]"Ohm's per square" usually refers specifically to measurements of the resistance from one side of a square sheet to the opposite side, ignoring any contact resistance, and is also referred to as the sheet resistance. That value can be derived from the Van der Pauw method used at the corners by dividing that result by a factor of $2 \ln 2/\pi$. For more information about 4-probe resistance measurements using alternate geometries, see Miccoli et al. (2015).

Problems

1. Find the Thevenin equivalent for the circuit of Fig. 2.P1, including all of the components *except* the 2k resistor on the right.

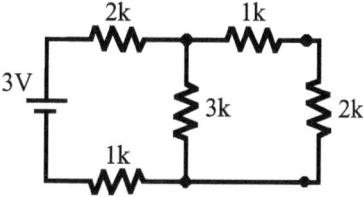

Fig. 2.P1 Problem 1

2. (a) What is the Thevenin equivalent for the piece of a circuit shown in Fig. 2.P2a? (b) What values should be chosen for resistors R_1, R_2, and R_3 so that the piece of a circuit shown in Fig. 2.P2b is equivalent to the circuit of part a?

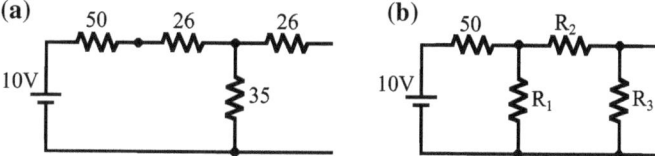

Fig. 2.P2 Problem 2

3. A less than ideal voltage source is connected to a voltmeter (that can be taken to be ideal) and the output voltage is measured to be 9.0 V. A variable resistor is then placed across the line, in parallel with the voltmeter. When the resistor is adjusted to 600 Ω the reading on the voltmeter is exactly half of the first measurement. If the voltage source is to be modeled using a Thevenin equivalent, what are appropriate values for V_{th} and R_{th}?

4. (a) What is the equivalent resistance for the infinite array of equal valued resistors shown in Fig. 2.P4a? (b) Compare the result from (a) to that of the finite array of resistors in Fig. 2.P4b.

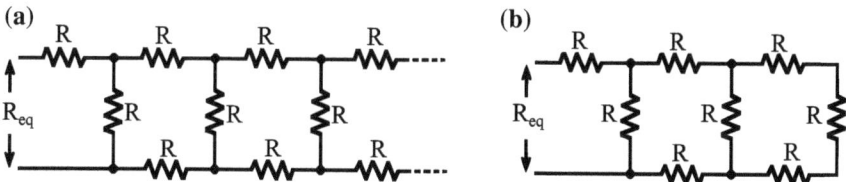

Fig. 2.P4 Problem 4

5. Demonstrate reciprocity with a current source and voltage measurement by computing the voltage, V, for the two circuits in Fig. 2.P5.

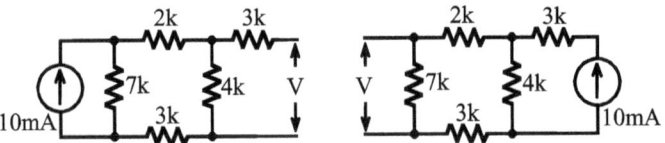

Fig. 2.P5 Problem 5

6. A clever student decided to use reciprocity to find the current I_5 in the bridge circuit shown in Fig. 2.P6a. First, the location of the source and the location where the current is measured are swapped, as in Fig. 2.P6b. Then the circuit is redrawn as shown in Fig. 2.P6c. The student correctly solves for the current $I = (72/85)$ mA and concludes that must also equal the current I_5 from the original bridge circuit. However, that answer is exactly a factor of 12 too large. What did the student do incorrectly and where does the factor of 12 come from?

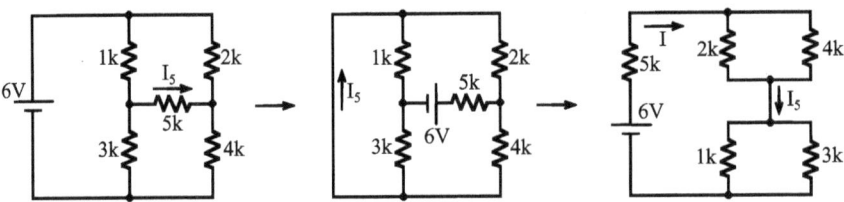

Fig. 2.P6 Problem 6

7. What is the Thevenin equivalent for the circuit of Fig. 2.P7?

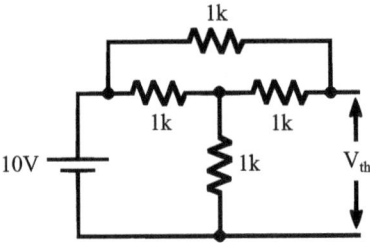

Fig. 2.P7 Problem 7

Problems

8. (Challenge Problem) Use symmetry and circuit theorems to determine the effective resistance for the following: (a) The circuit of Fig. 2.P8a, a square array with diagonal connections, measured across adjacent corners, (b) the circuit of Fig. 2.P8b, an infinite array of the squares from A. (Hint: if you end up using reams of paper for either of these problems, you are not thinking like a scientist).

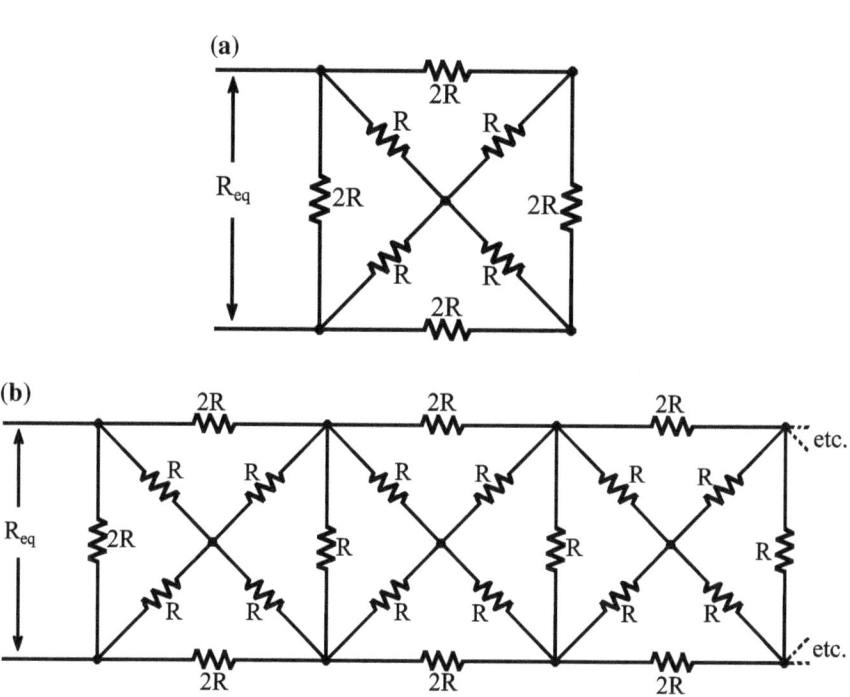

Fig. 2.P8 (a) Problem 8a, (b) Problem 8b

References

D. Atkinson, F.J. Van Steenwijk, Infinite resistive lattices. Am. J. Phys. **67**, 486–492 (1999)
I. Miccoli, et al., The 100th anniversary of the four-point probe technique: the role of probe geometries in isotropic and anisotropic systems. J. Phys.: Condens. Matter **27**, 223201 (2015)
L.W. Shapiro, An electrical lemma. Math. Mag. **60**(1), 36–38 (1987)
L.J. van der Pauw, *Philips Research Reports*, vol 13, No. 1, pp. 1–9. Philips Tech. Rev. **20**(8), 220–224 (1958/59)

Chapter 3
Complex Impedances

With the basic theorems in place for the simplest devices—ideal power sources, resistors and wires—some additional linear devices are now considered. In particular, capacitors, inductors, and time-dependent sources are introduced. In some cases, it will prove most convenient to describe voltage-current relationships using imaginary numbers. First, what constitutes a linear device is defined. Then it is shown how imaginary values might arise, why such values may be useful, and how they are used. Since, by definition, nothing measured is ever imaginary, it is important to understand how to interpret these imaginary values for real world results, as well as to know when they will not work.

What Is a Linear Device?

A device is considered linear if its current-voltage characteristic is linear in the same way that linear is defined for mathematical functions (and operators). This does not necessarily mean that the current-voltage characteristic can be plotted as a simple straight line.

Assume the current through a (two-lead) device can be described as a function of the voltage across the device. That is, it is the case that $I = I(V)$. If the device is linear then the function $I(V)$ always obeys

$$I(aV_1 + bV_2) = aI(V_1) + bI(V_2), \qquad (3.1)$$

where a and b are any (unitless) real numbers and V_1 and V_2 are any two voltages.

For example, it is clearly true that a resistor is linear. For a resistor, $I(V) = V/R$, and

$$I(aV_1 + bV_2) = \frac{1}{R}(aV_1 + bV_2) = a\frac{V_1}{R} + b\frac{V_2}{R} = aI(V_1) + bI(V_2). \quad (3.2)$$

Some other devices are also linear. Recall from electricity and magnetism that for a capacitor, C, with charge Q,

$$I(V) = \frac{dQ}{dt} = \frac{d(CV)}{dt} = C\frac{dV}{dt}, \quad (3.3)$$

so

$$I(aV_1 + bV_2) = C\frac{d}{dt}(aV_1 + bV_2) = aC\frac{dV_1}{dt} + bC\frac{dV_2}{dt} = aI(V_1) + bI(V_2), \quad (3.4)$$

and therefore the capacitor is a linear device. Similarly, an inductor is a linear device. Suppose, however, there were a device where

$$I(V) = D \exp(\alpha V), \quad (3.5)$$

where D and α are constants with appropriate units. Then

$$\begin{aligned} I(aV_1 + bV_2) &= D \exp(\alpha(aV_1 + bV_2)) = D \exp(\alpha a V_1) \exp(\alpha b V_2) \\ &= D(\exp(\alpha V_1))^a (\exp(\alpha V_2))^b \\ &= (I(V_1))^a (I(V_2))^b / D \neq aI(V_1) + bI(V_2), \end{aligned} \quad (3.6)$$

and this is *not* a linear device. Note that if $I(V)$ is a linear function, then $V(I)$ will also be a linear function (if the function exists) and vice versa.

Some Vocabulary

For success in any endeavor it is most important to know what you are talking about. Thus, before proceeding, it is useful to take a little space to define a number of terms and abbreviations that will occur throughout electronics and the rest of this text. The definitions here represent common use of these terms within the context of electronics. These are:

- **D.C.**: An abbreviation for "direct current," also written as DC, d.c., dc, etc. DC is used to indicate voltages and currents that can be considered constant in time.
- **VDC**: An abbreviation for "Volts DC," that is, a constant voltage.
- **A.C.**: An abbreviation for "alternating current," also written AC, a.c., ac, etc. A.C. is used to indicate voltages and currents that vary in time. Often it is implied that the time variation is periodic, with a time average of zero, and also that it is sinusoidal.

Some Vocabulary

- **VAC**: An abbreviation for "Volts AC," that is, a time dependent (usually sinusoidal) voltage.
- **Frequency, f**: Used for periodic signals, the frequency, in hertz (Hz), is the number of signal repeats, that is "cycles," per second. In older literature, the units may be written "cycles per second" or cps, rather than hertz.
- **Angular frequency, ω**: Principally used for sinusoidal periodic signals, angular frequency is the frequency expressed in radians per second, where there are 2π radians for each full repeat cycle. Hence, $\omega = 2\pi f$.
- **Period**: the repeat time for a periodic signal, often symbolized using "T." Related to the **frequency** in hertz, f, by $T = 1/f$, and the angular frequency, ω, by $T = 2\pi/\omega$.
- **RMS**: An abbreviation for "Root Mean Square." Also written "rms." The root mean square is the square root of the average of a series of values squared (or a function squared). This is a useful way to characterize AC signals since, for example, power depends on the voltage or current squared. For a sinusoidal signal, the RMS voltage is the amplitude of the sine wave divided by the square root of 2.
- **VRMS**: An abbreviation for "Volts RMS," also written as VRMS, Vrms, etc. This is the usual unit of Volts for a situation where the RMS value is being reported. That is, 115 VRMS would mean that one has an a.c. signal where the time average gives a root mean square value of 115 V. It is used (sometimes) to make sure that the reader knows the value is an RMS value. In electronics, "RMS" is often left off, but is implied, when a.c. values are reported. For example, 115 VAC usually implies that 115 is the RMS value.
- **Passive Device**: A device that does not require a separate power supply for normal operation.[1]

Passive Linear Circuit Elements with Two Leads

The simple ideal passive linear devices in electronics are the resistor, capacitor and inductor. Real devices are usually modeled using one or more of these three. The rules (i.e., the current-voltage characteristics) for these devices are summarized in Table 3.1. For all of these devices, the sign convention is such that a positive voltage corresponds to a decrease in the electric potential (or EMF) in the direction of positive current, illustrated in Fig. 3.1. That is, positive values are defined such

---

[1]There are several "definitions" of passive vs. active devices. None are particularly rigorous and none actually work in all cases. However, all agree that the resistor, capacitor, and inductor are passive.

Table 3.1 Passive linear devices

Schematic symbol	Name	V(I)	I(V)
—W—	Resistor	$V = IR$	$I = V/R$
		R is a constant and will be in ohms (Ω)	
—∣⊢—	Capacitor	$V = V_0 + \frac{1}{C}\int_0^t I(t)dt$	$I = C\frac{dV}{dt}$
		C is a constant and will be in farads (F)	
—⌒⌒⌒—	Inductor	$V = L\frac{dI}{dt}$	$I = I_0 + \frac{1}{L}\int_0^t V(t)dt$
		L is a constant and will be in henries (H)	

Fig. 3.1 For two-lead passive devices, a positive current is taken to be in the direction from higher to lower voltage

that there is a "voltage drop" in the direction of the current. The voltage-current relationships in Table 3.1 are for ideal devices and are approximations for the behavior of real devices.

Idealized Sources

Idealized power sources are linear devices with a very simple relationship between voltage and current. The ideal battery is one such device and was already presented. Real power sources are often modeled using one or more ideal sources combined with one or more passive linear devices. A summary of some ideal sources is in Table 3.2.

There is considerable variation for the specific symbols used for sources. For example, one might see the symbol shown for "DC Voltage Source" also used for AC voltage sources. The reader is often expected to be able to determine which it is from context.

Table 3.2 A variety of ideal linear sources

Schematic	Name	V(I) or I(V)
	Battery	$V = V_0$ A constant value, no matter what current
	DC voltage source	$V = V_0$ A constant value, no matter what current
	AC voltage source	$V = V_0 \cos(\omega t + \varphi)$ Where the voltage specified, V, is often the rms voltage, V_{rms}, $V_0 = V_{rms}\sqrt{2}$
	DC current source	$I = I_0$ A constant value, no matter what voltage
	AC current source	$I = I_0 \cos(\omega t + \varphi)$ Where the current specified, I, is often the rms current, I_{rms} $I_o = I_{rms}\sqrt{2}$

RC and L/R Time Constants

(This first section should be a review of material from introductory electricity and magnetism)

For both the capacitor and inductor, something interesting will happen only if the voltage and/or current is changing in time. If neither change with time, the derivatives (see Table 3.1) are zero and the current-voltage relationship is trivial. In this chapter, two simple cases involving a time dependence are considered: first, a time dependence due to a switch being opened or closed, and then the time dependence in the presence of sinusoidal signals. The more general case can be quite complicated and will be considered briefly in Chap. 5.

Consider the two circuits in Fig. 3.2 where there is a switch that will close (make contact) at a certain time, t_0. Assume the switch closes instantaneously. Before the switch is closed, it must be that $I = 0$ since the circuit is not complete—it is an open circuit. Also assume that the switch has been open a "long time" so that the voltage across the capacitor and the inductor are both zero. Long is a relative word and what constitutes a long time is an important question to ask, and the answer to which, for this case, should be evident later. The more immediate question to be answered is "what happens after the switch is closed?"

After the switch is closed, apply Kirchhoff's voltage law (KVL) along with the device rules, around the loops to get

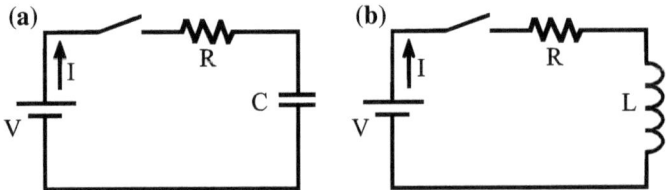

Fig. 3.2 Simple circuits to illustrate (**a**) RC and (**b**) L/R time constants

$$V - IR - V_C = 0; \quad V - IR - V_L = 0$$
$$-R\frac{dI}{dt} - \frac{1}{C}I = 0; \quad V - IR - L\frac{dI}{dt} = 0, \quad (3.7)$$

where, for the capacitor circuit, a derivative was taken on both sides of the equation. Both of these equations can be written in the form

$$\frac{dI}{dt} = -\frac{1}{\tau}(I - I_\infty). \quad (3.8)$$

The constant, τ (a lower-case Greek tau), will have units of seconds and is known as the "time constant." The constant I_∞ is the current after a very long time—a "long time" after the switch was closed things stop changing and hence the derivatives, that is, the rates of change, become zero. It should be evident that for the capacitor circuit shown, $\tau = RC$ and $I_\infty = 0$, and for the inductor circuit, $\tau = L/R$ and $I_\infty = V/R$.

A general solution to the differential equation (3.8) is given by

$$I(t) = I_1 e^{-(t-t_0)/\tau} + I_2, \quad (3.9)$$

where I_1 and I_2 are constants. The constant t_0 is the starting time, which is often taken to define $t = 0$. In fact, the solutions to all "time constant problems" will look just like (3.9). It is possible to skip directly to this step for virtually all future time constant problems. To determine the values of the constants, make the solution match the problem at certain times, such as at $t = t_0$ and $t \to \infty$ (though other times can be used).

Inductors will tend to keep the *current* through them constant. Hence for the inductor circuit above the initial current must be zero since it was zero before the switch was closed. To get the solution to do that, it must be that $I_2 = -I_1$. After a long time, an inductor simply looks like a wire so $I_2 = I_\infty = V/R$. So, for the inductor circuit at a time t after the switch is closed, $I(t) = V(1 - e^{-t/\tau})/R$.

Capacitors will tend to keep the *voltage* across them constant. Hence the voltage across the capacitor in the corresponding circuit above will start off at zero, which is what it was before the switch was closed. Hence initially the capacitor looks like a wire (or more generally a battery, and in this case a 0 V battery), and so the initial current is $(I_1 + I_2) = V/R$. After a long time, the current through the capacitor is zero (the capacitor becomes "charged up" and no further charge is entering or

RC and L/R Time Constants

leaving). Hence, $I_2 = 0$ and one gets that for the capacitor circuit shown with the switch closed at $t = 0$: $I(t) = V(e^{-t/\tau})/R$.

For all of these time constant problems, look at the initial state of the system and the final (long time) state of the system. Figure out the appropriate time constant (or effective time constant) and then use the general solution above to generate a solution that changes exponentially between the initial and final values. Examining these two examples, it should be clear that the time constant, τ, sets the time scale. What constitutes a long and a short time corresponds to those times that are long and short compared to τ.

RC Time Constant Example

In the circuit shown in Fig. 3.3, the switch has been open a long time and is then closed at $t = 0$. What is the voltage across the capacitor and the current, I, in the circuit after the switch is closed?

Initial Values First, look at the circuit just before the switch is closed. The battery is disconnected and there is no source of power. The capacitor will have discharged (through the 3k resistor) and so the voltages across each of the two resistors, and across the capacitor, will be zero. Just after the switch is closed, the capacitor will try to keep the voltage constant so the initial voltage across the capacitor, and hence also across the 3k resistor, will stay zero (its starting value) immediately after the switch is closed. So at $t = 0$ the entire 9 V is across the 2k resistor, and the initial current is $I = 9\ \text{V}/2\text{k} = 4.5$ mA.

Final Values Now look at the circuit a long time after the switch is closed. After a long time, the capacitor will be fully charged to its "final" value, the current through the capacitor will be zero, and the current travels through the two resistors in series. The current after a long time is $I = 9\ \text{V}/(2 + 3)\text{k} = 1.8$ mA. This means that the voltage across the 3k resistor is 1.8 mA $\times$ 3 kΩ = 5.4 V, and that is, of course, equal to the (final) voltage across the capacitor.

Time Constant For this problem, the time constant when the switch is closed is required. Any fixed power sources will have no effect on an RC time constant. Fixed voltage sources can be replaced with wires (a "0 V source") and fixed current sources with open circuits (a "0 current source"—see "superposition" in Chap. 1) and this will not change the result for this part of the calculation. Now ask, what is

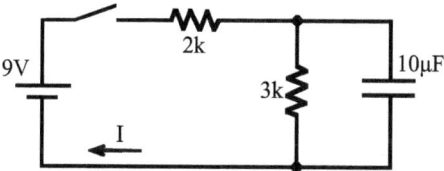

Fig. 3.3 Circuit used for an RC time constant example

the equivalent resistance *as seen by the capacitor?* In this case the capacitor "sees" the two resistors in parallel giving an effective resistance of (6/5) kΩ, so the RC time constant is τ = (6/5) kΩ × 10 μF = 12 ms. Make sure to keep track of the powers of ten carefully and correctly.

Now put it all together The current for this problem will look like $I(t) = I_1 e^{-(t-t_0)/\tau} + I_2$ with τ = 12 ms, t_0 = 0, $I(0)$ = 4.5 mA and $I(t \to \infty)$ = 1.8 mA. So it must be that I_2 = 1.8 mA and I_1 = (4.5 − 1.8) mA = 2.7 mA.

For time constant problems, the voltage is found using a similar approach. The voltage across the capacitor will look like $V_C(t) = V_1 e^{-t/\tau} + V_2$, and for this problem $V_C(0)$ = 0 and $V_C(t \to \infty)$ = 5.4 V. Hence, V_2 = 5.4 V and so V_1 = −5.4 V.

Solutions: $I(t) = \left(2.7e^{-t/12\text{ms}} + 1.8\right)$ mA; $V_C(t) = 5.4\left(1 - e^{-t/12\text{ms}}\right)$ V.

Now try to work the problem with the same circuit, but where the switch was *closed* for a long time and then *opened* at t = 0.

Capacitors and Inductors with Sinusoidal Sources

Recall that Fourier's theorem and superposition say that any time-dependent source can be represented as a sum of sinusoidal sources. The sum of the individual solutions is the solution to the sum. Knowing what happens to each single sinusoidal contribution by itself is sufficient knowledge to be able to treat all time-dependent sources, so a treatment based solely on sinusoidal time dependence is surely warranted. To learn more about Fourier's theorem, see the Appendix.

Consider a circuit consisting of a capacitor and an a.c. voltage source (Fig. 3.4a), where the current through the capacitor is easily computed to be

$$I(V) = C\frac{dV}{dt} = \omega C V_0 \cos(\omega t) = \omega C V_0 \sin(\omega t + 90°). \quad (3.10)$$

The current looks like a constant (ωC) multiplied by the original voltage, but with an additional "phase shift" of 90°. Such a phase shift is equivalent to a fixed shift in time. In this case, the voltage seems to be behind, or "lagging," the current—when t = 0, the argument inside the sine function is zero for the voltage, but is already larger than zero for the current.

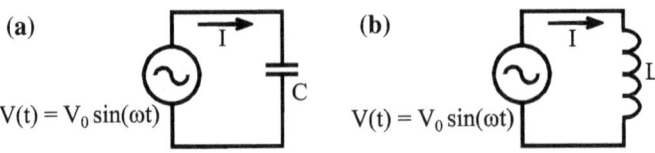

Fig. 3.4 The simplest circuits with a sinusoidal a.c. source for (**a**) a capacitor and (**b**) an inductor

Capacitors and Inductors with Sinusoidal Sources

Similarly, for an inductor connected directly to an a.c. voltage source (Fig. 3.4b), the current is found to be

$$I(V) = \frac{-1}{\omega L}\cos(\omega t) = \frac{1}{\omega L}\sin(\omega t - 90°), \qquad (3.11)$$

where the voltage is ahead of, or "leads," the current by 90°—when the argument inside the sine for the current is zero, it must be the case that $\omega t > 0$, so the argument inside the sine used for the voltage is already positive.

Hence, the amplitude of a sinusoidal voltage across (V) and the corresponding sinusoidal current (I) through the capacitor (C) and inductor (L), can be symbolically described as

$$V_C = \frac{1}{\omega C} I_C \angle -90°$$
$$V_L = \omega L I_L \angle 90°, \qquad (3.12)$$

where the sinusoidal function is not shown, but implied, and "$\angle$" means to shift the sinusoid by the phase angle that follows. With this notation, Ohm's law for resistors might be written $V_R = RI_R \angle 0°$, however since a shift by 0° is to do nothing, "$\angle 0°$" is usually omitted when it occurs.

Phase angles are always relative to some reference; they are phase *differences*. A phase angle is used to characterize any shifts in time of sinusoidal signals relative to one another. A signal that is delayed in time ("lagging") compared to a reference has a negative phase angle compared to the reference. Since there are 360° (or 2π radians) in one complete period, a phase angle of 90° (or $\pi/2$ radians) corresponds to a shift of 1/4th of a period, or a time equal to T/4. The general case is illustrated in Fig. 3.5.

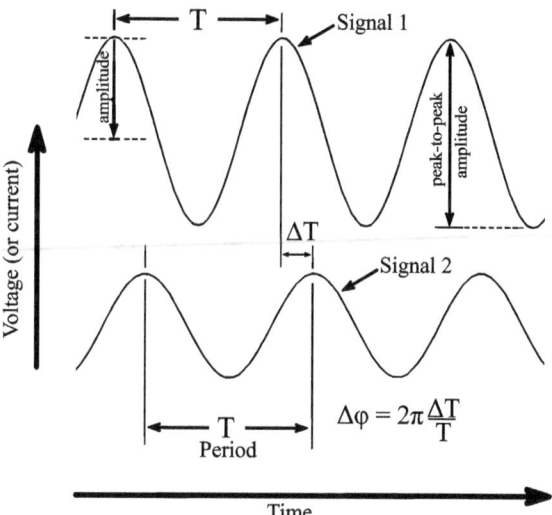

Fig. 3.5 Showing the offset in time, ΔT, associated with a phase shift $\Delta \varphi$. The two signals have been offset in the vertical direction for clarity

Now the question remains as to what to do with these relationships for circuits that are more complicated. That is, how are these results used in circuit analysis and how are the calculations performed? To answer this, first consider the following approach to solve a simple circuit.

Superposition and Complex Impedances

Complex numbers are those that involve a sum of real and imaginary numbers. Imaginary numbers are those that involve the square root of minus one. If needed, a review of complex numbers and complex number arithmetic is included in the Appendix.

It will be most convenient to consider complex numbers and complex arithmetic for a.c. circuits—in particular for circuits where the source(s) are sinusoidal. There is no such thing as an imaginary voltage or an imaginary current, however, so this approach should be greeted with some skepticism. The purpose here is to show why this works and to illustrate the appropriate interpretation for the complex numbers. Throughout this text the convention $i = \sqrt{-1}$ is used. In the literature, particularly for electronics, sometimes "j" may be used in place of "i".

Why and how complex values might be useful is illustrated using an example. Consider the circuit in Fig. 3.6a which has a sinusoidal voltage source of frequency $f = 10$ Hz (so $\omega = 2\pi f = 62.8$ rad/s). Using the mathematical identity,

$$\cos(\omega t) = \frac{1}{2}\left(e^{i\omega t} + e^{-i\omega t}\right), \qquad (3.13)$$

one can, *on paper*, draw the circuit with two sources, each of which produces a complex voltage (though the sum is always a real value). See Fig. 3.6b. Using the superposition principle, the total current in the circuit can be found by considering each of these voltage sources separately, with the other "off," and then summing the results, even though each source, by itself, can never actually exist.

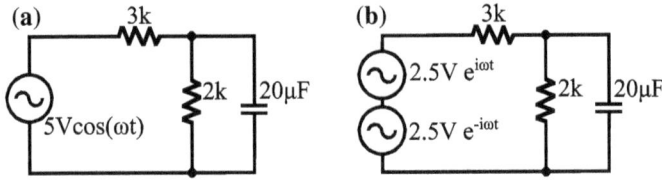

Fig. 3.6 The circuit shown in (**a**) is solved using mathematical identity to split the source into two sources, as shown in (**b**)

Superposition and Complex Impedances

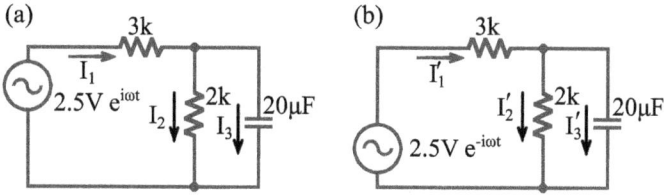

Fig. 3.7 The circuit of Fig. 3.6a is solved considering one source at a time. The two solutions will be identical except for the replacement of "i" with "$-i$"

Start by looking at the circuit of Fig. 3.7a, where three branch currents have been labeled. Writing down equations obtained by applying Kirchhoff's laws to this circuit,

$$I_1 = I_2 + I_3$$
$$2.5e^{i\omega t} - 3I_1 - 2I_2 = 0$$
$$2.5e^{i\omega t} - 3I_1 - \frac{1}{20\,\mu F} \int_{t \to -\infty}^{t} I_3 dt = 0, \qquad (3.14)$$

where voltages are in volts, and currents in mA. The integral yields the total charge that has accumulated on the capacitor. Now if the sinusoidal source has been on for a long time, all the currents will also be sinusoidal with *the same frequency*.[2] So write $I_k(t) = I_{k0} e^{i\omega t}$, $k = 1, 2, 3$, and these equations become

$$I_{10} = I_{20} + I_{30}$$
$$2.5 - 3I_{10} - 2I_{20} = 0$$
$$2.5 - 3I_{10} - \frac{1}{(i\omega)20\,\mu F} I_{30} = 0, \qquad (3.15)$$

where the capacitor has been assumed to be initially (i.e., at $t \to -\infty$) uncharged. Note that there is no longer any time dependence in these equations. All the time dependence has been factored out. For this example, $\omega = 2\pi f = 62.8$ s^{-1}, so the last equation becomes

$$2.5 - 3I_{10} + 0.8i\,I_{30} = 0. \qquad (3.16)$$

---

[2]There is no other frequency that it could be.

Solving these equations yields (in mA)

$$\begin{aligned} I_{10} &= (0.73 + 0.15i) = 0.75e^{+0.21i} \\ I_{20} &= (0.15 - 0.23i) = 0.28e^{-0.99i} \\ I_{30} &= (0.58 + 0.39i) = 0.69e^{+0.59i}, \end{aligned} \qquad (3.17)$$

and so the currents from this first (complex) voltage source are

$$\begin{aligned} I_1(t) &= 0.75e^{+i(\omega t + 0.21)} \\ I_2(t) &= 0.28e^{+i(\omega t - 0.99)} \\ I_3(t) &= 0.69e^{+i(\omega t + 0.59)}. \end{aligned} \qquad (3.18)$$

Now look at the second source by itself with the first source "off," shown in Fig. 3.7b, and analyze this circuit in the same way. The only difference from what is above is equivalent to replacing i with $-i$ everywhere it appears. This means that the result will be the same as the previous result, but with each i replaced with $-i$. Thus, for this second source

$$\begin{aligned} I_1(t) &= 0.75e^{-i(\omega t + 0.21)} \\ I_2(t) &= 0.28e^{-i(\omega t - 0.99)} \\ I_3(t) &= 0.69e^{-i(\omega t + 0.59)}. \end{aligned} \qquad (3.19)$$

To find the solution to the original problem, add these two solutions together to get

$$\begin{aligned} I_1(t) &= 0.75(e^{+i(\omega t + 0.21)} + e^{-i(\omega t + 0.21)}) = 1.50\cos(\omega t + 0.21) \\ I_2(t) &= 0.56\cos(\omega t - 0.99) \\ I_3(t) &= 1.38\cos(\omega t + 0.59). \end{aligned} \qquad (3.20)$$

Using the abbreviated notation from above, the solutions can be written (again in mA),

$$I_1 = 1.50\angle 12°, \quad I_2 = 0.56\angle -57°, \quad I_3 = 1.38\angle 34°. \qquad (3.21)$$

where the phase shifts have been converted from radians to degrees.[3] The phase shift here is relative to the original voltage source. Any current that can be measured will always have a real value, and these solutions are indeed real valued.

The second solution did not need to be found to know what the result would be. Doing the computation for one or the other of the complex sources provides all the information that is required. The other solution is found by replacing all i's with $-i$'s, but no new information is obtained. Henceforth, there is no need to waste time

---

[3]To convert an angle in radians to an angle in degrees, multiply by $(180/\pi) = 57.2958$.

considering that second solution. Notice how the amplitude and phase shift of the final (real-valued) result is already contained in the result for a (hypothetical) single complex source.

The solution above illustrates why using complex values for sinusoidal voltages and currents might be useful and why the method works to find changes in magnitude and phases, but *it is not the way to solve these problems*.

The procedure can be simplified for any sinusoidal source. Complex numbers can be used to describe an impedance, Z, between any two points in a circuit. Then a generalized version of Ohm's law, V = IZ, is used where the impedance describes not only the relative magnitudes but also any phase shifts that may occur. The impedance is no longer a single value since it must describe two things simultaneously: a magnitude and a phase shift. The phase angle of the impedance represents the phase difference between a sinusoidal applied voltage and the resulting sinusoidal current through the device.

The complex impedance for a capacitor and for an inductor are found using the general relationships between current and voltage and assuming the time dependence to be proportional to $e^{i\omega t}$. Designating the (time-independent) amplitudes with a subscript "0," for capacitors

$$C\frac{d}{dt}\left(V_0 e^{i\omega t}\right) = \left(I_0 e^{i\omega t}\right) \rightarrow i\omega V_0 = \frac{1}{C}I_0 \rightarrow V_0 = \left(\frac{1}{i\omega C}\right)I_0 \rightarrow Z_C = \left(\frac{1}{i\omega C}\right), \tag{3.22}$$

and for inductors:

$$\left(V_0 e^{i\omega t}\right) = L\frac{d}{dt}\left(I_0 e^{i\omega t}\right) \rightarrow V_0 = i\omega L I_0 \rightarrow Z_L = i\omega L. \tag{3.23}$$

Comparing these results to those above (3.12), it is clear that multiplying by "i" is equivalent to a 90° phase shift, and dividing by "i" (which is the same as multiplying by "$-i$") is a –90° phase shift. More generally, for this electronics application using sinusoidal sources, multiplication by $e^{i\theta}$ is equivalent to "$\angle\theta$." The arithmetic of the phase shifts is the same as the arithmetic of these exponentials. Some examples follow.

Examples

$$(6.3\angle 35°) \times (2.1\angle -22°) \Rightarrow \left(6.3 e^{i35°}\right) \times \left(2.1 e^{i-22°}\right)$$
$$= \left(6.3 \times 2.1 e^{i(35°-22°)}\right) \Rightarrow 13.23\angle 13° \tag{3.24}$$
$$\frac{9.6\angle 85°}{3\angle 25°} \Rightarrow \frac{9.6 e^{i85°}}{3 3 e^{i25°}} = \frac{9.6}{3} e^{i(85°-25°)} = 3.2 e^{i60°} \Rightarrow 3.2\angle 60°.$$

Analyzing time-dependent signals by considering the individual frequency components (i.e., individual sinusoidal contributions to a more complicated signal) is to be analyzing the signal "in the frequency domain" rather than in the "time domain." For capacitors and inductors, the complex impedance is also frequency dependent. Hence, if a complicated signal is written as a sum of sinusoidal signals then, in some sense, each sinusoid will experience a different impedance.

The use of complex numbers for problems involving any type of sinusoidal motion or signal is very common and is a very useful tool, whether the problem be from electronics, electricity and magnetism, mechanics, seismology, optics or any number of other fields. As long as there is a linear response, and hence superposition, complex numbers might be used to advantage when describing a time-dependent problem.

Series and Parallel Capacitors and Inductors

All the basic results presented for resistors apply more generally to complex impedances. For example, you can easily derive the following equivalences directly from the results for parallel and series resistors:

- Inductors in series:

$$Z_{eq} = Z_1 + Z_2 = i\omega(L_1 + L_2) = i\omega L_{eq} \rightarrow L_{eq} = L_1 + L_2 \qquad (3.25)$$

- Inductors in parallel:

$$\frac{1}{Z_{eq}} = \frac{1}{Z_1} + \frac{1}{Z_2} = \frac{1}{i\omega}\left(\frac{1}{L_1} + \frac{1}{L_2}\right) = \frac{1}{i\omega L_{eq}} \rightarrow \frac{1}{L_{eq}} = \frac{1}{L_1} + \frac{1}{L_2} \qquad (3.26)$$

- Capacitors in series:

$$Z_{eq} = Z_1 + Z_2 = \frac{1}{i\omega}\left(\frac{1}{C_1} + \frac{1}{C_2}\right) = \frac{1}{i\omega C_{eq}} \rightarrow \frac{1}{C_{eq}} = \frac{1}{C_1} + \frac{1}{C_2} \qquad (3.27)$$

- Capacitors in parallel:

$$\frac{1}{Z_{eq}} = \frac{1}{Z_1} + \frac{1}{Z_2} = i\omega(C_1 + C_2) = i\omega C_{eq} \rightarrow C_{eq} = C_1 + C_2 \qquad (3.28)$$

Of course, these same results can also be derived directly from the time domain relationships between voltage and current.

As an example, compare the infinite array of capacitors, Fig. 3.8a, to the infinite array of resistors solved previously. The translational symmetry is used as before (Fig. 3.8b). Then

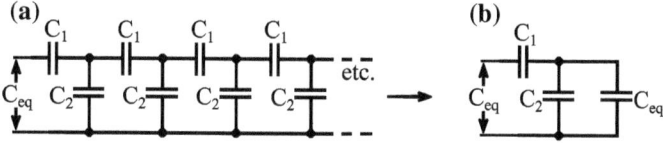

Fig. 3.8 Using complex impedances, an infinite array of capacitors can be solved in the same manner as an infinite array of resistors

$$\frac{1}{C_{eq}} = \frac{1}{C_1} + \frac{1}{C_2 + C_{eq}} \rightarrow C_1(C_{eq} + C_2) = C_{eq}(C_{eq} + C_1 + C_2), \quad (3.29)$$

so

$$C_{eq}^2 + C_2 C_{eq} - C_1 C_2 = 0, \quad (3.30)$$

and then, using the quadratic formula,

$$C_{eq} = \frac{1}{2}\left(-C_2 \pm \sqrt{C_2^2 + 4C_1 C_2}\right), \quad (3.31)$$

where the plus sign must be chosen to remain physical. For the special case where $C_1 = C_2 = C$, this becomes

$$C_{eq} = C\left(\frac{\sqrt{5}-1}{2}\right). \quad (3.32)$$

Comments About Complex Arithmetic

Complex values can be represented using real and imaginary parts and/or a magnitude and a phase angle. That is, a complex value can be represented using two real values, A and B, or two real values C and φ where

$$A + iB = Ce^{i\varphi}; \quad C = \sqrt{A^2 + B^2}; \quad \tan(\varphi) = B/A. \quad (3.33)$$

There is a direct analogy to vector notation in mechanics, where x-y coordinates or polar coordinates might be used. When adding or subtracting complex values, it is most useful to use real and imaginary parts (A and B). When multiplying or dividing, it is often most useful to use magnitudes and phases (C and φ). Hence being able to convert between the two notations is essential.

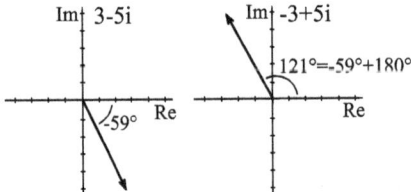

Fig. 3.9 Care must be taken when converting complex numbers to a magnitude and an angle since the arctangent function will return results in only two of the four quadrants. Representing the values graphically helps to identify the correct quadrant

An arctangent is often used when determining the phase angle from real and imaginary parts, however a calculator or computer will provide answers in only two quadrants. The result must be checked to make sure it is in the correct quadrant. For example, when computing the phase angles for $3 - 5i$ and $-3 + 5i$ using the arctangent, a calculator returns the same answer, but the angles are really 180° apart. For most calculators, if the real part is negative you need to add 180° to (or subtract 180° from) the result. For example, for the values above

$$\varphi_1 = \tan^{-1}\left(\frac{\text{Im Part}}{\text{Re Part}}\right) \to \tan^{-1}\left(\frac{-5}{3}\right) = -59°;$$
$$\varphi_2 = \tan^{-1}\left(\frac{5}{-3}\right) = -59° + 180° = 121°. \tag{3.34}$$

To check results, it helps to plot the values in the complex plane, such as in Fig. 3.9. The phase angle is the angle "up from the real axis." "Up" in this case would be counterclockwise.

For electronics, phase angles are normally reported between −180° and 180° unless there is a particular reason to extend them outside that range. If angles are outside that range, an appropriate multiple of 360° can be added or subtracted to get them back in that range.

Solving Circuits Using Complex Impedances

Here it is shown, by example, how to solve circuits with sinusoidal sources. The approach is identical to what was used for resistors, however the arithmetic becomes more complicated due to the addition of phase angles (i.e., complex numbers).

Example 1 It is desired to find the currents in the circuit of Fig. 3.10a. For this problem, the solution is shown using complex impedances and circuit reduction.

Solving Circuits Using Complex Impedances

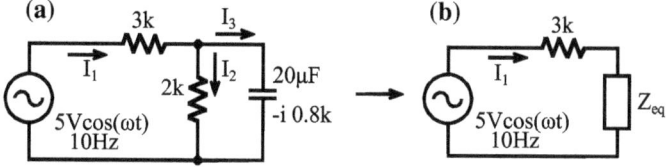

Fig. 3.10 The circuit shown in (a) is solved using complex impedances and circuit reduction

Kirchhoff's laws and equations could be used instead, if that is desired. The first step for either is to compute the impedance for the capacitor. Aside from forgetting to make the computation altogether, a common mistake when computing the impedance is to use the frequency, f, rather than the angular frequency, ω, in the computation. The difference is almost a factor of ten! The impedance of the capacitor for this example is

$$\frac{1}{i\omega C} = \frac{1}{i(2 \cdot \pi \cdot 10\,\text{Hz}) \cdot 20 \times 10^{-6}} = -800i = -i0.8\text{k}. \tag{3.35}$$

The 2k resistor and the capacitor are in parallel, so create an equivalent impedance

$$\begin{aligned} Z_{eq} &= \frac{Z_{2k}Z_C}{Z_{2k}+Z_C} = \frac{(2)(-0.8i)}{2-0.8i}\text{k} = \frac{-1.6i}{2-0.8i} \cdot \left(\frac{2+0.8i}{2+0.8i}\right)\text{k} \\ &= \frac{1.28 - 3.2i}{4.64}\text{k} = (0.28 - 0.70i)\text{k}. \end{aligned} \tag{3.36}$$

As seen by the voltage source and the 3k resistor, the circuit looks the same as Fig. 3.10b. Now the 3k resistor and Z_{eq} are in series. Hence, create another equivalent impedance for these series components,

$$\begin{aligned} Z'_{eq} &= 3\text{k} + Z_{eq} = (3.28 - 0.70i)\text{k} \\ &= \sqrt{3.28^2 + 0.70^2}\text{k}\angle \tan^{-1}(-0.70/3.28) = 3.35\text{k}\angle -12°. \end{aligned} \tag{3.37}$$

Hence,

$$I_1 = \frac{5\,\text{V}}{3.35\text{k}\angle -12°} = \left(\frac{5}{3.35}\right)\text{mA}\angle +12° = (1.45 + 0.31i)\,\text{mA}. \tag{3.38}$$

Now in the original circuit, this current was split two ways, and the current divider equation can be used to get

$$I_2 = I_1 \frac{Z_C}{2k + Z_C} = \frac{I_1 (2k)(Z_C)}{2k \; 2k + Z_C} = I_1 \frac{Z_{eq}}{2k}$$
$$= 1.5\angle 12° \left(\frac{0.28 - 0.70i}{2} \right) \text{mA} \qquad (3.39)$$
$$= (1.46 + 0.31i)(0.14 - 0.35i) \text{ mA}$$
$$= (0.31 - 0.47i) \text{ mA} = 0.56 \text{ mA} \angle 56°.$$

and a similar calculation is done for I_3. Alternatively, use $I_1 = I_2 + I_3$ (i.e., KCL) to get

$$I_3 = I_1 - I_2$$
$$= (1.46 + 0.31i) \text{ mA} - (0.31 - 0.47i) \text{ mA} \qquad (3.40)$$
$$= (1.15 + 0.78i) \text{ mA} = 1.4 \text{ mA} \angle 34°.$$

It is good form at this point to do some simple checks. One possible check is to make sure the voltages across the 2k resistor and across the capacitor are indeed equal. That is, compute

$$V_R = I_2(2k) = 1.1 \text{ V} \angle -56°$$
$$V_C = I_3 Z_C = (1.4 \angle 34°)(0.8 \angle -90°) = 1.1 \text{ V} \angle -56°, \qquad (3.41)$$

and the results do check out. Remember that when two complex values are multiplied, the magnitudes multiply but the angles *add*.

Example 2 Here it is desired to find the Thevenin equivalent for all of the circuit of Fig. 3.11a except for the 1k resistor. All of the theorems shown for resistors will work for a.c. analysis with linear devices. Hence, follow the same procedure used before except now with complex impedances. The first step is to compute the impedance of the capacitor. The angular frequency is $\omega = 2\pi(60 \text{ Hz}) = 377 \text{ s}^{-1}$. Then

$$Z_C = \frac{1}{i\omega C} = \frac{-i}{(377)(1 \times 10^{-6})} \Omega = -i2.7 \text{ k}\Omega. \qquad (3.42)$$

The problem is to find the Thevenin equivalent for the portion of the circuit shown in Fig. 3.11b.

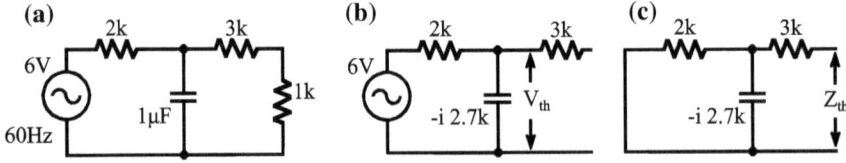

Fig. 3.11 The procedure to find the Thevenin equivalent of the circuit shown in (**a**) is the same as was used for resistors, except complex impedances are used

Solving Circuits Using Complex Impedances

First, find the Thevenin equivalent voltage, V_{th}, by finding the voltage between the two leads on the right when nothing is connected. With nothing connected, there is no current through the 3k resistor and hence there is no voltage change across the 3k resistor. Thus, the voltage across the capacitor equals V_{th}.

With nothing connected, the 2k resistor and the capacitor are effectively in series so you can use the voltage divider equation to find the voltage across the capacitor. That is,

$$V_{th} = \frac{Z_C}{Z_{2k} + Z_C} 6\,V = \frac{-i2.7k}{2k - i2.7k} 6\,V$$
$$= \frac{2.7k\angle-90°}{3.36k\angle-53°} 6\,V = 4.8\,V\angle-37°. \quad (3.43)$$

Second, find the Thevenin equivalent impedance by turning off the voltage source (making it a wire) and then look back into the circuit from the right. See Fig. 3.11c. Hence,

$$Z_{th} = 3k + 2k\|(-i2.7k) = 3k + \frac{(2k)(-i2.7k)}{2k - i2.7k}$$
$$= \frac{3k(2k - i2.7k) + 2k(-i2.7k)}{2k - i2.7k} \quad (3.44)$$
$$= \frac{6 - i13.7}{2 - i2.7} k = \frac{15\angle-66°}{3.36\angle-53°} k = 4.5k\angle-13°.$$

The Thevenin equivalent is then constructed using these complex values for the voltage and impedance. In each case, the phase shift is referenced to that of the original voltage source.

A.C. Power

The instantaneous electrical power, P, being delivered to, or absorbed by, a device is given by $P = VI$ (see Chap. 1). The time-averaged power is usually of most interest for a.c. devices. For example, the brightness of a light bulb, the power of an electric motor, or the loudness of a sound from a loudspeaker, all depend on the time-averaged power.

For a simple sinusoidal signal, the time-averaged power can be computed using a single cycle (all other cycles are the same). Hence, if the voltage across a device and current through the device are given by

$$V(t) = V_0 \cos(\omega t + \alpha), \quad I(t) = I_0 \cos(\omega t + \beta), \quad (3.45)$$

then the time-averaged power is given by

$$\bar{P} = \frac{1}{T}\int_0^T V(t)I(t)dt = \frac{V_0 I_0}{T}\int_0^T \cos(\omega t + \alpha)\cos(\omega t + \beta)dt \qquad (3.46)$$

$$= \frac{V_0 I_0}{2\pi}\int_0^{2\pi} \cos(\varphi + \alpha)\cos(\varphi + \beta)d\varphi = \frac{V_0 I_0}{2}\cos(\beta - \alpha),$$

where the period, $T = 2\pi/\omega$ and the substitution $\varphi = \omega t$ was used for integration. Similarly, the time-averages of $V^2(t)$ and $I^2(t)$ are

$$\overline{V^2} = \frac{V_0^2}{2}, \quad \overline{I^2} = \frac{I_0^2}{2}. \qquad (3.47)$$

In cases where the time averages of $V(t)$ and $I(t)$ are zero, it is often useful to characterize their time averages using the root mean square (rms) values.[4] Hence,

$$V_{rms} = \sqrt{\overline{V^2}} = \frac{V_0}{\sqrt{2}}, \quad I_{rms} = \sqrt{\overline{I^2}} = \frac{I_0}{\sqrt{2}}, \qquad (3.48)$$

where in the far right of each equation, the divisor of $\sqrt{2}$ is only valid for a sinusoidal time dependence.

Using these rms values, the time-averaged power for a sinusoidal signal is then

$$\bar{P} = V_{rms}I_{rms}\cos\phi = V_{rms} \cdot I_{rms} \qquad (3.49)$$

where ϕ is the difference between the phase angles. The vector dot product notation is used as if V and I were vectors that differ by an angle ϕ. When it is clear that sinusoidal signals are present, the over-bar and subscripts are often dropped but are understood to be present.

Alternatively, the problem can be approached using ideas from superposition. That is, write

$$V(t) = \frac{V_0}{2}e^{i\omega t}, \quad I(t) = \frac{I_0}{2}e^{i\omega t}, \qquad (3.50)$$

where

---

[4]In electronics, the phrase "average of a sinusoidal signal" may refer to the average of the absolute value of the signal. This somewhat sloppy use of the word average seems to have arisen because some analog a.c./d.c. meters use the interaction between two electromagnets to provide a needle movement. That interaction is, of course, unchanged when the sign of the current changes and so the meter reads absolute values. Due to the needle's mechanical inertia, the average value results. For a sinusoidal signal of amplitude A, the average of the absolute value is $2A/\pi$, which is slightly smaller than the rms value.

A.C. Power

V_0 and I_0 are complex values. The appropriate calculation is:

$$\bar{P} = Re\left[\frac{1}{T}\int_0^T V(t)I^*(t)dt\right]. \tag{3.51}$$

The form in brackets often appears when using complex numbers to do computations using sinusoidal functions. It can be regarded as a kind of dot product between two complex functions.

For a.c. signals, the rms voltages and currents are usually specified unless it is otherwise stated. Thus, as an example, for an a.c. voltage of 115 V and a corresponding a.c. current of 0.75 A with a relative phase angle of 32°, the time-averaged power is (115 V)(0.75 A)cos(32°) = 73 W. The result is quite simple, though there is a lot of math behind where it came from. The cosine of the phase difference, in this example cos(32°), is referred to as the "power factor" for the circuit.

The real part ("Re") used above for the power gives the actual power in watts. If power is treated as a complex value, the imaginary part is called the "reactive power." Reactive power is delivered to a device during one-half of the cycle, and is returned from the device during the other half. Any time the values of a voltage and a current are multiplied, the result will have units of volts times amps. In terms of basic units, 1 V · A = 1 W. However, the watt is usually reserved for actual power dissipated or delivered, which takes into account the power factor. This is why some devices are specified using V · A instead of watts. For example, V · A is often seen for higher power electric motors, that may have the current and voltage out of phase under some conditions.

Condenser Microphones

In some older literature, capacitors are sometimes referred to as "condensers." That historical name is still used on occasion. For example, a "condenser microphone" is a capacitor designed so that the capacitance depends on changes in air pressure. The capacitor is made part of an RC circuit and the changing voltage associated with the changing capacitance gives rise to an audio signal. Some external voltage source is necessary to charge the capacitor—usually supplied using a battery or using "phantom power" from an audio control board (48 V or so is typical).

A circuit with an impedance that depends on time can prove to be difficult to solve in general. Often it is the case that the time scales involved allow a simpler approximate solution. One example is that of a condenser microphone.

The basic setup is shown in the Fig. 3.12. The electrically conducting diaphragm responds to changes in air pressure, changing the spacing of the capacitor plates. The RC time constant is made long enough so that the charge on the sensing plates

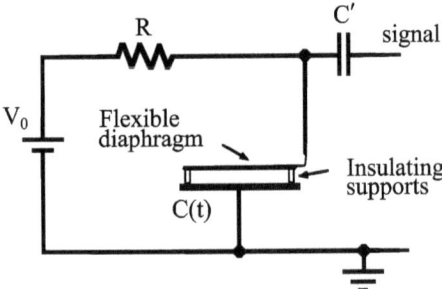

Fig. 3.12 A condenser microphone includes a time-dependent capacitance in an *RC* circuit. Finding the solution to this circuit is more difficult that it may first appear

is roughly constant in time. Hence, the voltage will change with time as the capacitor plate spacing changes. Assuming a long *RC* time constant, and using the simple parallel plate model for capacitors,

$$V_C(t) = \frac{Q}{C(t)} \approx \frac{Qd(t)}{\varepsilon_0 A} = \frac{Q[d_0 + \Delta d(t)]}{\varepsilon_0 A} = V_0 + \frac{Q\Delta d(t)}{\varepsilon_0 A}, \qquad (3.52)$$

where A is the area of the plates, d is the spacing between the plates, d_0 is the equilibrium spacing between the plates, and $\varepsilon_0 = 8.854 \times 10^{-12}$ F/m.

Try solving this circuit without making any approximations. Note that when the capacitance is time-dependent you cannot use the complex impedance derived above.

An electret microphone is also a kind of condenser microphone, though the plates include a highly insulating layer with embedded charge. That charge will stay present for a very long time (>100 yrs). Hence, the plates are effectively permanently charged. The signals from an electret microphone are low so a small amplifier is usually built into the device. Electret microphones require a power supply for that amplifier, typically a few volts, but none is needed for the capacitive sensor. Electret microphones can be very inexpensive and you will find them used for cell phones, personal computers, and lapel microphones. They tend to be somewhat noisier than the highest quality microphones so you will not usually see them used for the highest quality recordings.

Problems

1. For the circuit shown in Fig. 3.P1, the switch has been closed for a long time and then is opened at $t = 0$. What is (a) the voltage on the capacitor just after the switch is opened, (b) the time constant for changes in the capacitor's voltage, and (c) the voltage on the capacitor a long time after the switch is opened? Finally, (d) write an expression for the voltage on the capacitor as a function of time for $t > 0$.

Problems   75

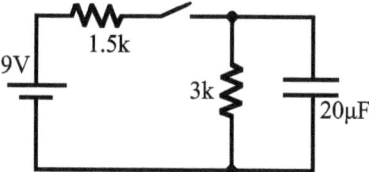

Fig. 3.P1 Problem 1

2. For the circuit of Fig. 3.P2, the switch is initially open (and has been for a long time, so the capacitor is uncharged), and at $t = 0$ the switch is suddenly closed. After 5 ms, the voltage across the capacitor is 2.0 V. What is C? (Hint: replace the voltage source and resistors with their Thevenin equivalent).

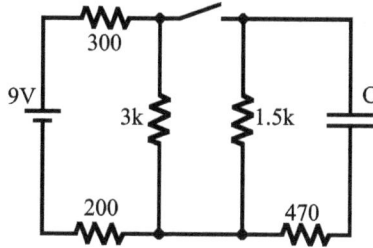

Fig. 3.P2 Problem 2

3. For the circuit of Fig. 3.P3, the switch is initially closed (and has been closed for a long time) and at $t = 0$ the switch is suddenly opened. Obtain an expression for the voltage across the inductor as a function of time and plot the results.

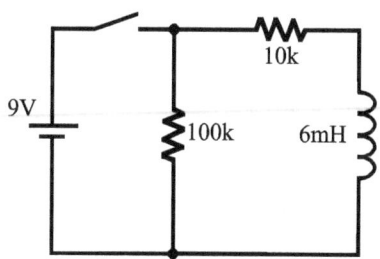

Fig. 3.P3 Problem 3

4. Compute the following for $A = 3 + 4i$, $B = 7 - 3i$ and $i = \sqrt{-1}$. Put the result in terms of a real and imaginary part and in terms of a magnitude and phase angle.

(a) $|A|$ (b) $A/|A|$ (c) $A - A^*$
(d) $A + B$ (e) $A \cdot B$ (f) A/B

5. What is the magnitude and phase shift of the current through the 1k resistor for the circuit of Fig. 3.P5?

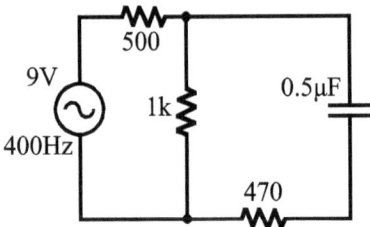

Fig. 3.P5 Problem 5

6. Find the Thevenin equivalent for the portion of the circuit in Fig. 3.P6.

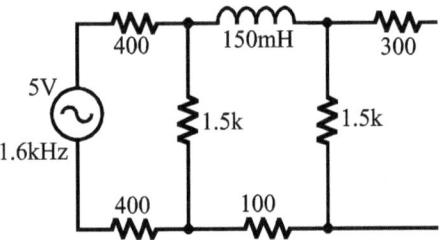

Fig. 3.P6 Problem 6

7. For the portion of a circuit in Fig. 3.P7, what is the magnitude of the Thevenin equivalent voltage?

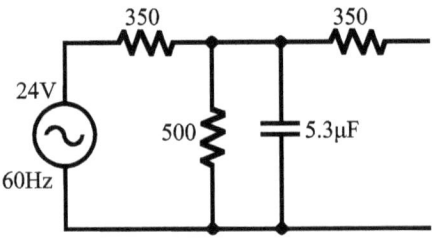

Fig. 3.P7 Problems 7 and 8

Problems

8. For the circuit of Fig. 3.P7, what is the magnitude and phase of the Thevenin equivalent impedance?

9. (Challenge Problem) Suppose you have a voltage that is periodic in time but not sinusoidal. You know it can always be expressed as a sum of sinusoidal functions due to Fourier's theorem, but can you treat the time-averaged power as being the sum of the power from each of the separate sinusoids? That is, if

$$V(t) = \sum_i V_i \cos(\omega_i t + \alpha_i), \quad I(t) = \sum_i I_i \cos(\omega_i t + \beta_i),$$

will the time-averaged power be given by $\bar{P} = \frac{1}{2}\sum_i V_i I_i \cos(\beta_i - \alpha_i)$? Prove or disprove.

10. (Challenge Problem) The voltage across a capacitor, V_C, is related to the magnitude of the net charge on the plates of the capacitor Q by the relationship $V_C = Q/C$, where C is the capacitance. The charge is taken to be $+Q$ on one plate and $-Q$ on the other, so the total net charge on the capacitor is zero. What happens if the charges on the two plates do not balance out? In particular, would such a charge imbalance be of concern when analyzing electronic circuits?

Chapter 4
More on Capacitors and Inductors

Real capacitors and inductors are often less ideal than real resistors. The non-ideal behavior can lead to some inconvenience that must be dealt with when capacitors and inductors are measured and used. In addition, some interesting new applications and results are presented using capacitors and inductors.

Real Capacitors and Inductors

Real inductors and capacitors often have a significant resistive component. At very high frequencies, inductors can also have a capacitive component, and vice versa. In general, the values used to model a real component may depend on the conditions (e.g., frequency, etc.). Any time that these imperfections are an issue, a real inductor or capacitor can be modeled using a number of ideal components. Two very simple models for real components, that include some resistance, are the parallel and the series models shown in Fig. 4.1.

A "quality factor," Q, is defined at (angular) frequency ω analogous to what is done for the driven harmonic oscillator. As long as Q is not too small, for these models Q is given approximately as

$$Q_L \approx \frac{R_p}{\omega L} \approx \frac{\omega L}{R_s}; \quad Q_C \approx \omega R_p C \approx \frac{1}{\omega R_s C}. \qquad (4.1)$$

Q is a measure of the magnitude of the reactive to resistive impedance. A larger value of Q means the component is closer to being ideal—it is, in some sense, of higher quality. These approximate equations and the simple parallel and series models work well if $Q \gg 1$. Some devices are specified using a "dissipation factor," which is $1/Q$, often expressed in percent, or through the use of a dissipation angle, δ, where $\tan\delta = 1/Q$.

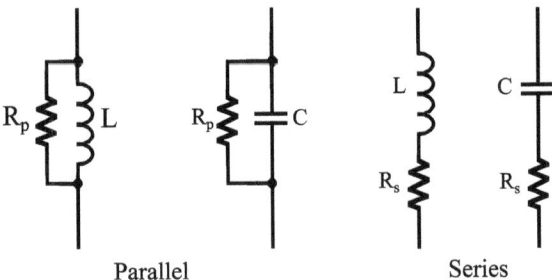

Fig. 4.1 Parallel and series models used to model the resistance for real inductors and capacitors

For the parallel model, a more ideal component will have a very large parallel resistance, $R_p \gg \omega L$. For the series model, on the other hand, a more ideal component will have a very small resistance, $R_s \ll \omega L$. Whether the parallel or series model will work best depends on the specific circumstances. In extreme cases, more elaborate models, for example including both a parallel and a series resistance, may be appropriate.

A general trend is that inductors and capacitors that have larger values tend to have a lower Q. This trend is due to the compromises that may be necessary to create the larger values and still have a device that is usable and of reasonable size. Practically speaking, for components often found in electronics, the presence of resistance is often more of a concern for inductors than for capacitors.

Measuring Capacitors and Inductors

Perhaps the simplest methods to determine a capacitance or inductance is to measure a time constant. The component to be determined is placed in series or in parallel with a known resistance and the time constant is determined. Results from the previous chapters are used to extract the component value. If the resistance in the device is not negligible, multiple measurements using different known resistors may be necessary.

If the device is to be used in an a.c. circuit, it is perhaps best to determine the values using an a.c. circuit operating near the intended operating frequency. The more precise methods use some sort of bridge circuit where the unknown components are compared to known values. A generalized version of the Wheatstone bridge is one obvious choice for such measurements. Some possible configurations are illustrated in Fig. 4.2.

Capacitive Position Sensors

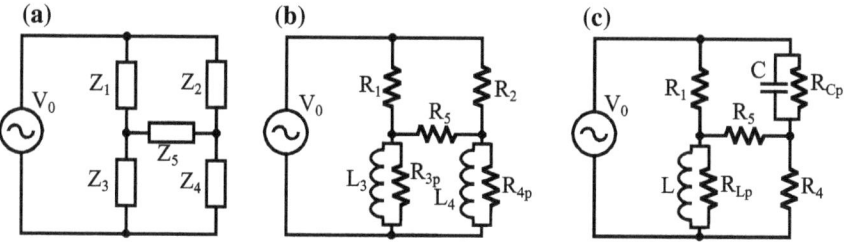

Fig. 4.2 A generic bridge circuit to measure a complex impedance is shown in (**a**). Specific implementations to measure an inductor and/or a capacitor are shown in (**b**) and (**c**).

Capacitive Position Sensors

A simple position sensor can be made using a bridge circuit and variable capacitors. While there are many ways to do this, one way is to use a capacitive sensor consisting of three parallel plates, illustrated in Fig. 4.3a.[1] The central plate moves back and forth as shown, and the outer two are fixed in position.

These plates can be connected in a Wheatstone bridge configuration such as shown in Fig. 4.3b. The meter "M" provides a reading. When the center plate (2a and 2b) of the capacitor is equidistant between the outer two plates (1 and 3), the meter reads zero due to the symmetry of the circuit.

Using results from Electricity and Magnetism ($C = \varepsilon_0 A/d$), the fractional imbalance in the capacitance when the center plate is a small distance Δx from the center will be $\Delta x/d$, where d is the spacing between the outer plates. Since d can be quite small, such sensors can be used to see very small changes in position. Commercial capacitance sensors, and associated electronics, are available that have a nominal resolution smaller than the size of an atom.

Position sensors are also useful as a major component in force bridges, pressure sensors, or accelerometers. The center plate is attached to a spring or cantilever and is deflected by an applied force F_1 or by a fictitious force if you subject the entire device to an acceleration. This is the principle used for many micro-electromechanical systems (MEMS) based acceleration detectors used in laptop computers, some game controllers, and as sensors for air bag deployment. In some cases, provisions are included to supply a known adjustable force F_2 to keep the inner plate centered and thus the force F_1 is determined directly by knowing F_2. Devices can be constructed to apply the appropriate F_2 automatically using "feedback." Feedback will be discussed later.

---

[1]For more information about capacitance sensors, see Jones and Richards (1973).

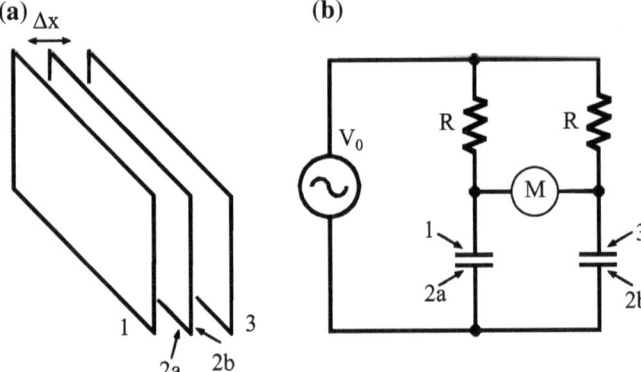

Fig. 4.3 A sensitive capacitive position sensor can be constructed with two fixed outer plates and a movable inner plate. The imbalance as the inner plate moves is measured using a bridge circuit

A Simple Circuit for Measuring Inductors

Since the resistive part of an inductor is often non-negligible, any determination of the value of an inductance might need to take into account that resistive component. That is, there are two values to measure—an inductance and a resistance. For a bridge circuit, such as the Wheatstone bridge mentioned earlier, there will need to be (at least) two variable components to obtain a good null on the detector. As an alternative to the Wheatstone bridge, consider the bridge circuit shown in Fig. 4.4a, where the impedance, Z_x, is an unknown inductor, including its resistive component. Here the inductor's resistance has been modeled using the simple parallel model.

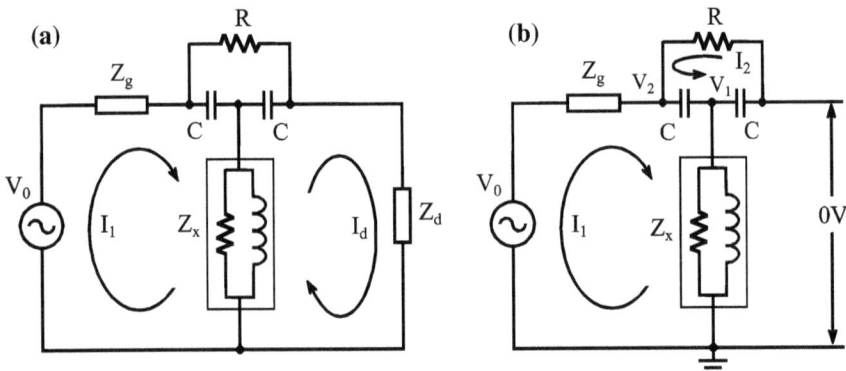

Fig. 4.4 A bridge circuit for measuring an inductor (**a**) and an equivalent (**b**) when the circuit has been balanced to have no current through the detector, Z_d

A Simple Circuit for Measuring Inductors

The resistor (R) and the two matched capacitors (C) are calibrated and adjustable. The other two impedances will not be of consequence for the balance condition, though will be important when determining the sensitivity of the measurement. The voltage source is sinusoidal at an angular frequency ω.

To determine Z_x, the resistor and the two matched capacitors are adjusted so that there is no current through the detector ($I_d = 0$). When that condition is met, there is no voltage drop across the detector. For that condition (only) the situation is as shown in Fig. 4.4b. Two intermediate voltages, labeled V_1 and V_2, and a ground symbol have been added. The ground symbol at the bottom indicates the position used for the 0 V reference. Starting from the upper right (which is also at 0 V due to the match condition) the voltage drops lead to

$$V_1 = \frac{1}{i\omega C} I_2$$
$$V_2 = -R I_2 \tag{4.2}$$
$$(V_1 - V_2) = -\frac{1}{i\omega C}(I_1 + I_2) = \left(R + \frac{1}{i\omega C}\right) I_2,$$

which can be solved to give $I_1 = -(2 - i\omega RC) I_2$.

Now find V_1 by starting at the bottom and going up through the unknown, Z_x,

$$V_1 = I_1 Z_x = -I_2(2 + i\omega RC) Z_x = \frac{1}{i\omega C} I_2. \tag{4.3}$$

The far right-hand side comes from the relationship found above. Using these results

$$Z_x = \frac{-1}{i\omega C}\frac{1}{(2 + i\omega RC)} = \frac{1}{(\omega^2 RC^2 - 2i\omega C)}. \tag{4.4}$$

The parallel model for the inductor yields

$$Z_x = (i\omega L_x) \| R_x = \frac{1}{1/R_x - i/(\omega L_x)}. \tag{4.5}$$

Comparing to the result above for the circuit under the balance condition, the corresponding model components are

$$R_x = \frac{1}{\omega^2 RC^2}; \quad L_x = \frac{1}{2\omega^2 C}, \tag{4.6}$$

and the quality factor is given by $Q = 1/(2\omega RC)$. That is, once the match condition is achieved, the component values of the inductor model are easily determined from the (presumed known) values of the bridge resistor and the two matched capacitors.

Switched Capacitor Methods

There are a number of interesting circuits developed around the idea of a "switched capacitor." Only the basic principle is shown here. In practice, the switching operation is often achieved using transistor circuits and may occur at frequencies well above 1 kHz.

Consider the circuit of Fig. 4.5a. The switches S_1 and S_2 will be alternately opened and closed, taking care so they are never both closed at the same time. Assume that the voltage across C_2 is initially V_2 and so the initial charge on C_2 is $Q_2 = C_2 V_2$.

Now S_1 is closed (S_2 is open) so capacitor C_1 is charged to a voltage V_0. The charge on C_1 is then $Q_1 = C_1 V_0$. Next, S_1 is opened and then S_2 is closed. Since the two capacitors are now in parallel, the total effective capacitance is $C_1 + C_2$ and the total net charge on that capacitance is $Q_{tot} = Q_1 + Q_2 = C_1 V_0 + C_2 V_2$. That charge will rearrange so that the potentials (voltages) across the two capacitors are equal. The voltage across the effective capacitance is then

$$\frac{Q_{tot}}{C_1 + C_2} = \frac{C_1 V_0 + C_2 V_2}{C_1 + C_2} = \frac{C_1}{C_1 + C_2} V_0 + \frac{C_2}{C_1 + C_2} V_2$$
$$= V_2 + \frac{C_1}{C_1 + C_2}(V_0 - V_2) = V_2 + \Delta V_2. \tag{4.7}$$

After S_2 is again opened, this is the new value for V_2. If this entire switching process occurs over a time Δt, then the average change in the charge on C_2 per unit time is given by

$$\frac{\Delta Q_2}{\Delta t} = C_2 \frac{\Delta V_2}{\Delta t} = \frac{C_1 C_2}{C_1 + C_2} \frac{1}{\Delta t}(V_0 - V_2). \tag{4.8}$$

Compare that to a circuit with a resistor and capacitor, Fig. 4.5b, where at any time $I = dQ_2/dt = (V_0 - V_2)/R$. On time scales long compared to Δt, the switched capacitor circuit looks the same with an effective resistance

$$R_{eff} = \frac{C_1 + C_2}{C_1 C_2} \Delta t, \tag{4.9}$$

even though there are no resistors in the circuit. The effective resistance can be controlled by changing the switching time rather than by changing physical components.

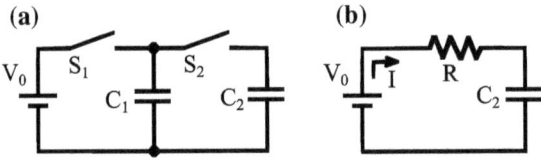

Fig. 4.5 The switching capacitor arrangement in (**a**) can mimic the behavior of a simple RC circuit, such as shown in (**b**)

Switched Capacitor Methods

Such a circuit has many uses, one of which is to measure small capacitance values. If C_1 is the unknown capacitance then a measurement of the charging time of the known capacitor C_2 determines R from which you can determine C_1. The value of Δt can be adjusted to get convenient values. Another use is illustrated by Van Den Akker and Webb (1936), who show how to use a switched capacitor "resistor" as part of a Wheatstone bridge in order to measure very large resistance values.

Charging a Capacitor Efficiently

If you charge a capacitor from 0 V to a voltage V_0 using a simple voltage source and resistor (Fig. 4.6a), the total energy change of the capacitor is $\frac{1}{2}CV_0^2$. During the process, the energy lost in the resistor as heat is found to be the same value. This can be seen by integrating the power dissipated in the resistor ($I^2R = V^2/R$) over time,

$$E = \frac{V_0^2}{R} \int_0^\infty e^{-2t/RC} dt = \frac{1}{2}CV_0^2. \quad (4.10)$$

While this result may seem purely academic, much of the power consumption that occurs in modern digital electronics arises during what is, effectively, the charging and discharging of capacitors. If this simple process could be made more efficient, the electronics could be packed that much tighter without overheating and/or batteries would last longer.[2]

One scheme that can reduce the energy loss to almost nothing is to charge the capacitor in many small steps using gradually increasing battery voltages. It is straightforward to show that if this is done using n equal steps of size V_0/n, the total energy loss is given by $E = CV_0^2/2n$. By making n large, the loss is reduced significantly. Such a procedure may be difficult to implement, but at least it shows that energy loss is not inevitable and depends on the process. There are many similar examples of such situations from the field of thermodynamics.

Consider another approach, which is to charge the capacitor using a current source from time $t = 0$ (Fig. 4.6b) until it reaches the required final voltage, V_0. That is, charge for a time T, such that

$$V_0 = \frac{1}{C} \int_0^T I_0 dt \quad \text{or} \quad T = CV_0/I_0. \quad (4.11)$$

---

[2]For more detailed considerations, see, for example, Paul et al. (2000), and references therein.

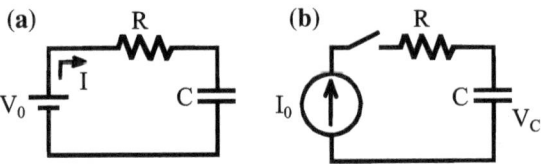

Fig. 4.6 Charging a capacitor with the circuit (**a**) loses half the energy. Using the circuit of (**b**) and a slow charge, less energy is lost

The energy lost in the resistor during this time is $I_0^2 RT = CV_0^2(RC/T) = CV_0^2(I_0R/V_0)$. The loss can be made very small by taking a long time compared to RC, which is equivalent to saying that the voltage across the resistor, I_0R, is always very small compared to V_0.[3]

Mutual Inductance and Transformers

Lenz's law, from Introductory Electricity and Magnetism, says that a time-changing magnetic field will induce an electromotive force (a voltage) around a closed path,

$$V = -\frac{d\Phi}{dt}, \qquad (4.12)$$

where Φ is the magnetic flux through any area bounded by the path and the minus sign is somewhat symbolic, indicating that the induced voltage is in a direction to create currents which, in turn, create their own magnetic field in a direction that opposes the original time-changing magnetic field.

Any current carrying circuit will create magnetic fields. If the current changes, those magnetic fields change, and hence a change in the magnetic flux through the circuit that created the field in the first place. This gives rise to "self inductance." For fixed geometry and in the absence of magnetic materials, or for small enough changes in the presence of magnetic materials, the changes in the magnetic flux are proportional to the changes in the current. That proportionality constant is the self inductance, usually signified with the symbol L, and

$$V = -L\frac{dI}{dt}. \qquad (4.13)$$

On the other hand, if you have two circuits near each other, the changing magnetic field from one of them can create a changing flux in the other. Of course, the largest contributions arise from pieces of the circuit that are designed to create

---

[3]To learn more about this problem see Heinrich (1986), and related articles.

Mutual Inductance and Transformers

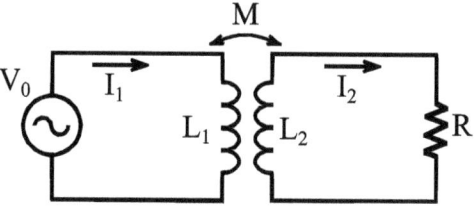

Fig. 4.7 Two nearby inductors can interact to make a transformer

large magnetic fields (e.g., inductors). Consider two circuits that are magnetically coupled, as shown in Fig. 4.7. That is, there are two inductors that can "feel" each other's magnetic field. If the geometry is fixed, the field felt by circuit 2 on the right is proportional to the current in circuit 1 on the left, and so

$$V_2 = M_{12} \frac{dI_1}{dt}, \quad (4.14)$$

where the proportionality constant M_{12} is called the "mutual inductance." Similarly, if there is a changing current in circuit 2 it will create a voltage in circuit 1,

$$V_1 = M_{21} \frac{dI_2}{dt}. \quad (4.15)$$

For any two circuits, no matter how they are constructed, it will always be the case that $M_{21} = M_{12} \equiv M$, a result of the reciprocity theorem. There is still some arbitrariness about the minus signs. In many practical circuits, those minus signs are of no consequence. However, when they do matter, there is a "dot convention." A small dot is placed on each of the two inductors that allows the minus signs to be determined accurately. For the moment, this is ignored and hence the resulting current I_2 may be off by a minus sign. The dot convention will be described later.

Now, to analyze the circuit above, use Kirchhoff's laws around the two circuits,

$$\begin{aligned} V - i\omega L_1 I_1 - i\omega M I_2 &= 0 \\ i\omega L_2 I_2 + i\omega M I_1 + R I_2 &= 0. \end{aligned} \quad (4.16)$$

These equations can be readily solved to give

$$\begin{aligned} I_1 &= \frac{R + Z_2}{RZ_1 + Z_1 Z_2 - Z_M Z_M} V \\ I_2 &= \frac{Z_M}{RZ_1 + Z_1 Z_2 - Z_M Z_M} V, \end{aligned} \quad (4.17)$$

where $Z_1 = i\omega L_1$, $Z_2 = i\omega L_2$, and $Z_M = i\omega M$.

Now the total power being delivered by the voltage source is VI_1 and the total power absorbed in the resistor is $I_2^2 R$. Ideally, those will be the same. Now,

$$VI_1 = \frac{R+Z_2}{RZ_1 + Z_1 Z_2 - Z_M Z_M} V^2 = \frac{(R+Z_2)(RZ_1 + Z_1 Z_2 - Z_M Z_M)}{(RZ_1 + Z_1 Z_2 - Z_M Z_M)^2} V^2$$

$$I_2^2 R = \frac{Z_M^2 R}{(RZ_1 + Z_1 Z_2 - Z_M Z_M)^2} V^2,$$

(4.18)

and so these will be the same if the numerators are equal, which requires

$$Z_M^2 (2R + Z_2) = Z_1 (R + Z_2), \quad (4.19)$$

which can be accomplished if $|Z_2| \gg R$ and $Z_M = \sqrt{Z_1 Z_2}$. That is, $M = \sqrt{L_1 L_2}$. Note that M is always taken to be positive. Such a power-conserving transformer is the so-called "ideal transformer." Using the result for M and the condition $|Z_2| \gg R$ in the solutions above, yields

$$I_1 = \frac{V}{R} \frac{L_1}{L_2}; \quad I_2 = \frac{V}{R} \sqrt{\frac{L_2}{L_1}}. \quad (4.20)$$

Most often, this is expressed using the "turns ratio." That is, if coil 1 is wound with n_1 turns of wire, and coil 2 with n_2, and remembering that all other things being equal, the inductance is proportional to the number of turns squared, the following relations can be obtained:

$$\frac{V}{I_1} = \left(\frac{n_1}{n_2}\right)^2 R, \quad I_2 R = \left(\frac{n_2}{n_1}\right) V, \quad \text{and} \quad I_2 = \left(\frac{n_1}{n_2}\right) I_1. \quad (4.21)$$

In words, the effective impedance seen by the voltage source is the resistance, R, "transformed" by the turns ratio squared. The voltage seen by the resistor is the voltage supplied transformed by the inverse of the turns ratio, and the current seen by the resistor is the current supplied transformed by the turns ratio. The ideal transformer changes the ratio of the voltages and currents in a circuit, but not the power (i.e., the product of voltage and current is preserved). In a real transformer, some power loss is expected.

Geometrically, the ideal transformer requires that all of the magnetic field lines created by circuit 1 go through the area bounded by circuit 2 (and vice versa). In addition, the inductive impedances of the two circuits must be large compared to any load resistance placed on them.

The Dot Convention for Transformers

In situations where the relative signs of the currents on the two sides of a transformer are important, the "dot convention" or "dot rule" is often used. When you see a transformer (two inductors next to each other) with dots, such as in Fig. 4.8, then when you assign currents and apply Kirchhoff's laws, do the following:

> If both currents enter or both leave the transformer at the dotted terminal, the sign of the M terms is the same as that of the corresponding L terms, otherwise the signs of the two terms are opposite.

In the example above, the signs of the M terms were made the same as those of the L terms somewhat arbitrarily. If dots had been assigned in this case, and the solution had been done correctly, the inductors would have had one dot on the top and one on the bottom, as shown on the right of Fig. 4.8. Note that when using dots to get the minus signs correct, what matters is which direction is defined to be the positive direction for current—i.e., which way you draw your arrow—and not the direction the current actually flows. The actual direction will come out in the solution to the problem.

Inductive Position Sensors

A simple position sensor can be constructed using coils set up so that the mutual inductance depends on a position. Two examples are shown in Fig. 4.9. In both cases, two "sense" coils are used and one excitation coil. The coils are wound so that the induced voltage measured across the detector, M (not to be confused with the mutual inductance, M), is zero when the moving portion is centered.

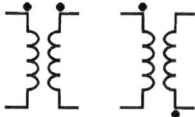

Fig. 4.8 Examples of the use of the dot convention for transformer windings

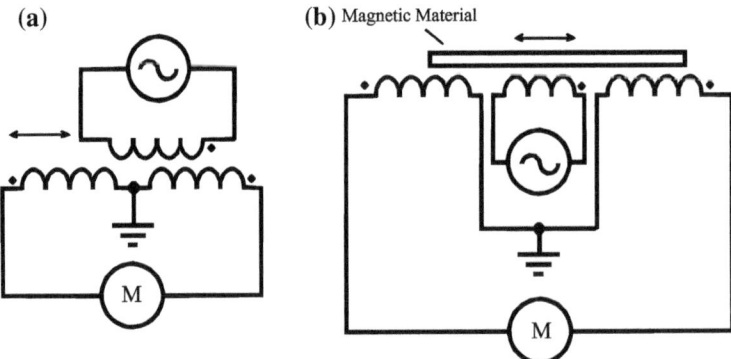

Fig. 4.9 Two examples of an inductive position sensor

RLC Circuits

There are a number of combinations of inductors and capacitors that form a tuned circuit. Of course, any real circuit will also have some resistance, but it is useful to look at the ideal case as a starting point.

Consider the two simple circuits of Fig. 4.10. For these idealized series and parallel LC circuits, the (sinusoidal) voltage source sees an effective impedance,

$$Z_{series} = Z_C + Z_L = \frac{1}{i\omega C} + i\omega L = \frac{1 - \omega^2 LC}{i\omega C}$$
$$Z_{parallel} = \left(\frac{1}{Z_C} + \frac{1}{Z_L}\right)^{-1} = \left(i\omega C + \frac{1}{i\omega L}\right)^{-1} = \frac{i\omega L}{1 - \omega^2 LC},$$
(4.22)

which have interesting behavior when $\omega^2 LC = 1$. In particular, the series impedance becomes zero and the parallel impedance approaches infinity. Looking at the behavior for very large and very small frequencies ($\omega^2 LC \gg 1$ and $\omega^2 LC \ll 1$ respectively),

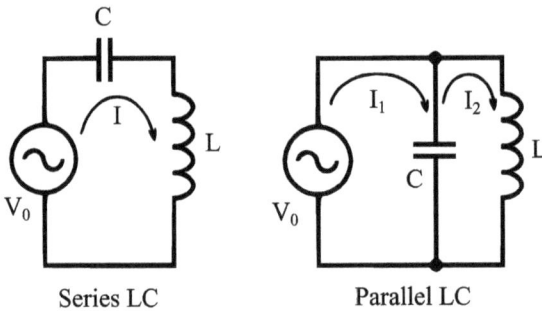

Fig. 4.10 Basic series and parallel LC circuits

RLC Circuits

Table 4.1 Limiting behavior for simple LC circuits

	Z_{series}	$Z_{parallel}$
$\omega^2 \to 0$	$\to \infty \angle -90°$	$\to 0 \angle +90°$
$\omega^2 \to \infty$	$\to \infty \angle +90°$	$\to 0 \angle -90°$

the limiting behavior summarized in Table 4.1 is obtained. For real circuits replace "0" with "small" and "∞" with "large" for the impedance values in the table.

Now consider a more realistic series circuit that includes some resistance. In fact, as an example consider a series circuit with two resistors added (Fig. 4.11): one, R, that represents a "load" where the signal is delivered (this might be a speaker in a stereo system, for example) and the other, r, represents all the other resistance in the circuit, including the resistive losses in the inductor. What is desired is to compute is the total (time-averaged) power delivered to the load, R, for the circuit, which is given by the time average of $P_R = |I|^2 R$. The current can be found by creating an effective impedance for the series combination, Z_{ser}, and then $I = V_0/Z_{ser}$. The results are

$$Z_{ser} = r + i\omega L + \frac{1}{i\omega C} + R = \frac{(1 - \omega^2 LC) + i\omega C(R+r)}{i\omega C}$$

$$I = \frac{V_0}{Z_{ser}} = V_0 \frac{i\omega C}{(1 - \omega^2 LC) + i\omega C(R+r)} \cdot \frac{(R+r)}{(R+r)} \quad (4.23)$$

$$P_R = |I|^2 R = I^* I R = \frac{V_0^2 R}{(R+r)} \cdot \frac{\omega^2 C^2 (R+r)^2}{(1 - \omega^2 LC)^2 + \omega^2 C^2 (R+r)^2}.$$

Defining $\omega_0 = 1/\sqrt{LC}$, the power dissipated in the load can be written

$$P_R = \frac{V_0^2 R}{(R+r)} \cdot \frac{(\omega/\omega_0)^2/Q^2}{\left(1 - (\omega/\omega_0)^2\right)^2 + 1/Q^2}, \quad (4.24)$$

where a "quality factor for the circuit," Q, has been defined for the series combination as

$$Q = \frac{\omega_0 L}{(R+r)}. \quad (4.25)$$

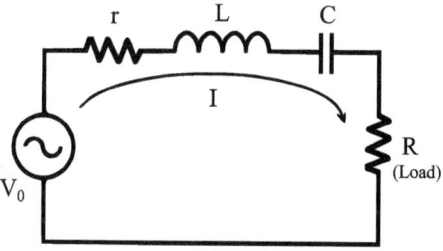

Fig. 4.11 A series RLC circuit

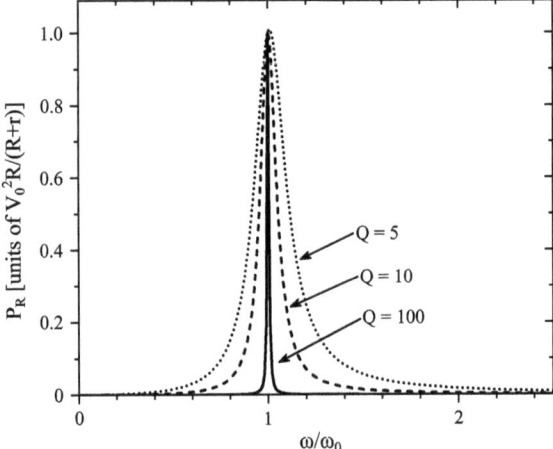

Fig. 4.12 The power delivered to a load for a RLC circuit as a function of frequency for different quality factors. The higher the quality factor, the narrower is the range of response

This result for P_R is shown graphically in Fig. 4.12 for several different values of Q. Aside from some name changes, the behavior is the same resonant behavior exhibited by a driven, damped, harmonic oscillator.

Such a circuit can be used as a bandpass filter—that is, only signals with a frequency within a certain range of frequencies—a "band of frequencies"—are allowed through to the load while signals with frequencies outside the band are blocked. As a guide, the width of the "pass band" (the range of frequencies that "get through") is roughly $\Delta\omega = \omega_0/Q$. A more precise definition requires a precise statement for a cut-off between frequencies that "make it through" and those that do not.

As an exercise, try computing the power delivered to R_2 for the circuit in Fig. 4.13. The result should look very similar to the result above for the series circuit. An appropriate definition for the "Q of the circuit" here would be $Q = (R_1 \| R_2)/(\omega L)$.

Low-pass, high-pass, and more complicated band-pass filters can be created using different arrangements and multiple LC resonant circuits. There are standard tables available that can be used to help determine values to use. In general, the more components, the more control one has over the filter properties.

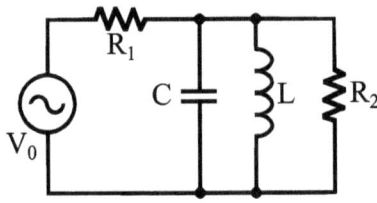

Fig. 4.13 A parallel RLC resonant circuit

Cable Models

Cables, or more generally "transmission lines," consist of two (or more) wires in close proximity. For example, coaxial cables commonly used to connect laboratory equipment (e.g., with BNC connectors) have an outer jacket acting as one wire, and an inner conductor as the other. When cables carry electrical signals, both electric and magnetic fields are involved—hence there must be both capacitance and inductance. For a cable, these are spread out or "distributed" along the length of the cable and so it is convenient to describe these quantities "per unit length." That is, a cable will have resistance per unit length, r, capacitance per unit length, c, and inductance per unit length, l.

One of the first models used for cables came from Lord Kelvin as a model for early trans-Atlantic cables. He included the resistance and capacitance but took the inductance to be negligible in comparison. That is reasonable only at low enough frequencies. A discretized model for the cable looks like the circuit of Fig. 4.14a, where R and C are the resistance and capacitance for some arbitrarily chosen finite length of the cable, Δx, and so $R = r\Delta x$ and $C = c\Delta x$. Following the procedure used previously for a similar array made entirely of resistors (see Chap. 2), the first section is separated and by translational symmetry the remaining (infinite) cable must have the same effective impedance, as in Fig. 4.14b.

For sinusoidal signals, the equivalent impedance must satisfy

$$Z_{eq} = R + \left(\frac{1}{i\omega C} \| Z_{eq}\right). \tag{4.26}$$

It is left as an exercise to show that the solution is

$$Z_{eq} = \frac{R}{2}\left(1 + \sqrt{1 + 4/(i\omega RC)}\right). \tag{4.27}$$

Now to complete the model for a continuous cable, take the limit as Δx becomes very small. To do that, write $R = r\Delta x$, and $C = c\Delta x$, where r and c are the resistance per unit length and the capacitance per unit length respectively. Then take the limit that Δx becomes infinitesimally small (but r and c stay constant). In that limit the complex equivalent impedance of the cable is proportional to $(i\omega)^{-1/2}$, i.e.,

$$Z_{eq} = \frac{r\Delta x}{2}\left(1 + \sqrt{1 + \frac{4}{i\omega rc\Delta x^2}}\right) = \frac{r}{2}\left(\Delta x + \sqrt{\Delta x + \frac{4}{i\omega rc}}\right), \tag{4.28}$$

$$\lim_{\Delta x \to 0} Z_{eq} = \frac{r}{2}\sqrt{\frac{4}{i\omega rc}} = \sqrt{\frac{r}{c}}(i\omega)^{-\frac{1}{2}}. \tag{4.29}$$

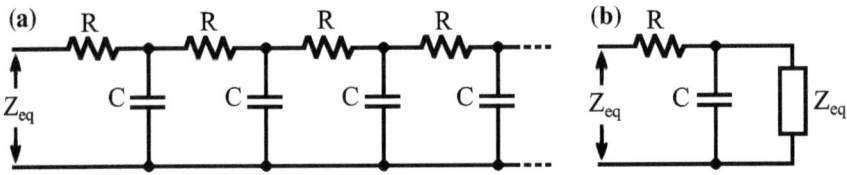

Fig. 4.14 (a) A simple RC cable model and (b) an equivalent due to symmetry

Some care is necessary with the square root of $1/i$. It is easy to see that

$$\left[\pm\left(\frac{1-i}{\sqrt{2}}\right)\right]^2 = \frac{1-2i-1}{2} = -i = \frac{1}{i}, \quad (4.30)$$

and so

$$\lim_{\Delta x \to 0} Z_{eq} = \pm \frac{1-i}{\sqrt{2}} \frac{1}{\omega^{1/2}} \sqrt{\frac{r}{c}}, \quad (4.31)$$

and only the solution with the "+" sign can be valid for a real cable (why?). This equivalent impedance has a phase angle of −45° for all frequencies and is, in some sense, half-way between a resistor (always 0°) and a capacitor (always −90°).

This model has found use in the study of electronic devices made with Carbon nanotubes[4] indicating that distributed resistance and distributed capacitance are important even in much smaller devices. It is also an interesting model to motivate fractional derivatives.

Because it is interesting, a little sidetrack to consider fractional derivations seems appropriate. The basic idea behind fractional derivatives can be seen by noting that

$$\frac{d^1}{dt^1}e^{i\omega t} = (i\omega)^1 e^{i\omega t}; \quad \frac{d^2}{dt^2}e^{i\omega t} = (i\omega)^2 e^{i\omega t}; \quad \frac{d^\alpha}{dt^\alpha}e^{i\omega t} = (i\omega)^\alpha e^{i\omega t}, \quad (4.32)$$

and that any function encountered in physics can be written as a Fourier series. Generalizing this result for terms such as $(i\omega)^{1/2} e^{i\omega t}$ suggests $\alpha = \frac{1}{2}$, that is, a half derivative. A half derivative applied twice is, of course, the same as a first derivative. Since the half derivative can be applied to every term in a Fourier series, it can be applied to any function that can be written as a Fourier series! Of course one need not stop at half derivatives, why not make α any real number? You could have the "square root of two-th" derivative if you set $\alpha = \sqrt{2}$. Note also that if $\alpha < 0$ then the derivative is equivalent to an integral—the inverse operation to a derivative.

Using this notation, the relationship between voltage and current for the infinite cable model described above can be written in terms of half derivatives:

---

[4]For example, see Esen et al. (2007).

Cable Models

$$V(t) = \sqrt{\frac{r}{c}} \frac{d^{-1/2}}{dt^{-1/2}} I(t) \text{ or } I(t) = \sqrt{\frac{c}{r}} \frac{d^{1/2}}{dt^{1/2}} V(t). \quad (4.33)$$

For more information about fractional derivatives, their various definitions and some of their potential uses, see Sokolov et al. (2002) or do an internet search for "fractional derivatives." This ends the sidetrack.

Cable Impedance

For higher frequency applications the inductance in a cable becomes important. As a starting point to model cables, such as the coax cables commonly used for experimentation, it is useful to look at the inductance and capacitance and presume the resistance is negligible. To model such a cable, assumed uniform along its length, consider the infinite array in Fig. 4.15a, where each segment of length Δx is modeled using an inductance $L = l\,\Delta x$ and capacitance $C = c\,\Delta x$.

To find the equivalent impedance, the infinite array to the right of the first segment is replaced with its (as yet unknown) equivalent impedance (Fig. 4.15b). Then

$$Z_{eq} = i\omega L + \frac{1}{i\omega C} \| Z_{eq} = \frac{i\omega C}{i\omega C} \cdot \left(\frac{Z_{eq}/i\omega C}{Z_{eq}+1/i\omega C} + \frac{i\omega L(Z_{eq}+1/i\omega C)}{Z_{eq}+1/i\omega C} \right), \quad (4.34)$$

or $Z_{eq}(1+i\omega C Z_{eq}) = Z_{eq} + i\omega L(1+i\omega C Z_{eq}),$

so

$$\begin{aligned} Z_{eq}(1+i\omega c \Delta x Z_{eq}) &= Z_{eq} + i\omega l \Delta x(1+i\omega c \Delta x Z_{eq}) \\ i\omega Z_{eq}^2 &= -\omega^2 l c \Delta x Z_{eq} + i\omega l, \end{aligned} \quad (4.35)$$

and in the limit $\Delta x \to 0$, the simple result is that $Z_{eq} = \pm\sqrt{l/c}$, where only the + sign is valid here (why?). Note that this equivalent impedance is a real, frequency independent value. That is, this infinite cable looks like a resistor even though no resistance was in the model. This is reasonable since when you put power in it never comes back, which looks the same as if the power had been converted to heat in a resistor.

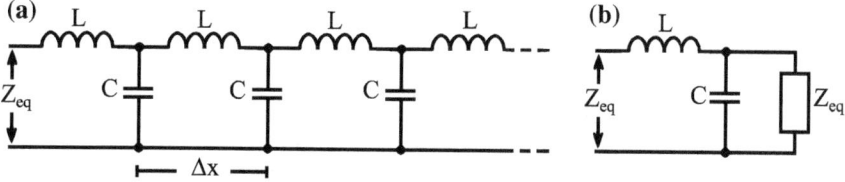

Fig. 4.15 (a) A cable model for higher frequencies and (b) an equivalent due to symmetry

Often a cable will be referred to by its equivalent impedance. For example, a coaxial cable often used with scientific equipment might be referred to as "50 Ω coax cable" and another similar cable (often used for cable TV) is "75 Ω coax cable." Note, however, that the impedance specified is for an infinite length. You cannot expect to measure that value for any finite length of cable. What will be measured depends on what is connected at the other end, as will be seen shortly.

Signal Speed in a Cable

To compute the voltage and current inside the cable, start by considering a spot in the cable, which will be labeled $x = 0$, where there is a sinusoidal voltage with angular frequency, ω. Referring to Fig. 4.16 and using the discrete model, the voltage a small distance Δx away can be computed using the voltage divider equation,

$$V(\Delta x) = \frac{(Z_{eq} \| 1/i\omega C)}{i\omega L + (Z_{eq} \| 1/i\omega C)} V(0) = \frac{Z_{eq}}{Z_{eq}(1 - \omega^2 LC) + i\omega L} V(0) \qquad (4.36)$$
$$= \frac{Z_{eq}}{Z_{eq}(1 - \omega^2 lc(\Delta x^2)) + i\omega(l\Delta x)} V(0).$$

To take the limit as $\Delta x \to 0$, use the approximation

$$\frac{1}{1+\varepsilon} \approx 1 - \varepsilon, \qquad (4.37)$$

valid for $\varepsilon \ll 1$, and drop the Δx^2 terms compared to the Δx terms to get

$$V(\Delta x) = \left(1 - \frac{i\omega l}{Z_{eq}} \Delta x\right) V(0). \qquad (4.38)$$

This can be rearranged to get

$$\frac{V(\Delta x) - V(0)}{\Delta x} = \frac{i\omega l}{Z_{eq}}. \qquad (4.39)$$

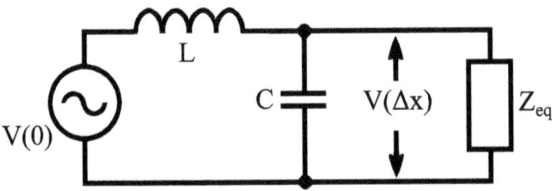

Fig. 4.16 The discretized model for the transmission line used to derive the signal along the cable

Cable Models

Taking the limit $\Delta x \to 0$, including the limiting value for Z_{eq} found previously, and using the fact that the location of $x = 0$ was arbitrary, this equation becomes

$$\frac{dV}{dx} = -\left(i\omega\sqrt{lc}\right)V, \tag{4.40}$$

that has as its solution $V(x) = V_0 \, e^{-ikx}$, where V_0 is a constant and $k = \omega\sqrt{lc}$. That is, the solution is also sinusoidal with *position* along the cable, with wavelength $\lambda = 2\pi/\sqrt{\omega^2 lc}$. Since $f = 2\pi\omega$, $f\lambda = 1/\sqrt{lc}$. Hence, the wave is traveling with wave speed[5] $1/\sqrt{lc}$. For typical cables, the wave speed is a good fraction of the speed of light.

Impedance of Finite Cables

Of course, real cables are not infinite. Now compute the effective impedance seen at one end of a cable when an impedance Z, the "load," is attached at the other end, illustrated in Fig. 4.17. The characteristic cable impedance (Z_{eq} above, in the limit $\Delta x \to 0$) is taken to be Z_0.

For signals at frequency ω, the voltage and current in the finite cable can be described as the superposition of two traveling waves, one traveling toward the load and the other away from the load. That is

$$V(x) = V_R e^{i(\omega t - kx)} + V_L e^{i(\omega t + kx)}. \tag{4.41}$$

The current will then be given by

$$I(x) = \frac{V_R}{Z_0} e^{i(\omega t - kx)} - \frac{V_L}{Z_0} e^{i(\omega t + kx)}, \tag{4.42}$$

where the sign convention used has $I(x) > 0$ corresponding to a total current traveling to the right and a minus sign is added for current traveling to the left. Note that in general, $V(x)/I(x) \neq Z_0$.

The two unknown values, V_R and V_L, can be determined by the conditions

$$V(0) = V_s e^{i\omega t} \tag{4.43}$$

And

$$\frac{V(l)}{I(l)} = Z. \tag{4.44}$$

---

[5]This speed is often designated using a "c," however "c" here is the capacitance per unit length.

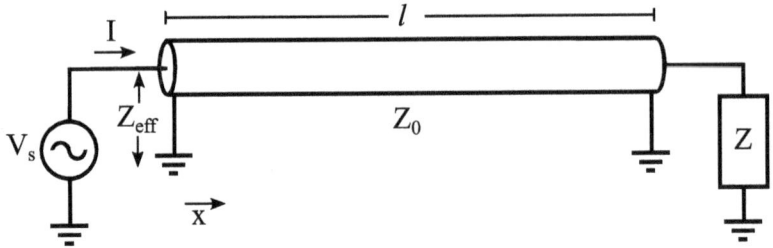

Fig. 4.17 An impedance at one end of a cable may appear to be completely different if measured at the other end

Applying the condition at $x = 0$ gives

$$V_S = V_R + V_L. \tag{4.45}$$

The condition at $x = l$ gives

$$Z = Z_0 \frac{V_R e^{-ikl} + V_L e^{+ikl}}{V_R e^{-ikl} - V_L e^{+ikl}}, \tag{4.46}$$

or, solving this for V_L,

$$V_L = V_R e^{2ikl} \frac{Z - Z_0}{Z + Z_0}. \tag{4.47}$$

Putting that result back into the condition at $x = 0$ gives

$$V_R = V_S \left(1 + \frac{Z - Z_0}{Z + Z_0} e^{2ikl}\right)^{-1}, \tag{4.48}$$

and then

$$\begin{aligned} I(0) &= (V_R - V_L)/Z_0 = (2V_R - V_S)/Z_0 \\ &= \frac{V_S}{Z_0} \left(\frac{Z_0 \cos(kl) - iZ \sin(kl)}{Z \cos(kl) - iZ_0 \sin(kl)}\right). \end{aligned} \tag{4.49}$$

The "effective impedance," Z_{eff}, seen by the signal source is then

$$Z_{\mathit{eff}} = \frac{V_S}{I(0)} = Z_0 \left(\frac{Z \cos(kl) - iZ_0 \sin(kl)}{Z_0 \cos(kl) - iZ \sin(kl)}\right) = Z_0 \left(\frac{Z - iZ_0 \tan(kl)}{Z_0 - iZ \tan(kl)}\right). \tag{4.50}$$

In addition to computerized solutions, a special graphic known as a Smith chart may be useful to those for whom such computations occur often. If $Z = Z_0$, then

Cable Models

$Z_{eff} = Z_0$ and the cable is "impedance matched" to the load.[6] One other result that is easy to show is that if $kl = n\pi$, where n is any integer (that is, the transmission line length is a multiple of $\lambda/2$), then $Z_{eff} = Z$ regardless of the value of Z_0.

Of interest for some applications are transmission lines (cables) where $kl = (n + \frac{1}{2})\pi$, the shortest being a quarter wavelength ($n = 0$). It is easy to show that for a quarter-wavelength transmission line

$$Z_{eff} = \frac{Z_0^2}{Z}, \tag{4.51}$$

and in particular, if Z is a short circuit, Z_{eff} corresponds to an open circuit, and if Z is an open circuit, Z_{eff} is that of a short circuit. In addition, if Z corresponds to a capacitance, Z_{eff} looks like an inductance and vice versa. The effect of the cable length is very far from negligible in this case.

In practice, these results show that if the cable is very short compared to a wavelength (i.e., the signal's wavelength within the cable) then the effects of the cable can be neglected. On the other hand, once the cable becomes comparable to a fair fraction of the wavelength, the cable can have a huge effect. Remember that the speed of light is roughly 1 foot (25 cm) per nanosecond, and modern electronics can have clock speeds over 1 GHz (that is, less than 1 ns per clock pulse), then even a relatively short cable may need to be treated carefully.

For most commercially available cables the wave speed is about 0.5–0.8 times the speed of light in a vacuum. Hence, minimum cable lengths that may become of concern can be estimated from those values. For frequencies below 1 MHz, a cable 10 m long will not be much different from a short cable, however at 100 MHz, even a 10 cm cable may be significant.

Capacitor and Inductor Labels

It is common practice that numerical identifiers on capacitors are usually either in pF or μF. If the value does not include units, then if the value shown is below 1.0, it is (probably) μF and if the value is larger than 1, it is (probably) pF. Increasingly nF and mF are also being used, though generally not without units displayed. Hence, a capacitor labeled 0.1 would probably be 0.1 μF and a capacitor labeled 33 would be 33 pF. For values in pF, a third number is often used for a power of ten. For example, a capacitor labeled "104" would be 10×10^4 pF, which is the same as 10^5 pF = 0.1 μF, and a capacitor labeled "561" would be 560 pF. If the third digit is a zero, however, it is probably a third digit. Thus, a label 150 would (probably)

---

[6] On some equipment, the input/output connectors may be labeled with a resistance value such as "50 Ω" or "1 MΩ." These values are the equivalent input/output impedance of the device. When so marked, it is usually the expectation that the characteristic impedance for cables which make those connections match that value.

be 150 pF, the same as a label 151. As for resistors, an "R" is sometimes used as a decimal point for smaller capacitors. For example, 4R7 would be 4.7 pF.

On many capacitors, a single uppercase letter follows the value of the capacitor. This is usually not a multiplier but a tolerance code, a temperature coefficient, or some other identifier.

Numerical identifiers on inductors are highly variable and depend much more strongly on the manufacturer and intended use than is seen for resistors and capacitors.

Duality

In electronics, and more generally in electricity and magnetism, there is a concept called duality. For electronics, if two things are related by the interchange of voltage and current, then they are called duals. Voltage sources and current sources are obvious duals. Ohm's law is usually written $V = IZ$, where Z is the (complex) impedance. Defining the (complex) admittance, $Y \equiv 1/Z$, Ohm's law can equally well be written $I = YV$. This has the same form as the previous version, but the roles of current and voltage have been switched. Hence, impedance and admittance are duals.

Inductors and capacitors are also duals. The basic relationships are

$$I_C = C\frac{dV_C}{dt} \text{ and } V_L = -L\frac{dI_L}{dt}, \tag{4.52}$$

which are of the same form except the roles of current and voltage have been switched. Likewise, series resistors (that make a voltage divider) and parallel resistors (that make a current divider) are duals. Defining the conductance, $G \equiv 1/R$, the voltage divider written for series resistors and the current divider written for parallel conductances are[7]

$$V_k = V_o \frac{R_k}{\sum_{all} R_i} \text{ and } I_k = I_0 \frac{G_k}{\sum_{all} G_i}, \tag{4.53}$$

which are of the same form except the roles of current and voltage have been switched.

Entire circuits can have a dual circuit—one with equations to solve that are the same except that voltage and current have been swapped. Duals will not play a significant role in this text beyond this brief introduction to them. It should be noted, however, that sometimes when solving a circuit, and perhaps even understanding a circuit, the task might be made easier by considering a dual.

---

[7]These dividers were first introduced in Chap. 1.

Problems

1. Consider the two circuits in Fig. 4.P1 for the special case when $R = \omega L = 1/\omega C$, where ω is the angular frequency of the voltage source. Hence $Z_L = iR$ and $Z_C = -iR$. (a) Compute I_0 and V_{out} for both circuits. (b) Now compute I_0 for both circuits for the case where the output has been shorted to ground (so $V_{out} = 0$).

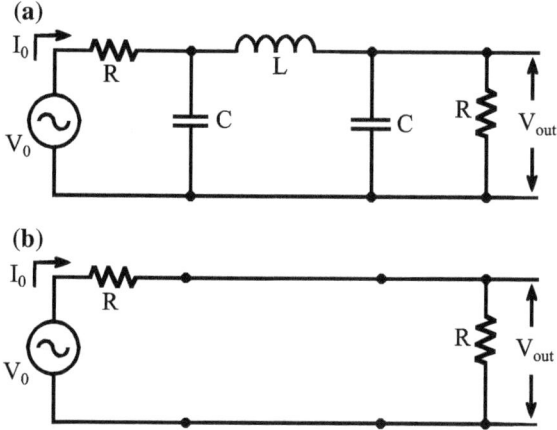

Fig. 4.P1 Problem 1

2. The bridge circuit in Fig. 4.P2 is to be used to determine L and its resistance using the parallel model. A known capacitance, C, and resistance R_2, are varied to find a null condition across R_5. Once the null condition is found, what is the value of the unknown inductance and its parallel resistance, R_{LP}, in terms of R_1, C, and R_2?

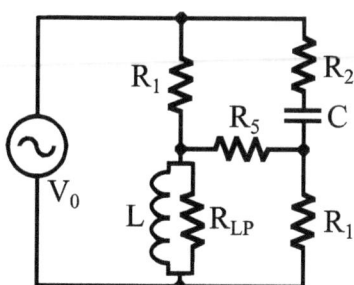

Fig. 4.P2 Problem 2

3. The switching circuit in Fig. 4.P3 has two pairs of switches. The switches operate together within each pair, as indicated by the dashed line. That is, the pair of switches S_1 and S_2 are either both on or both off, and similarly for the pair S_3 and S_4. In operation, the two pairs are switched back and forth such that only one pair is on at any given time. If this switching is done very rapidly compared to $R_L C_2$, what is V_{out}?

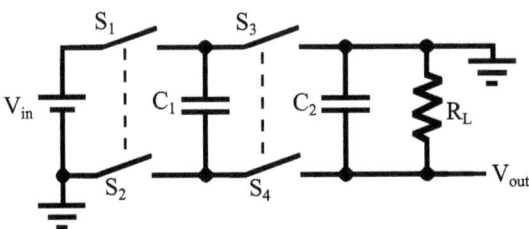

Fig. 4.P3 Problem 3

4. One common coaxial cable used with scientific instrumentation is referred to as "RG-58." While there are several varieties, the cable impedance will be close to 50 Ω and will have a capacitance per unit length of about 100 pF/m. As a fraction of the speed of light, how fast do electric signals propagate in such a cable?

5. Assuming a force that is sinusoidal in time with angular frequency ω, and using imaginary numbers as was done in Chap. 3 for capacitors and inductors, derive the relationship between force and velocity corresponding to Newton's second law ($F = ma$) and between the force and velocity for an ideal spring ($F = -kx$). For what value of ω will those forces be equal?

6. (Challenge Problem) A capacitor is charged through an inductor, as shown in Fig. 4.P6. The switch is closed at $t = 0$ at which time the current begins to rise and then to oscillate. When the current next becomes zero, the switch is re-opened. After this process, what is the voltage across the capacitor? Assume the quality factor, $Q = \omega L/r$, is much larger than 1. Show that the energy lost due to the resistance is proportional to $1/Q$, and hence can be much less than the energy stored in the capacitor.

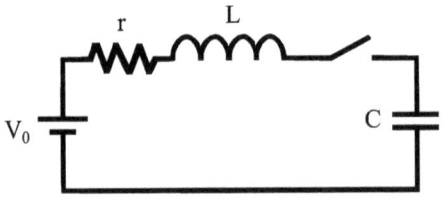

Fig. 4.P6 Problem 6

References

G. Esen et al., Transmission line impedance of carbon nanotube thin films for chemical sensing. Appl. Phys. Lett. **90**, 123510 (2007)

F. Heinrich, Entropy change when charging a capacitor. Am. J. Phys. **54**, 742–744 (1986)

R.V. Jones, J.C.S. Richards, The design and some applications of sensitive capacitance micrometers. J. Phys. E **6**, 589–600 (1973)

S. Paul, A.M. Schlaffer, J.A. Nossek, Optimal charging of capacitors. IEEE Trans. Circ. Syst. I **47**, 1009 (2000)

I.M. Sokolov, J. Klafter, A. Blumen, Fractional kinetics. Phys. Today **55**(11), 48–55 (2002)

J.A. Van Den Akker, G.M. Webb, A method for measuring high resistance. Rev. Sci. Instruments **7**, 44–46 (1936)

Chapter 5
The Laplace Transform

The Laplace transform is a linear operation that can be performed on functions of a single variable. For electronics, the transformation is taken from the time domain (the single variable is t) to the "s-domain" where s is a generalized complex frequency. The transform is useful for solving systems of linear differential equations such as those found for some electronic circuits. An understanding of the transform can lead to a description of circuit behavior expressed in terms of poles and zeros in the s-domain.

As a word of warning, this chapter is somewhat mathematically intense. The mathematics is not particularly advanced, at least for those who have made it through calculus, however there is a lot of it.

A mathematical transform is a way of rearranging all the information contained within a function. An invertible transform loses no information so you can get back to where you started. It is like rearranging the books on a bookshelf—an arrangement of the books based on alphabetic order by title can be transformed to an arrangement based on alphabetic order by author without losing any books. The problem of finding all the books by a certain author is certainly easier if the books are in order by author rather than by title. The mathematics is much more abstract, but that is, in essence, what is happening here.

The Transform

One form of the Laplace transform, $F(s)$, of a function $f(t)$, is given by[1]

---

[1]The Laplace Transform is named after the French mathematician Pierre Simon, Marquis de Laplace, (1749–1827), who is credited with its development as well as with many other discoveries in mathematics, physics, and astronomy.

© Springer Nature Switzerland AG 2020
B. H. Suits, *Electronics for Physicists*, Undergraduate Lecture Notes in Physics, https://doi.org/10.1007/978-3-030-39088-4_5

$$\mathcal{L}(f(t)) = \mathbf{F}(s) = \int_{0+}^{\infty} f(t)e^{-st}dt, \quad (5.1)$$

where s is a complex number and "0+" means the lower limit should be interpreted to be "just after $t = 0$." For the discussion here, t is time and the function is transformed to the "s-domain." The inverse transform can be written as a complex integral along a particular path in the complex domain and will not be reproduced here.

It should be noted that the Fourier transform can be considered to be the special case of the Laplace transform where s is restricted to be a purely imaginary number. That is, in general write $s = \sigma + i\omega$ and for the Fourier transform include only the special case where $\sigma = 0$. Note also that in general the frequency, ω, can be any real number including both positive and negative—that is, for this use, negative frequencies are taken as distinct from positive frequencies.

The Laplace transform is invertible. That is, there is uniqueness so that there is a one-to-one correspondence between a function and the function that is its transform. They appear in pairs. To execute the appropriate integral for the inverse is not always a simple thing to do. Fortunately, for electronics applications, one rarely needs to compute these integrals. Instead, a table of Laplace Transform Pairs is used. That is, one simply looks up the answers. A selection of such pairs is included in Table 5.1.

Table 5.1 Some Laplace transform Pairs

	$f(t)$	$F(s)$
1.	1	$1/s$
2.	t	$1/s^2$
3.	e^{-at}	$1/(s+a)$
4.	te^{-at}	$1/(s+a)^2$
5.	$\frac{t^{n-1}}{(n-1)!}e^{-at}$	$1/(s+a)^n = (s+a)^{-n}$
6.	$\sin \omega t$	$\omega/(s^2+\omega^2)$
7.	$\cos \omega t$	$s/(s^2+\omega^2)$
8.	$e^{-at}\sin \omega t$	$\omega/\left[(s+a)^2+\omega^2\right]$
9.	$e^{-at}\cos \omega t$	$(s+a)/\left[(s+a)^2+\omega^2\right]$
10.	$\sqrt{c^2+d^2}e^{-at}\cos(\omega t - \tan^{-1}(d/c))$	$\dfrac{c(s+a)+d\omega}{(s+a)^2+\omega^2}$
11.	$\sinh \omega t$	$\omega/(s^2-\omega^2)$
12.	$\cosh \omega t$	$s/(s^2-\omega^2)$
	Special transforms	
13.	$\frac{df}{dt}$	$s\mathbf{F}(s) - f(0^+)$
14.	$\int_0^t f(\tau)d\tau$	$\mathbf{F}(s)/s$

(continued)

Table 5.1 (continued)

	$f(t)$	$F(s)$
15.	$c_1 f_1(t) + c_2 f_2(t)$	$c_1 \mathbf{F}_1(s) + c_2 \mathbf{F}_2(s)$
16.	$\int_0^t f_1(t) f_2(t-\tau) d\tau$	$\mathbf{F}_1(s)\mathbf{F}_2(s)$
17.	$\delta(t-T)$ (delta function, $T > 0$)	e^{-sT}
18.	$\frac{d^n}{dt^n}(\delta(t))$ (n-th derivative of delta function)	s^n
19.	$\delta^{(-n)}(t)$ (n-th integral of delta function)	s^{-n}

Comments on the Transform Table
(A) The transform variable, s, is complex. The Fourier transform is the special case of this more general transform where s is restricted to be pure imaginary. Thus, such a table can also be used for Fourier transforms.
(B) In the table above, a, c, and d are considered real constants, though many of the transforms are still valid if they are complex or pure imaginary. Consult a more general table for those cases.
(C) 1–4 are special cases of 5.
(D) 1, 3, and 6–9 are special cases of 10.
(E) 6 and 7 are related to 11 and 12. If you replace ω with $i\omega$ in one pair, you generate the other.
(F) 15 is a statement of the linearity of the transform.
(G) 14 is a special case of 16 (where one of the functions is constant).
(H) 16 is known as the "convolution theorem."
(I) 18 and 19 are particularly useful if you have an equation with a delta-function in it and need to take a derivative. This occurs for some idealized models of real systems (e.g., In E&M, current through a thin wire is often modeled using a delta function for the current density.).

An advantage of the Laplace transform is that operations and functions such as derivatives, delta functions, derivatives of delta functions, etc., become nicely behaved in the s-domain making solutions, in some sense, very easy. The major work becomes transforming the problem and then transforming back after the easy solution is found.

Using transforms of various kinds is a standard practice in Physics. A problem is transformed into another one that is easier to solve, then transformed back. One of the simplest ways this is accomplished is to transform from one coordinate system to another. Indeed, the Laplace transform is a generalization of that idea where the t-coordinate axis is transformed into an s-coordinate axis.

Since the transform is an integral, which is a linear operation, the transform itself is linear. That is for any two constants a and b and any two functions $f(t)$ and $g(t)$:

$$\mathcal{L}(af(t) + bg(t)) = a\mathcal{L}(f(t)) + b\mathcal{L}(g(t)) \qquad (5.2)$$

This linearity is important for finding solutions by reverse look-up.

The use of the Laplace transform will be illustrated using the concrete examples that follow. These examples from electronics use a look-up table to do the transforms and inverse transforms. The difficulty will be putting the inverse problem into a form that appears in the look-up table.

Laplace Transform Example 1

Consider the circuit in Fig. 5.1, where the switch is open for $t < 0$ and closed for $t > 0$. The previous methods (simple L/R time constants and/or complex impedances) will not work for this circuit because it has a sinusoidal source *and* a switch.

Using Kirchhoff's voltage law around the loop (after the switch is closed),

$$V_0 \sin(\omega_0 t) - L\frac{dI}{dt} - IR = 0, \tag{5.3}$$

and the objective is to find $I(t)$. Note that here ω_0 is the frequency of the source and is a constant, and is not the same as the variable ω, which is the imaginary part of s. Now transform each term in the equation (e.g., using a look-up table) to get

$$V_0 \frac{\omega_0}{s^2 + \omega_0^2} - L(s\mathbf{I}(s) - I(0)) - \mathbf{I}(s)R = 0, \tag{5.4}$$

which has transformed the immediate problem to that of finding $\mathbf{I}(s)$. Note that $I(0)$ refers to the current at $t = 0$ and for this problem $I(0) = 0$ (the inductor keeps the current constant, as best it can, and before the switch was closed the current was zero). Solving for the current in the s-domain is now straightforward, and

$$\mathbf{I}(s) = \frac{V_0}{L}\frac{\omega_0}{s^2 + \omega_0^2}\frac{1}{s + R/L} = \frac{V_0}{L}\frac{\omega_0}{(s+i\omega_0)(s-i\omega_0)(s+R/L)} \tag{5.5}$$

Note that a strategy being used is that if possible, the result is written using products of terms where the coefficient in front of each "s^n" term is one. The second part of the strategy is to try to write it as a product involving terms such as $(s^n + a)$, where a is some (possibly complex, possibly zero) constant. For this example,

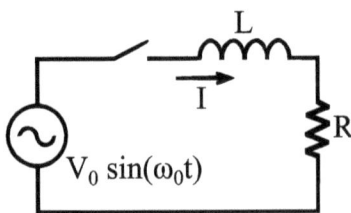

Fig. 5.1 A simple LR circuit with a switch used as an example

Laplace Transform Example 1

$n = 2$ or $n = 1$ for all the terms in the denominator. This is done because such terms appear in the look-up table. Also, this makes the denominator look like the denominator obtained when several fractions are added using a common denominator. Now the hardest part of the problem is to transform back to find $I(t)$.

The expression on the right above does not appear in the look-up table. One way to express it using terms that do show up is to use the method of partial fractions, illustrated below.

Method I

Noting that the individual terms in the denominator are in the table, guess a solution using unknown constant values A, B, and D, and then require that[2]

$$\frac{A}{s+i\omega_0} + \frac{B}{s-i\omega_0} + \frac{D}{s+R/L} = \frac{\omega_0}{(s+i\omega_0)(s-i\omega_0)(s+R/L)}. \quad (5.6)$$

The constant out front (V_0/L) is dropped for now, and will simply be reintroduced at the end. The problem is to find constants A, B, and D that make this guess work. Creating a common denominator and adding, the left side becomes

$$\text{(left side)} = \frac{A(s-i\omega_0)(s+R/L) + B(s+i\omega_0)(s+R/L) + D(s+i\omega_0)(s-i\omega_0)}{(s+i\omega_0)(s-i\omega_0)(s+R/L)}. \quad (5.7)$$

This will equal the right side only if the numerators are the same *for all s*. To get the numerators to be the same for all s, the coefficients for *each* power of s must be equal. In this example, the only power of s in the numerator on the right side that has a non-zero coefficient is s^0 (i.e., the constant term, $\omega_0 = \omega_0 s^0$). Now expand the numerator on the left side and make the coefficient for each power of s have the same coefficient as on the right. That is

$$\begin{aligned} s^2 \text{ terms: } & (A+B+D) = 0 \\ s^1 \text{ terms: } & A(R/L - i\omega_0) + B(R/L + i\omega_0) = 0 \\ s^0 \text{ terms: } & A(-i\omega_0 R/L) + B(i\omega_0 R/L) + D\omega_0^2 = \omega_0, \end{aligned} \quad (5.8)$$

which now need to be solved for A, B, and D. If a unique solution cannot be found, the initial guess was bad.

---

[2] "C" is skipped here so that there is no confusion with any capacitance values.

For this example, after some algebra, and defining $\tau = L/R$,

$$A = -\frac{\tau}{2i}\left(\frac{1+i\omega_0\tau}{1+\omega_0^2\tau^2}\right); \quad B = A^*; \quad D = \frac{1}{\omega_0}\frac{\omega_0^2\tau^2}{1+\omega_0^2\tau^2}. \tag{5.9}$$

Note that A, B, and D are now all constants with known values. Now write (reintroducing the constant out front, V_0/L),

$$\mathbf{I}(s) = \frac{V_0}{L}\left(\frac{A}{s+i\omega_0} + \frac{B}{s-i\omega_0} + \frac{D}{s+R/L}\right) \tag{5.10}$$

and all these terms are in the form $1/(s + a)$ which is in the Laplace transform tables. When transformed back such terms result in exponentials. Hence, using the table

$$\begin{aligned}I(t) &= \frac{V_0}{L}\left(Ae^{-i\omega_0 t} + Be^{i\omega_0 t} + De^{-t/\tau}\right) \\ &= \frac{V_0\tau}{L}\left(\frac{1}{1+\omega_0^2\tau^2}\right)\left(\frac{1}{2i}(1-i\omega_0\tau)e^{i\omega_0 t} - \frac{1}{2i}(1+i\omega_0\tau)e^{-i\omega_0 t} + \omega_0\tau e^{-t/\tau}\right) \\ &= \frac{V_0}{R}\left(\frac{1}{\sqrt{1+\omega_0^2\tau^2}}\sin(\omega_0 t - \varphi) + \left(\frac{\omega_0}{1+\omega_0^2\tau^2}\right)e^{-t/\tau}\right),\end{aligned}$$

$$\tag{5.11}$$

where $\phi = \tan^{-1}(\omega_0\tau)$. Note that for long times the solution is the same as obtained from the usual a.c. analysis using complex impedances—that is, eventually one cannot tell that a switch was closed a long time ago. The term on the far right is a transient that is present just after the switch is closed and the transient will persist for times of order $\tau = L/R$. Hence in this case "a long time" means long compared to L/R.

Method II

Multiplying the first two terms in the denominator of $\mathbf{I}(s)$,

$$\mathbf{I}(s) = \frac{V_0}{L}\frac{\omega_0}{(s^2+\omega_0^2)(s+R/L)}, \tag{5.12}$$

each of the terms in the denominator appears in the table separately. Hence, using constants A, B, and D (none of which are assumed to be the same values as found in Method I), try

Laplace Transform Example 1

$$\frac{As+B}{\left(s^2+\omega_0^2\right)} + \frac{D}{(s+R/L)} = \frac{\omega_0}{\left(s^2+\omega_0^2\right)(s+R/L)}, \tag{5.13}$$

noting that the numerators in our guess should be one power less than their corresponding denominator. Creating a common denominator and adding on the left side

$$\frac{As^2 + AsR/L + Bs + BR/L + Ds^2 + D\omega_0^2}{\left(s^2+\omega_0^2\right)(s+R/L)} = \frac{\omega_0}{\left(s^2+\omega_0^2\right)(s+R/L)}, \tag{5.14}$$

and once again, the coefficients in front of all the powers of s must match:

$$\begin{aligned} s^2 \text{ terms: } & A + D = 0 \\ s^1 \text{ terms: } & AR/L + B = 0 \\ s^0 \text{ terms: } & BR/L + D\omega_0^2 = \omega_0, \end{aligned} \tag{5.15}$$

which yields

$$A = -B\frac{L}{R} = -D = \frac{\omega_0}{\omega_0^2 + R^2 L^2}. \tag{5.16}$$

Once again defining $\tau = L/R$ and putting these known constants back into the initial guess,

$$\mathbf{I}(s) = \frac{V_0}{\omega_0 L} \left(\frac{\omega_0^2 \tau^2}{1+\omega_0^2 \tau^2}\right) \left(\frac{-s}{s^2+\omega_0^2} + \frac{1/\tau}{s^2+\omega_0^2} + \frac{1}{s+1/\tau}\right), \tag{5.17}$$

which can easily be inverted using Table 5.1 (entries 3, 6, and 7) to give

$$I(t) = \frac{V_0}{R} \left(\frac{1}{1+\omega_0^2 \tau^2}\right) \left(-\omega_0 \tau \cos \omega_0 t + \sin \omega_0 t + \omega_0 \tau e^{-t/\tau}\right), \tag{5.18}$$

which can be made to match the form of the final result from Method I through use of the identity[3]

$$p \cos z + q \sin z = \sqrt{p^2+q^2} \sin(z+\phi); \quad \phi = \tan^{-1}(p/q). \tag{5.19}$$

Note also that while Kirchhoff's voltage law produced a differential equation, the Laplace Transform eliminated the derivative making the problem easy to solve in the s-domain using simple algebra. The Laplace transform may also be a useful tool

---

[3] As was the case for complex phase angles, the arctangent must be used with care.

to solve many other differential equations, including equations with delta functions, derivatives of delta functions, non-linear terms, and possibly some other pathological terms.

Laplace Transform Example 2

Consider the circuit of Fig. 5.2, where the switch has been open for a long time and is closed at $t = 0$ (the capacitor is initially uncharged). Now solve for the currents in the circuit for $t > 0$. This circuit can be solved by finding the appropriate time constant, writing an exponential solution, and applying the appropriate initial and "long time" values. On the other hand, if the voltage source were sinusoidal, or even more complicated, such a solution is not valid. Consider solving this with the Laplace transform.

For $t > 0$ Kirchhoff's voltage law around the two loops shown yields two equations,

$$V_0 - I_1 R_1 - (I_1 - I_2)R_2 = 0$$
$$(I_1 - I_2)R_2 - I_2 R_3 - \frac{1}{C} \int_0^t I_2 \, dt = 0, \quad (5.20)$$

where the integral yields the net charge accumulated on the capacitor since $t = 0$. Since the charge started at zero for this example, this integral also gives the total charge on the capacitor.

Applying the Laplace transform to both sides of both equations, indicating the transformed function with bold, and with a little rearrangement,

$$\mathbf{I}_1(R_1 + R_2) - \mathbf{I}_2 R_2 = V_0/s$$
$$\mathbf{I}_1 R_2 - \mathbf{I}_2(R_2 + R_3 + 1/(sC)) = 0. \quad (5.21)$$

Solving the second of these to get $\mathbf{I}_2$ in terms of $\mathbf{I}_1$, and putting that result back into the first yields solutions for $\mathbf{I}_1$ and $\mathbf{I}_2$:

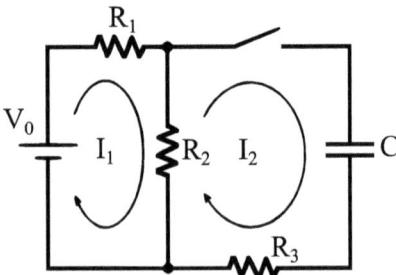

Fig. 5.2 A two-loop RC circuit used as an example

Laplace Transform Example 2

$$\mathbf{I_2} = \frac{R_2}{R_2 + R_3 + 1/(sC)} \mathbf{I_1}$$

$$\mathbf{I_1}\left(R_1 + R_2 - R_2 \frac{R_2}{R_2 + R_3 + 1/(sC)}\right) = \frac{V_0}{s}.$$

(5.22)

Now, in an attempt to simplify the notation, define the constants

$$\alpha = (R_2 + R_3)C; \quad \beta = (R_1 R_2 + R_1 R_3 + R_2 R_3)C; \quad \gamma = (R_1 + R_2)/\beta, \quad (5.23)$$

and then

$$\mathbf{I_1} = V_0 \frac{\alpha}{\beta} \frac{s + 1/\alpha}{s(s+\gamma)} = V_0 \frac{\alpha}{\beta}\left[\frac{1}{s+\gamma} + \frac{1/\alpha}{s(s+\gamma)}\right]$$

$$\mathbf{I_2} = \frac{sR_2 C/\alpha}{s + 1/\alpha} \mathbf{I_1} = V_0 \frac{R_2 C}{\beta} \frac{1}{s+\gamma}.$$

(5.24)

Now these currents must be transformed back to the time domain. From the look-up table:

$$\frac{1}{s} \to 1; \quad \frac{1}{s+\gamma} \to e^{-\gamma t}; \quad \frac{1}{s + 1/\alpha} \to e^{-t/\alpha}.$$

(5.25)

The products in the denominator for $\mathbf{I_1}$ suggests that the method of partial fractions might work. That is, for $\mathbf{I_1}$ find constants A and B such that

$$\frac{A}{s} + \frac{B}{s+\gamma} = \frac{A(s+\gamma) + Bs}{s(s+\gamma)} = \frac{1/\alpha}{s(s+\gamma)},$$

(5.26)

which will be true if the coefficients in front of each power of s in the numerator are the same on both sides. That is,

$$A + B = 0 \text{ and } A\gamma = 1/\alpha,$$

(5.27)

which gives

$$A = \frac{1}{\alpha\gamma}, \quad B = \frac{-1}{\alpha\gamma},$$

(5.28)

so the solution can be written

$$I_1 = V_0 \frac{\alpha}{\beta}\left(e^{-\gamma t} + \frac{1}{\alpha\gamma}(1 - e^{-\gamma t})\right)$$

$$I_2 = V_0 \frac{R_2 C}{\beta} e^{-\gamma t}.$$

(5.29)

Substituting in values for α, β, and γ, yields the final result.

Laplace Transform Example 3

Consider the circuit in Fig. 5.3 where initially (at $t = 0$) the current is zero and the charge on the capacitor is Q_0 with positive charge as defined in the schematic. This problem is analogous to a damped harmonic oscillator starting at rest that is struck at $t = 0$.

Writing Kirchhoff's voltage law around this loop,

$$\frac{Q}{C} + IR + L\frac{dI}{dt} = 0, \tag{5.30}$$

where the charge on the capacitor at time t (t > 0) is given by

$$Q = Q_0 + \int_0^t I\, dt. \tag{5.31}$$

Using the Laplace transform on both sides of the equation,

$$\frac{Q_0}{sC} + \frac{\mathbf{I}(s)}{sC} + R\mathbf{I}(s) + L(s\mathbf{I}(s) + I(0)) = 0, \tag{5.32}$$

and solving for $\mathbf{I}(s)$, using $I(t = 0) = 0$, as specified above, yields

$$\mathbf{I}(s) = Q_0 \frac{1}{sC(sL + R + 1/(sC))} = \frac{Q_0}{LC} \frac{1}{(s^2 + sR/L + 1/(LC))}. \tag{5.33}$$

For convenience, define $\omega_0 = \sqrt{1/(LC)}$ and $2\gamma = R/L$, so this can be written

$$\mathbf{I}(s) = \omega_0^2 Q_0 \frac{1}{(s^2 + 2\gamma s + \omega_0^2)}. \tag{5.34}$$

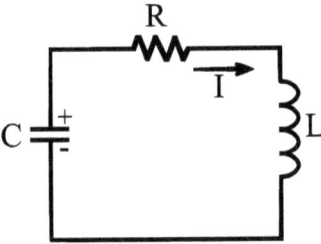

Fig. 5.3 An RLC circuit used as an example

Laplace Transform Example 3

Now the table does not include a generic quadratic but it does include a few entries that are close. Instead, use the quadratic formula to find the zeros, and then write the denominator as a product of two first order terms. The zeros of the quadratic are given by

$$s = \frac{1}{2}\left(-2\gamma \pm \sqrt{4\gamma^2 - 4\omega_0^2}\right) = \left(-\gamma \pm i\omega_0\sqrt{1 + \frac{\gamma^2}{\omega_0^2}}\right), \quad (5.35)$$

and defining $\omega' = \omega_0\sqrt{1 - (\gamma/\omega_0)^2}$, gives

$$\frac{1}{s^2 + 2\gamma s + \omega_0^2} = \frac{1}{s + (\gamma + i\omega')} \cdot \frac{1}{s + (\gamma - i\omega')}. \quad (5.36)$$

Using partial fractions with the constants A and B,

$$\frac{A}{s + (\gamma + i\omega')} + \frac{B}{s + (\gamma - i\omega')} = \frac{1}{s + (\gamma + i\omega')} \cdot \frac{1}{s + (\gamma - i\omega')}, \quad (5.37)$$

and creating a common denominator on the left, the two will be equal if A and B are chosen so that the numerators are equal. That is, for all s, A and B must satisfy

$$A(s + \gamma - i\omega') + B(s + \gamma + i\omega') = 1, \quad (5.38)$$

and as before, the solution can be found by equating all powers of s. That is

$$\begin{aligned} s^2 \text{ and higher} &\to \text{none.} \\ s^1 &\to A + B = 0 \\ s^0 &\to A(\gamma - i\omega') + B(\gamma + i\omega') = 1, \end{aligned} \quad (5.39)$$

which is easily solved to give

$$A = \frac{i}{2\omega'}; \quad B = -A. \quad (5.40)$$

Hence,

$$\mathbf{I}(s) = \omega_0^2 Q_0 \left(\frac{1}{2\omega'}\right)\left(\frac{1}{s + \gamma + i\omega'} - \frac{1}{s + \gamma - i\omega'}\right), \quad (5.41)$$

which can be readily transformed using the table to give

$$I(t) = \omega' \left(\frac{\omega_0}{\omega'}\right)^2 Q_0 \left(\frac{1}{2i}\right) \left(e^{-(\gamma - i\omega')} - e^{-(\gamma + i\omega')}\right)$$
$$= \omega' \left(\frac{\omega_0}{\omega'}\right)^2 Q_0 e^{-\gamma t} \sin \omega' t. \tag{5.42}$$

The solution looks like a damped sinusoid, as should be expected. However, one must be careful, as this conclusion is actually only valid if ω' is real and non-zero (the "underdamped case"), which was implicitly assumed but which is not necessarily true. The case where $\gamma/\omega_0 > 1$ (the "overdamped case") results in ω', as defined above, being pure imaginary. That case is easily treated using identities for the sine of a complex value, rather than starting over. That is, redefining $\omega' = \omega_0 \sqrt{(\gamma/\omega_0)^2 - 1}$, so that it is real and positive, the solution becomes

$$I(t) = i\omega' \left(\frac{\omega_0}{\omega'}\right)^2 Q_0 e^{-\gamma t} \sin(-i\omega' t) = \omega' \left(\frac{\omega_0}{\omega'}\right)^2 Q_0 e^{-\gamma t} \sinh(\omega' t)$$
$$= \frac{\omega'}{2} \left(\frac{\omega_0}{\omega'}\right)^2 Q_0 \left(e^{-(\gamma - \omega')t} - e^{-(\gamma + \omega')t}\right). \tag{5.43}$$

As a function of time, this solution grows to a maximum and then decays to zero. Consult a good set of math tables for the trigonometric identities used for the complex arguments.

Now if $\gamma/\omega_0 = 1$, or equivalently $\omega' = 0$, then the solution above is invalid because during the solution both sides of the equation were multiplied by zero. Rewriting $I(s)$ for this special case (known as "critical damping") one gets

$$I(s) = \omega_0^2 Q_0 \frac{1}{(s+\gamma)^2}. \tag{5.44}$$

which is found on line 4 of the table and so is easily inverted to give

$$I(t) = \omega_0^2 Q_0 t \, e^{-\gamma t}, \tag{5.45}$$

that also starts at zero, grows, and then decays to zero. Though it may not be obvious, this solution can also be obtained by taking the limit $\omega' \to 0$ using either of the previous solutions.

Comment on Partial Fractions

Whenever there is a polynomial in the denominator of degree n, it can be factored, as above, or the guess can include a polynomial in the numerator of degree $n-1$. For example, to invert

Comment on Partial Fractions

$$F(s) = \frac{1}{s(s^2+b^2)}, \tag{5.46}$$

try

$$\frac{A}{s} + \frac{Bs+D}{s^2+b^2} = \frac{1}{s(s^2+b^2)} \tag{5.47}$$

with A, B, and D unknown constants to be determined. Note that the total number of constants necessary will be the same no matter how the fraction is divided. Using this method only helps if the remaining expressions can be found in the Laplace transform look-up table. Example 1, Method II, uses this method, and Example 3 above could have been done using this method.

Whenever there is a degeneracy in the denominator, that is, the same quantity to a power larger than 1, then do the partial fractions including all powers up to and including the one in the denominator. That is, to use partial fractions to invert

$$F(s) = \frac{1}{s(s+b)^3}, \tag{5.48}$$

that has the root $-b$ appearing three times, start by using four unknown constants and write

$$\frac{A}{s} + \frac{B}{(s+b)} + \frac{C}{(s+b)^2} + \frac{D}{(s+b)^3} = \frac{1}{s(s+b)^3}. \tag{5.49}$$

This section on the partial fraction method was for review and should not be considered a complete treatment by any means. For more information about the method of partial fractions, consult an appropriate mathematics text.

Poles and Zeros

In general, the currents in a linear circuit can often be expressed in the s-domain in the form

$$\mathbf{I}(s) = A \frac{(s-z_1)(s-z_2)\cdots(s-z_n)}{(s-p_1)(s-p_2)\cdots(s-p_m)} \tag{5.50}$$

where z_i are "zeros" of the response (where the numerator is zero) and p_i are "poles" of the response (where the denominator is zero). Both z_i and p_i may be complex values, including pure real and pure imaginary values.

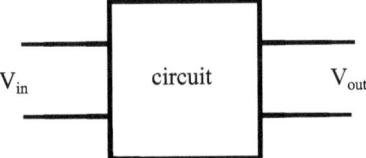

Fig. 5.4 A general representation of a circuit with an input and output

In the Laplace Transform Example 1 (see Method I) above there were no zeros and three poles ($i\omega$, $-i\omega$, and $-R/L$). In Example 2, I_1 has two poles and one zero and I_2 has one pole, and in Example 3 there were two complex poles.

More generally, one may wish to consider the response (a current or voltage) at one position in a circuit due to an excitation (a current or voltage source) at some other position in the circuit. For linear circuits, the ratio of the response to the excitation, known as the transfer function, can always be expressed in the form shown above. One place this occurs is in "input-output" devices (which includes filters and amplifiers) that look like the simplified circuit of Fig. 5.4.

Hence, if the circuit is linear, V_{out}/V_{in} can be put in the form shown above (for the s-domain) and can be described using poles and zeros. If the poles and zeros are specified, the behavior of the circuit, aside from a multiplicative constant, has also been specified. Thus, the difficult part of the problem, the inverse, can be avoided if one has an understanding of poles and zeros. The main reason for considering the Laplace transform here is, in fact, to get some of that understanding.

In many cases, the behavior of even a very complicated circuit, at least over the range where a circuit will actually be used, will be dominated by just a few poles and/or zeros. Not all of the poles and zeros need to be known. Hence the use of dominant poles and zeros is a convenient way to specify the important behavior of many circuits, or more generally, any system with behavior that can be described using linear differential equations.

Problems

1. Show that if $\mathbf{X}(s)$ is the Laplace transform of $x(t)$, and $y(t) = x(t)e^{-\alpha t}$, then $\mathbf{Y}(s) = \mathbf{X}(s + \alpha)$.
2. Show that if $x(t)$ is a square wave with period T, such that

$$x(t) = \begin{cases} 1, & nT \leq t < \left(n+\tfrac{1}{2}\right)T \\ -1, & \left(n+\tfrac{1}{2}\right)T \leq t < (n+1)T \end{cases}$$

Problems

where n is any integer, then the Laplace transform of $x(t)$ is given by

$$X(s) = \frac{1}{s} \cdot \frac{1 - e^{-sT/2}}{1 + e^{-sT/2}} = \frac{1}{s}\tanh(sT/4).$$

Recall that the sum for an infinite geometric series is

$$a + ar + ar^2 + ar^3 + \cdots = a/(1-r) \quad \left(|r|^2 < 1\right).$$

3. Using the result above for the square wave, derive the Laplace transform for a triangle wave by noting that a triangle wave can be written as a time integral of a square wave.

4. For the circuit of Fig. 5.P4, the source operates at a frequency of $f_0 = 50$ Hz and the switch is closed at $t = 0$. Use the Laplace transform method to solve for the current through the inductor in this circuit when $t > 0$.

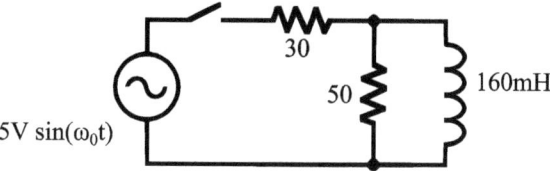

Fig. 5.P4 Problem 4

5. In the process of solving for the current in a circuit using the Laplace transform method, a student, Gunner Dufein, found that

$$\mathbf{I}(s) = \frac{3s}{(s+1)(s-2)} \text{ mA}.$$

Identify the poles and zeros associated with the circuit. What is this current as a function of time?

Chapter 6
Diodes

This chapter considers diodes—in particular, semiconductor diodes. Diodes are non-linear devices that pass current much more easily in one direction than in the opposite direction. The emphasis here is to learn techniques for simple analysis for circuits that contain diodes and to look at some simple and useful diode circuits. For a more detailed description of semiconductor diodes and how they work, or if more precise circuit analysis is desired, the reader should refer to a more advanced text.

Semiconductor Diodes

To understand how semiconductor diodes work, it is important to remember that an electron cannot share its state with another electron. Hence, when there are a number of possible electron states to fill, only one electron can go into each state. The lowest energy state for an N electron system, the so-called ground state, will have the lowest N energy states filled and the remainder of the states, all of which have higher energy, empty.

An atom will have discrete energy levels. When two such atoms are brought together those energy levels split into two due to the interactions between the electrons on each atom. A classical analog is the coupling of two harmonic oscillators or pendulums. With a weak coupling (weak compared to the other forces in the problem), what was a single resonant frequency for the oscillators becomes two distinct frequencies. The process continues as more atoms (oscillators) are brought together. Thus, if, say, 10^{23} atoms are brought near each other, such as in a solid, then the levels break up into 10^{23} sublevels. This large number of levels, that are almost of equal energy, are usually described using "band theory." That is, rather than talk about discrete levels as in an atom, one talks of a band of levels. While a discrete level is either occupied or empty, a band can be fully occupied, completely empty, or partially filled.

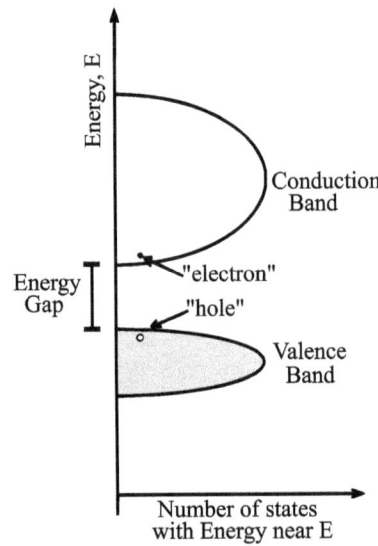

Fig. 6.1 Representation of the uppermost filled and lowermost empty energy bands in a pure semiconductor at absolute zero

The energy bands for a pure ("intrinsic") semiconducting material at absolute zero (see Fig. 6.1) will have an uppermost energy band that is completely filled, known as the valence band. Above that is a lowermost band that is completely empty, called the conduction band. In between is a region of energy where there are no allowed states—that is, there is an "energy gap" between the two bands. Completely filled or completely empty bands are "inert" in that they do not contribute to electrical conduction at normal operating voltages—in much the same way that filled electron shells make the noble gases inert to chemical reactions. In the case of an empty band, there are no charge carriers to carry a current. In the case of a full band there will always be as many electrons going in any given direction as are going in the opposite direction, and hence no net current.

A semiconducting material is one where the band gap is "not too large." When the gap is large, the material is an insulator. There is no sharp cut-off between semiconductors and insulators, but a practical value is that a band gap of ~0.3 to ~1 eV is desirable for a useful electronic device.[1]

At finite temperatures (such as room temperature) some electrons from the valence band will be excited into the conduction band leaving behind an empty state, a "hole," in the valence band. The "hole" is the absence of an electron. The empty state can move around and, compared to the inert state (completely full), it is positively charged.

---

[1] The electron-volt, or eV, is a non-SI unit of energy equal to the amount of energy gained by an electron when accelerated through 1 V. 1 eV = 1.60218×10^{-19} J.

Semiconductor Diodes

When referring to "electrons" in a semiconductor what is usually meant is "electrons in the conduction band" only. All the other electrons are inert and so need not be considered. Likewise, a "hole" refers only to empty states in the valence band. If an electron and hole "recombine," that is, an electron in the conduction band falls back to fill a hole in the valence band, one is left with the inert state, or "nothing." It is as if both the electron and hole disappeared.

Electrical current in a semiconductor can be carried by electrons in the conduction band and/or holes in the valence band. These charge carriers are free to move under the influence of an applied electric field—there are plenty of states available for them to occupy.

As a crude analogy, imagine a room filled with N chairs representing N possible states. If N people occupy the room and all are sitting, there will be no empty chairs. If the people are only allowed to change chairs one at a time, no one will be able to move in such a room. However, if the number of people is small compared to N, then they are pretty much free to move about. If it gets hot in the room, they all can move toward the window, and so on. On the other hand, if there are just a few less than N people in the room, there will be only a small number of empty chairs. The people next to those empty spots can choose to move into the empty spot, causing the empty spot to move in the opposite direction. If the room gets warm and the people try to move to the window, the empty spot will move away from the window. That empty spot is analogous to the hole in the semiconductor, but rather than talk about what fills the hole, it is easier to talk about the hole, in this case the absence of a person, as if it were the object of interest.

The conducting properties of the intrinsic ("i") semiconducting material can be modified by adding impurities. Adding one type of impurity effectively adds electrons to the system and is called n-type doping ("n" for the negative charge). Adding another type effectively removes electrons, the same as "adding holes," and is called p-type doping ("p" for the positive charge of a hole). The product of the electron density and the hole density is (roughly) constant, so with strong n-doping there will be very few holes around, and with p-doping there will be very few electrons available to carry current.

No net charge is added to the system with impurity doping. The compensating charge is contained in the immobile nucleus of the impurity atoms. That is, for n-doping, there will be extra protons present and with p-doping, a deficiency of protons when compared to the pure system (the "inert" state).

The basic semiconductor diode is the "p-n junction diode" and consists of a layer of p-doped semiconductor immediately adjacent to a layer of n-doped semiconductor (Fig. 6.2). The diode conducts from p to n relatively well, but does not conduct well in the reverse direction.

The electrons and holes will be moving around at speeds near the speed of light and can diffuse across the barrier. Of course, the "n" carriers will leave their extra proton behind and the "p" carriers their missing proton—a negative charge compared to the inert state. Hence there are forces trying to pull the electrons and holes back into their own region. An equilibrium is established between the tendency to diffuse and this electric field that pulls them back.

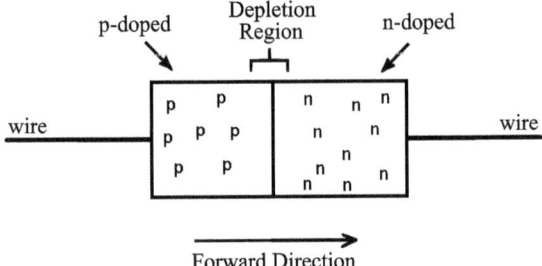

Fig. 6.2 A semiconductor diode is constructed from layers that are "doped" with impurities created an excess of holes ("p") in the valence band and/or an excess number of electronics ("n") in the conduction band. In equilibrium at the boundary, there will be a depletion region with no mobile charge carriers

Near the *p-n* junction is a "depletion region" where electrons have diffused across into the *p*-region and holes have diffused across to the *n*-region resulting in recombination, i.e., "nothing." The size of the depletion region varies considerably with conditions but is of the order of 1 μm across. Associated with the depletion region is an energy barrier to cross the junction that arises from the equilibrium electric field within the material. With no external electric field applied, the equilibrium is established so that the barrier is just barely high enough so that there is no net current. When an electric field is applied in one direction, the barrier is lowered and conduction occurs, when applied in the other direction the barrier is raised further and, of course, no conduction occurs since the barrier was already high enough to stop conduction.

The schematic symbol for a diode is shown in Fig. 6.3. The voltage across the diode is taken to be positive if the voltage at the anode is larger than the voltage at the cathode. The current through the diode is taken to be positive when it is in the forward direction. When the voltage and current are positive, the diode is said to be "forward biased." When the voltage is negative, trying to push the current backwards through the diode, the diode is said to be "reverse biased."

To understand electronic circuits containing diodes, the voltage-current relationship for the diode is all that really matters. Since that relationship is somewhat complicated for real diodes, simpler models are often used to describe the behavior. Which model is chosen will depend on the demands of the application being considered.

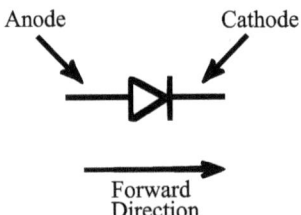

Fig. 6.3 The standard schematic for a diode. If the triangle is considered an arrowhead, it points in the forward direction, where conduction is easy. In the reverse direction, there is a wall to block the current

Diode Models

As stated above, a diode is a two-terminal device that conducts easily in one direction and poorly in the other. It is, therefore, a non-linear device. In general, the basic theorems for linear devices (superposition, etc.) cannot be expected to work for non-linear devices. Even simple equation solving methods can become quite cumbersome, if not impossible, to use when there are non-linear elements present. However, there are methods that can be used to get practical results.

Kirchhoff's laws must be obeyed in all circuits, and at the same time the current-voltage relationship for each device, the device rules, must be satisfied. To solve circuits with diodes, a current-voltage relationship for diodes is required. The most accurate "rule for the diode," that is, the relationship between the voltage across and current through the diode, is known only graphically based on measurements. In order to handle many non-linear devices, simplified models are used that approximate the real behavior of the devices. Those models may be based in theory or may simply be an ad hoc description of observed behavior. Each of those models will have a range of validity which must be verified. Determining which model to use depends on the situation, the accuracy required for the result, and the computational resources available.

A real semiconductor diode does not start conducting appreciably until a small "turn-on voltage" has been reached in the forward direction. This is typically 0.3–0.7 V. Beyond that there will be some resistance within the device. These behaviors are often approximated using simple models. The simplest models for calculations involving diodes are based on piece-wise linear models. For each linear region the problem can be treated as a linear circuit. The solution for each linear piece is then stitched together to find the total behavior.

Piece-Wise Linear Diode Models

Table 6.1 illustrates three of the simplest models for a diode, the piece-wise linear models. These are constructed using several straight line segments.

For the piece-wise linear models the circuit can often be solved by assuming that the solution lies on one of the linear pieces, solving the (now linear) circuit based on that assumption, and then checking to see if that assumption is borne out by the result. If the solution matches the assumption, then the problem is solved. If the solution does not match, another linear piece of the model must be used.

Alternatively, of course, a graphical solution can be used. In this case an equation is found for the circuit in question relating the voltage and current across the diode, V_d and I_d. These equations are obtained using Kirchhoff's laws (the general rules). The equations are plotted alongside the diode characteristics (the rules for the device). Where these two curves intersect is where both rules are

Table 6.1 Piecewise linear models for diodes

Model	Description	Graph of I versus V
"Ideal Diode" (The simplest model)	Either $V = 0$, $I > 0$ or $I = 0$, $V < 0$	
Ideal diode with "turn-on" voltage (Improves on the ideal diode if voltages encountered are less than or comparable to about 1 V)	Either $V = V_0$, $I > 0$ or $I = 0$, $V < V_0$ ($V_0 \approx 0.5$ V for typical semiconductor diodes, though somewhat larger for some special diodes such as LED's)	
Ideal diode with turn-on voltage and some resistance (Improves on the above by adding some diode resistance. Use if diode resistance is not negligible compared to others in the circuit)	Either $I = 0$, $V < V_0$ or $I = (V - V_0)/R_d$ if $V > V_0$ (R_d is an effective resistance for the diode. Its value depends on the "typical current" encountered)	

satisfied simultaneously and is the solution. Accurate numerical solutions obtained using a computer are often equivalent numerical implementations of such graphical solutions.

An Analytic Model for the Semiconductor Diode

An approximate expression relating the voltage across, V_d, and current, I_d, through a real semiconductor diode is

$$I_d = I_0(\exp(V_d/\eta V_T) - 1)$$
$$V_T = \frac{T}{11600} \frac{V}{K} = 0.026 \text{ V at } 300 \text{ K}, \quad (6.1)$$

where V_T is the "volt equivalent of temperature" (temperature, T, should be in kelvin). The parameter η (eta) will depend on the particular semiconductor and is approximately 1 for most silicon diodes.

The constant out front, I_0, is the "reverse saturation current," which is also somewhat temperature dependent. If the temperature does not stray too far from room temperature, it is not necessary to worry about that dependence here. The reverse saturation current is the magnitude of the current for the diode in the reverse

Diode Models

direction. For typical modern silicon diodes, I_0 can be as small as a few picoamps (pico- is 10^{-12}) and rarely exceeds a nanoamp (nano- is 10^{-9}). The reverse current for a real diode is not zero, but it is usually very small compared to most currents of concern.

The expression above is reasonably accurate provided the diode current and/or the voltage are not too large in magnitude—that is as long as one stays well within the ratings of the device—and if operation is not too far from room temperature. Even though the expression above is reasonably simple, when used to solve for currents in a circuit it is likely that either graphical or numerical techniques will be required to find solutions to the transcendental equation(s) that result. This expression will be useful later for analyzing an op-amp circuit that contains a diode. This expression also indicates there is some temperature dependence which may need to be taken into account and/or can be calibrated and used to make temperature measurements.

Solving Circuits with Diodes

The choice of solution method for a circuit with a diode depends on the desired accurately of the results. For most cases that are likely to be encountered, one of the simple piece-wise linear models is more than accurate enough. When the diode is simply used as an on/off device, that is, as a "rectifier," the ideal diode model is usually quite sufficient, at least to understand what the circuit does.

Since the diode is non-linear, some care is necessary when solving circuits. However, if a piece-wise linear model is used and conditions are such that the diode stays on one linear piece, the diode can be treated as a linear device. Sometimes it is not obvious ahead of time if this will be true. In such cases, a solution can be found by guessing. The guess is checked once the solution is found. If everything works out, then that solution is (probably) ok. If not, the guess was wrong and a new solution must be found starting with a different guess.

If more precision is required, in particular if the "turn-on" voltage of the diode is comparable to the accuracy needed, then it may be necessary to use graphical or numerical techniques.

The Ideal Diode

Many circuits can be treated using the ideal diode model. Those that require one of the other piece-wise linear models can be solved in the same way, since the piece-wise linear models are equivalent to an ideal diode with the addition of some linear components (e.g., resistors and/or batteries).

Consider the simple circuit shown in Fig. 6.4a. Assume V_0 is positive and enough larger than the diode's turn-on voltage so that the turn-on voltage can be

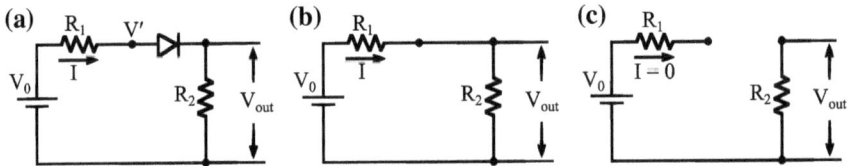

Fig. 6.4 The simple circuit in (**a**) is analyzed using the ideal diode model in (**b**) and (**c**). In (**b**) the diode is considered conducting and in (**c**) it is considered to be in reverse bias. Only one of (**b**) or (**c**) can be correct

neglected. Using the ideal diode model, the diode is either conducting as if it were a wire or it is an open circuit. These two possibilities are shown in Fig. 6.4b, c. For circuit (b), the resistors are in series so the current is easily found to be $I = V_0/(R_1 + R_2)$ in the direction shown by the arrow. For circuit (c), the current will be zero since there is no complete circuit. Now go back to check which of these is consistent with the ideal diode. The solution for (b) clearly works as the current found is indeed in the forward direction, as assumed. The circuit at (c) has $V' = 0$, and $V_{out} = 0$. Since it was assumed that $V_0 > 0$, the diode must be forward biased, which is not consistent with the open circuit behavior. Hence the solution is the value obtained from circuit (b) and not (c). For circuit (b) the series resistors form a voltage divider, so $V_{out} = V_0 R_2/(R_1 + R_2)$. In practice the measured value will be about 0.5 V less than this due to the turn-on voltage of real semiconductor diodes.

Diodes are often found in circuits involving time-dependent signals. They may be used as part of a signal processing application or in a power supply for rectification of incoming a.c. power (e.g., from a wall outlet) to create d.c. power. The circuit analysis is the same in either case. Some examples are presented next and are discussed using the ideal diode model.

Half-Wave Rectifier

The circuit in Fig. 6.5a is called a "half-wave rectifier." It is perhaps the simplest way to convert signals that may be both positive and negative into a signal that is only positive (or by turning the diode around, only negative). Using the ideal diode model as above, it is clear that when $V_{in} > 0$ the diode conducts and $V_0 = V_{in}$. When $V_{in} < 0$, the diode blocks the current and so with no current across R, it must be that $V_{out} = 0$. Hence, V_{out} as a function of time (solid) when V_{in} is sinusoidal (dotted) looks as shown in Fig. 6.6a.

The output represents only the positive half cycle of a sinusoidal input so this circuit has come to be referred to as a half-wave rectifier.

The output of this half-wave rectifier has a non-zero average, but the output still has a significant time dependence. To smooth that out, a low-pass filter can be used. Perhaps the simplest filter is to add a capacitor, C, as shown in Fig. 6.5b. When $V_{in} > V_{out}$ the capacitor charges, however when $V_{in} < V_{out}$, the capacitor can only

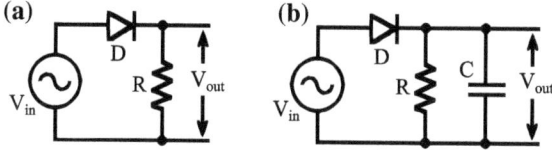

Fig. 6.5 (a) A simple half-wave rectifier, shown with a sinusoidal source, lets through only the positive part of the sine wave. In (b) the output is filtered by a capacitor so it will be closer to being constant in time

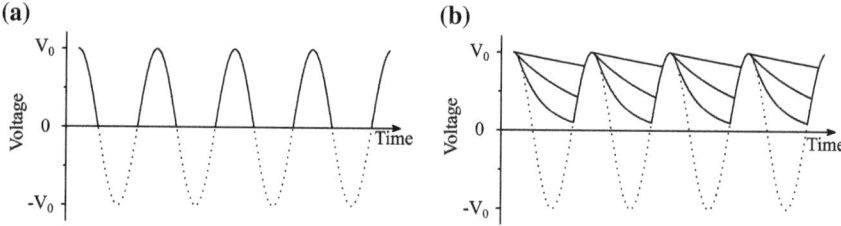

Fig. 6.6 The output of the circuits of Fig. 6.5a, b respectively. In (b) traces for three different capacitance values are shown. The larger the capacitance, the less droop will occur between maxima

discharge through the resistor. If the RC time constant is long compared to the period of the sinusoidal input, the output will smooth out and will appear as illustrated in Fig. 6.6b for three different capacitance values. Of the values shown, the upper curve corresponds to the largest capacitance and the lower to the smallest. When the capacitance becomes very, very large, the output will be essentially flat.

The half-wave rectifier finds use for more than just sinusoidal inputs. Such a circuit can be used to extract the maximum amplitude of any time-dependent signal. Another example is for use as protection against an externally supplied signal or power source. If the source is negative when it should be positive, this circuit blocks the negative signal from the remaining circuitry, thus preventing possible damage.

Diode Limiter

A limiter is useful in order to protect later circuitry from signals that are too large or too small. A simple diode limiter is shown in Fig. 6.7a, where $V_{max} > V_{min}$. When $V_{in} > V_{max}$ the upper diode is forward biased and the lower reverse biased. The ideal diode model would then predict behavior as shown in Fig. 6.7b, and hence $V_{out} = V_{max}$. On the other hand, if $V_{in} < V_{min}$ the behavior will be as shown in Fig. 6.7c and $V_{out} = V_{min}$. When $V_{min} < V_{in} < V_{max}$ both diodes are "off" and so, using the ideal diode model, the diodes each look like an open circuit. In that case $V_{out} = V_{in}$.

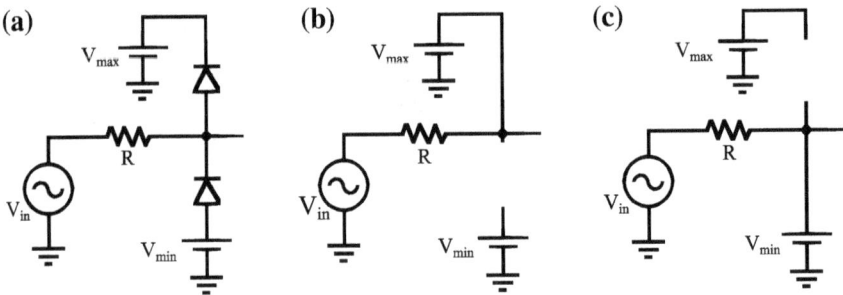

Fig. 6.7 A limiting circuit shown in (**a**) is analyzed using the ideal diode model in (**b**) and (**c**). If the input exceeds V_{max} or gets smaller than V_{min}, the corresponding diode conducts

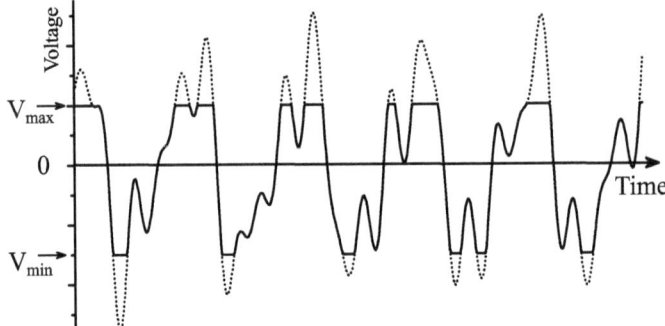

Fig. 6.8 An example showing the effects of a limiting circuit on a time-dependent signal. The original signal is shown dotted, while the output of the limiter is solid

An example of how this circuit functions for a time-dependent signal is illustrated in Fig. 6.8, where a hypothetical input signal, V_{in}, is shown with the dotted curve and the corresponding output voltage, V_{out}, is the solid curve. When a signal is simply truncated in this manner it is said to have been subjected to "hard limiting." Imagine an object moving in a room between a hard ceiling and a hard floor. "Soft limiting" would be more like an object moving vertically[2] in a room where pillows were spread across the floor and fastened to the ceiling.

Diode Clamp and Voltage Doubler

The circuit in Fig. 6.9a is referred to as a diode clamp. Using the ideal diode model, when the input is negative the equivalent circuit is as shown in Fig. 6.9b. The capacitor is charged to match the (negative) input voltage. When the voltage returns

---

[2]Imagine a child jumping on their bed.

Solving Circuits with Diodes 131

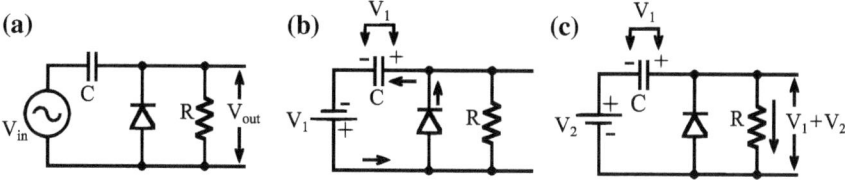

Fig. 6.9 The diode clamp circuit in (**a**) is analyzed in (**b**) and (**c**) for a negative input followed by a positive input respectively

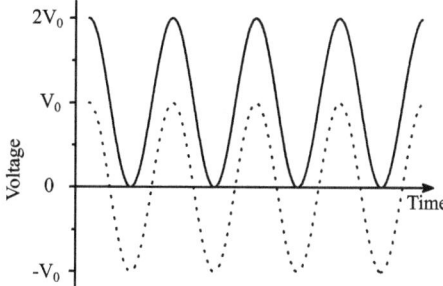

Fig. 6.10 The output from the diode clamp of Fig. 6.9a (solid) compared to the input (dotted) for a sinusoidal source

to positive, the capacitor cannot discharge through the diode. If the RC time constant is long compared to the period of the (sinusoidal) input, the capacitor will have comparatively very little time to discharge at all. Since the capacitor will ultimately charge to match the maximum negative voltage, when the input returns positive, that maximum negative value is added (positively) to the input. The net result is that the signal is translated so that what was the maximum negative voltage is now at zero. This is illustrated in Fig. 6.10 for a sinusoidal input, where the input is shown dotted and the output is the solid line.

If the amplitude of the input signal is changed, the minimum of the signal will adjust so that it is always at zero—it is "clamped" at zero. The maximum of the signal is then the peak-to-peak amplitude of the input. If the diode is reversed, the signal is clamped so that the maximum is always zero. If desired, a constant voltage source (e.g., a battery) can be added to the circuit to clamp the voltage at a non-zero value.

If the clamp circuit is followed by a half-wave rectifier with a capacitor, the result is similar to the filtered result found above for the half-wave rectifier by itself, but with the peak amplitude now doubled. Such a circuit is called a voltage doubler and would look like Fig. 6.11a. Additional diodes and capacitors can extend this idea to create extremely large (d.c.) output voltages.[3] Each stage of such a circuit

---

[3]Such a scheme was used in the 1930s by J. D. Crocket and E. T. S. Walton for their Nobel Prize winning particle accelerator, and the so-called Crocket-Walton multiplier circuit shown here is named for them.

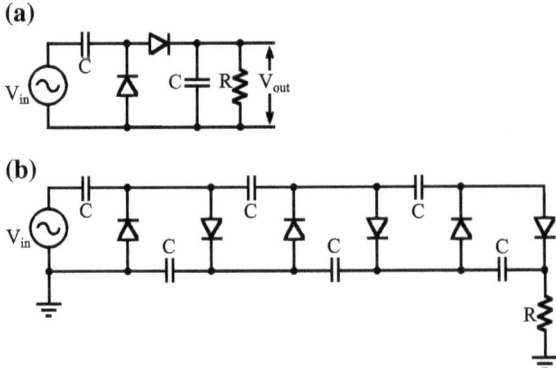

Fig. 6.11 A simple voltage doubler is shown in (**a**). The idea can be extended to achieve very high voltages, such as shown in (**b**)

consists of two diodes and two capacitors. The output voltage across R for an N stage multiplier, provided the capacitors are large enough, will be NV_{pp}, where V_{pp} is the peak-to-peak voltage of the input signal. For example, a 3-stage multiplier is shown in Fig. 6.11b. Such a circuit is sometimes referred to as a "voltage ladder."

Full-Wave Rectifier

The half-wave rectifier can be used to turn a.c. power into d.c. power, however half of the available signal is unused. The full-wave rectifier, or absolute value circuit, can be used instead. A simple full-wave rectifier based on a diode bridge is shown in Fig. 6.12a.

As long as the magnitude of V_{in} is significantly larger than the diode turn-on voltage, the behavior can be analyzed using the ideal diode model. When $V_{in} > 0$ the circuit behaves as shown in Fig. 6.12b, and when $V_{in} < 0$ the circuit behaves as shown in Fig. 6.12c. In each case, consider which way the input source is trying to push the current. If it is in the forward direction, the diode is replaced with a wire, if in the reverse, an open circuit. The result is that in all cases, the positive side of the input is connected to just one side of the resistor, the negative side to the other, and hence the output is always positive. For a sinusoidal input, the output as a function of time will look like what is shown in Fig. 6.13, where once again the input is shown dotted.

As was done for the half-wave rectifier, the full-wave circuit can be smoothed with the addition of a capacitor (Fig. 6.14a). In practice, it is important to note that for the circuit shown, the input source and the output cannot both have a ground connection. Remember that all grounds are connected to each other, even if that connection is not visible in the diagram. In power supply applications, the input power may come from a wall outlet at a voltage level different from what is desired. Both the grounding and voltage level can be addressed using a transformer on the

Solving Circuits with Diodes

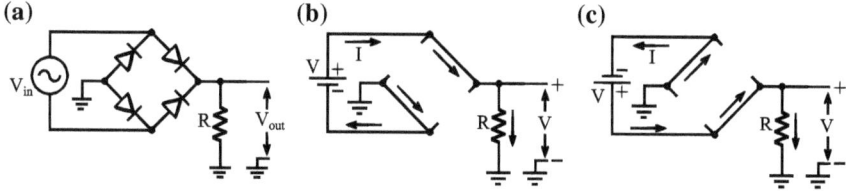

Fig. 6.12 The diode-bridge full-wave rectifier circuit shown at (**a**) is analyzed using the ideal diode model in (**b**) and (**c**)

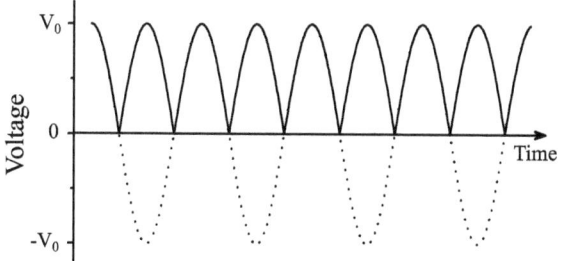

Fig. 6.13 The output from the circuit of Fig. 6.12a (solid) compared to the input (dotted)

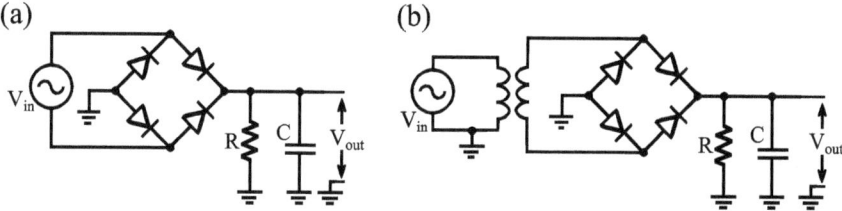

Fig. 6.14 (**a**) A capacitor is added to the full-wave rectifier to smooth the output. A common implementation uses a transformer on the input, as shown in (**b**), that can scale the voltage and eliminate potential problems with the ground connections

input (Fig. 6.14b). The transformer only passes the time-dependent signals and hence the time independent ground level is blocked. As discussed earlier, a transformer can also be used to change the amplitude of a time-dependent input signal. Such a circuit is common for some types of inexpensive d.c. power supplies or as the first stage of a more elaborate supply.

The diode bridge—four diodes wired appropriately for a full-wave rectifier—is available as a single device. Those devices designed for higher current applications include provisions that can be used to help dissipate the heat generated by real diodes.

Graphical Solutions

An important tool for understanding the behavior of all non-linear circuit elements is the use of graphical techniques. Remember that all circuits must obey Kirchhoff's laws and each element of the circuit must obey its own rule. The strategy is then to write down the conditions for each of those separately, graphing the two results on a common grid, and then finding the spot where both rules are satisfied.

Consider the circuit of Fig. 6.15 where an ideal voltmeter is used to measure V_{out}. First, write down the equations from Kirchhoff's laws,

$$I_1 = I_2 + I_3$$
$$V_1 - I_1 R_1 - I_2 R_2 = 0 \tag{6.2}$$
$$I_2 R_2 - V_d - I_3 R_3 = 0.$$

Note that the (as yet unknown) voltage across the diode is simply entered as V_d. The voltage across the diode is a function of the current through the diode, which in this case is $I_d = I_3$. There is also a relationship between V_d and I_d, which is the "rule for the diode." To get the results of Kirchhoff's laws into a form so they may be graphed on the same grid, they need to be put into a form where V_d is a function of $I_3 = I_d$.

For this circuit, one way to do this is to replace I_1 in the second equation above using the first equation. Now multiply the second equation by $R_2/(R_1 + R_2)$,

$$[V_1 - (I_2 + I_3)R_1 - I_2 R_2] \frac{R_2}{R_1 + R_2} = \frac{R_2}{R_1 + R_2}[V_1 - I_3 R_1] - I_2 R_2 = 0, \tag{6.3}$$

and add this to the third equation to eliminate I_2

$$\frac{R_2}{R_1 + R_2} V_1 - I_3 (R_1 \| R_2 + R_3) = V_d. \tag{6.4}$$

The result is a straight-line relationship between V_d and $I_3 = I_d$ that must be obeyed for this circuit in order to satisfy Kirchhoff's laws.

Now plot the straight line result from Kirchhoff's laws on a graph of $I_d (= I_3)$ as a function of V_d. That line can be plotted by finding any two points on that line and

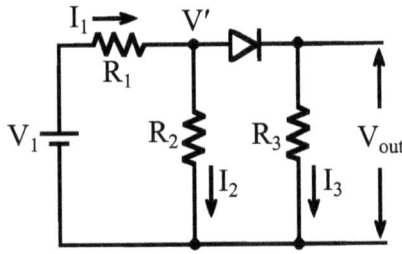

Fig. 6.15 A simple example circuit used as an example to illustrate graphical solutions

Solving Circuits with Diodes

connecting them. It is perhaps most convenient to use the intercepts with the axes. That is, put in $V_d = 0$ and solve for I_3 to get one point, then put in $I_3 = 0$ and solve for V_d to get another point. Now connect those points with a straight line. Put the device characteristics on the same graph. The point where the two curves intersect is the solution—both the general and device specific rules are satisfied there.

Example Graphical Solution

Consider a diode curve obtained from direct measurement of a real diode (solid curve in Fig. 6.16), and use the circuit above as an example, with $V_1 = 3$ V, and $R_1 = R_2 = R_3 = 1$k. Then the straight line from Kirchhoff's laws is

$$V_d = 1.5\,\text{V} - 1.5\text{k} \cdot I_3, \tag{6.5}$$

so if $V_d = 0$, I_3 is 1 mA, and if $I_3 = 0$, $V_d = 1.5$ V. The resulting line is shown dashed in Fig. 6.16.

The only spot that satisfies both the overall rules and the device rule will be the solution. The solution here is then $I_d = I_3 = 0.6$ mA, $V_d = 0.6$ V. These values are substituted into the Kirchhoff's law equations to get $I_1 = 1.8$ mA and $I_2 = 1.2$ mA.

This type of graphical method can be used with any two-terminal device, no matter how strange, if measured device characteristics are available.

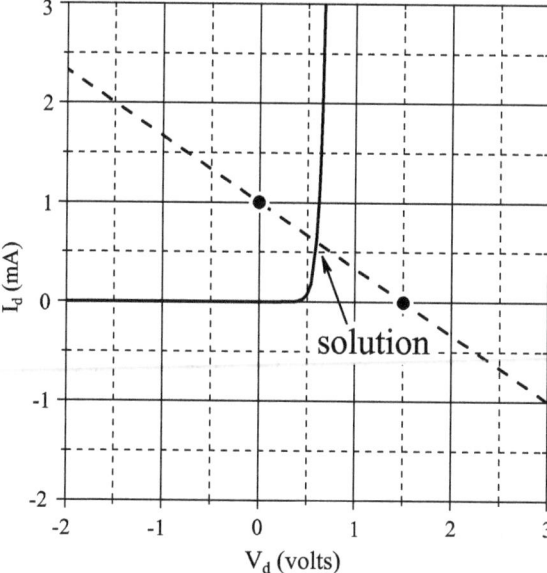

Fig. 6.16 To solve the circuit of Fig. 6.15, the diode characteristic (solid) and the results obtained from Kirchhoff's laws (dashed) are plotted together. Since both must be satisfied simultaneously, the only solution is at the intersection of these curves

Diode Ratings

The two diode ratings that are often important are the "peak inverse voltage (PIV) " and "maximum forward current." The PIV is the largest voltage one can expect to use in the reverse direction before breakdown occurs. The maximum forward current is the largest current one can expect to get through the diode before damaging the device (principally due to the heat generated). Sometimes a maximum forward voltage is also specified.

Diode Capacitance and Response Time

Since a semiconductor diode has adjacent layers with opposite charges (electrons and holes) separated by a non-conducing region (the depletion region), a diode can have a significant capacitance (a few pF). This capacitance, along with any resistance in the circuit that contains the diode, form an *RC* time constant. This *RC* behavior is one important factor that limits how quickly the diode can respond to a change in the input conditions. If speed is of concern, some capacitance may need to be added to the diode model to get accurate predictions from the circuit analysis.

Specialty and Other Diodes

There are various other diodes that have a more specialized purpose. A summary of many of them is included in Table 6.2. While specialized schematics are often seen for these diodes, the simple diode schematic (Fig. 6.3) is often used as well. Two of these will be considered in more detail below: Zener diodes and light-emitting diodes (LEDs).

All diodes have a reverse breakdown voltage. If the reverse voltage gets too large in magnitude, the diode begins to conduct. In many cases, this leads to destruction of the device. Zener diodes are designed to do this in a controlled way so that, in a sense, they have a very large turn-on voltage in the reverse direction. The reverse conduction can occur through a quantum tunneling process (the Zener effect) or through an avalanche process. The avalanche starts with one electron (or hole) breaking free, and that causes more to break free and so on. The reverse voltage where reverse conduction occurs is controlled during the manufacture. Whether the Zener effect or the avalanche dominates, the diodes are usually referred to as Zener diodes[4] and the voltage where it occurs is referred to as the Zener voltage of the device.

Zener diodes are specified by the (magnitude of the) reverse voltage where conduction begins. Examples of measured values for several Zener diodes are

---

[4]References to an "avalanche diode" may still be found from time to time.

Specialty and Other Diodes

Table 6.2 Some specialty diodes

Schematic	Name	Brief description
⟶▷⌿⟶	Zener diode	Designed to be used at the reverse breakdown voltage as a limiter or voltage reference. Reverse break-down current is limited by power considerations
⟶▷⟶ (with arrows)	Light emitting diode (LED)	A semiconductor diode optimized to emit light due to electron-hole recombination when the diode is conducting in the forward direction. Electrically similar to other semiconductor diodes, but usually with a larger turn-on voltage (Can be used in reverse as a light detector)
⟶▷⟶ (with arrows in)	Photodiode	A semiconductor diode optimized to detect incident light. The diode is used in series with a resistor and with reverse bias. Incident light creates electron-hole pairs and increases the conductivity of the diode
⟶▷⊦⟶	Varactor diode	Designed to use the inherent capacitance of a semiconductor diode due to the depletion region. By varying a DC biasing voltage in the reverse direction, the size of the depletion region, and hence the capacitance, can be adjusted
⟶▷⌐⟶	Schottky diode	Generally has a smaller turn-on voltage and a very rapid turn-on time compared to other semiconductor diodes. Used when either of those parameters is a significant issue. Relies on a metal-semiconductor junction. Also known as a barrier diode or a hot-carrier diode. Not to be confused with the four-layer Shockley diode
⟶▷⟶	Tunnel diode	Based on quantum mechanical tunneling, and has the unusual property that for some operating conditions the dynamic resistance is negative—that is, the current decreases when the voltage across it increases
⟶▷⟶ (with arrow)	Laser diode	As its name implies, a laser diode is constructed to produce laser light. Usually based on a p-i-n layer structure. The laser light arises from the area near the junction. Often represented in a circuit using the simple diode schematic, without the arrow
⟶(H⊏)⟶	Vacuum tube diode	A diode based on vacuum tube technology. Largely obsolete except for novelty use and for very high-voltage and/or very high power devices. Otherwise, used like other diodes

shown in Fig. 6.17. In the forward direction, they behave the same as a normal semiconductor diode. In the reverse direction they behave normally until the Zener voltage is reached.[5] The almost vertical behavior is then well-modeled with a constant voltage source (a battery).

---

[5]For most such diodes, the avalanche effect tends to dominate at higher Zener voltages, whereas the Zener effect dominates at lower Zener voltages. The Zener effect has a more gradual "turn-on" characteristic.

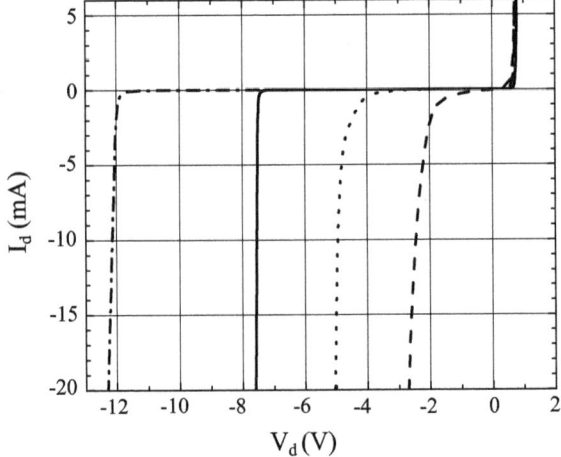

Fig. 6.17 The measured (d.c.) characteristic curves for 3.3 V (dashed), 5.1 V (dotted), 7.5 V (solid), and 12 V (dot-dash) Zener diodes

Zener diodes are used for a number of purposes. Two Zener diodes in series, but pointed in opposite directions, can be used to make a simple limiter. In addition, Zener diodes are often used as voltage references and, for lower power circuits, these diodes can be used as a simple voltage regulator. Note that the power dissipated in the diode is the product of the voltage and the current. Since the voltage can be large, the current is generally kept small.

When electrons and holes recombine, energy equal to the energy gap must go somewhere. The two likely places are into a phonon (a quantized vibrational state of the atoms, and ultimately heat) or into a photon of light. A diode optimized to produce light, encased in a transparent case, can be used as a light source. The color of the light depends on the size of the band gap.[6] Such diodes are referred to as LEDs, which is short for light emitting diodes. Within the normal operating range, the energy of the light emitted (or the number of photons) goes like the current through the device. Small signal LEDs are used with currents of 1–20 mA. Those used for higher power applications, such as home lighting and automobile brake lights, require much more current.

Electrically, an LED looks like a regular semiconductor diode, except with a larger turn-on voltage. Example characteristics are shown in Fig. 6.18 for several small LEDs.

If an LED is reverse biased, light can enter and, through the reverse process, create an electron-hole pair. That pair can then carry current. The reverse current is then a measure of the incident light intensity. Diodes optimized for such detection

---

[6]There are exceptions to this. For example, white LEDs may be created using fluorescent materials that change the color and so the observed light does not come directly from the electron-hole annihilation.

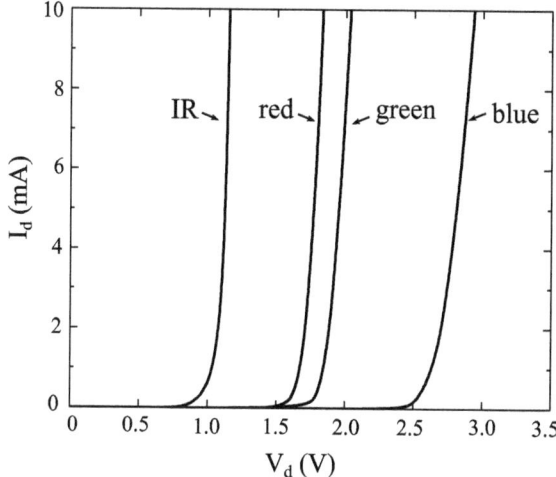

Fig. 6.18 The measured characteristic curves for several small LEDs which produce different wavelengths of light

are referred to as "photodiodes." When interaction with light is undesirable, the casing material should shield the diode from light.

Problems

1. A full-wave rectifier (Fig. 6.P1) is constructed for use as a power supply. Use the ideal diode model to predict the output which will be seen as a function of time on an oscilloscope, for each of the following individual faults. Assume the power from the wall is sinusoidal and when working correctly, the RC time constant at the output is *short* compared to the period of the sinusoidal input.

   (a) The transformer is not plugged in.
   (b) Diode D1 has failed and acts like a short circuit.
   (c) Diode D1 has failed and acts like an open circuit.
   (d) Diode D2 has failed and acts like a short circuit.
   (e) Diode D2 has failed and acts like an open circuit.

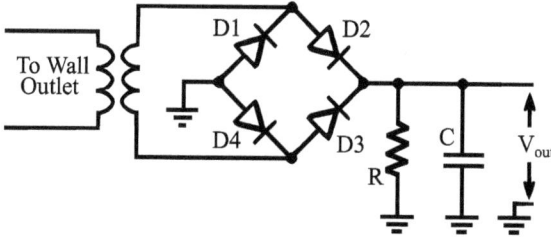

Fig. 6.P1 Problem 1

(f) Diode D2 was placed in the circuit backwards.
(g) All the diodes were placed in the circuit backwards.
(h) The resistor has burned out and is an open circuit.

2. A mystery device, "X," (Fig. 6.P2) that has the measured voltage-current relationship shown in Fig. 6.P2c, is placed in the circuit shown. What is the current, I, in the circuit?

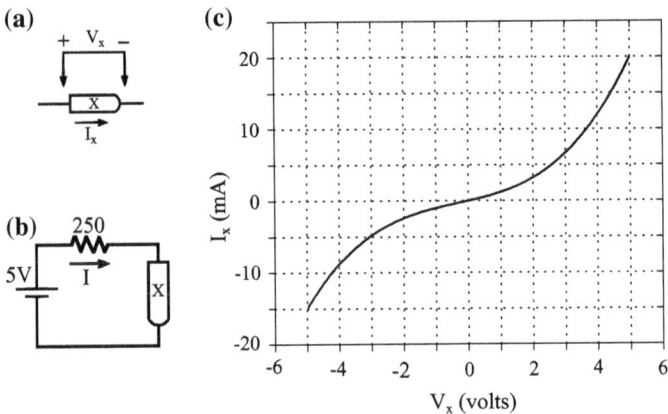

Fig. 6.P2 Problem 2

3. Use the analytic model for the diode to predict the voltage across a silicon diode when the current is 0.01, 0.1, and 1.0 mA. Assume room temperature, $I_0 = 1$ pA, and $\eta = 1$.
4. A 7.5 V Zener diode with a maximum power rating of 1 W is to be used in a circuit. What is the maximum (sustained) current that can go through the diode?
5. For the circuit of Fig. 6.P5, that contains three identical small LED's, describe the behavior before and after the switch is closed. See Fig. 6.18 for representative current-voltage relationships for LEDs.

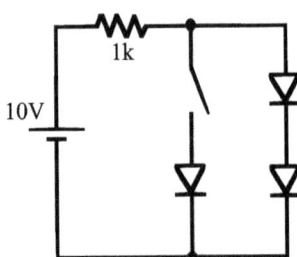

Fig. 6.P5 Problem 5

6. (Challenge Problem) Consider the circuit in Fig. 6.P6, where the capacitor is initially uncharged and the switch is closed at $t = 0$. Assuming ideal components, show that if L is large enough, the capacitor will become charged to close to $2V_0$ with virtually no loss of energy in the resistor. In the process, you need to determine and state clearly what is "large enough."

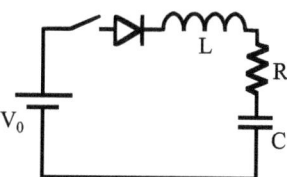

Fig. 6.P6 Problem 6

7. (Challenge Problem) How might you measure the reverse current vs. voltage characteristics for a diode using a real ammeter and voltmeter, given that the reverse current is so small?

Chapter 7
FETs

The circuit elements considered thus far were two-lead devices. When describing the behavior for those devices the relationship between just one current and one voltage was required. Here the discussion moves on to transistors that are three-lead non-linear devices. Field effect transistors will be considered in this chapter and bipolar junction transistors will be treated in the next. These two different types of transistors serve as examples as to how more complicated devices can be treated using linearized models.

Junction Field Effect Transistors

There are several different types of field effect transistors (FET) in use. When one simply refers to a(n) "FET," it is likely that one is referring to one based on a simple *p-n* junction, but with three wires instead of 2, as shown in Fig. 7.1a. A FET made with a simple junction is sometimes called a JFET ("J" is for "Junction").

FET is either pronounced as three separate letters (F-E-T) or as a word that rhymes with "bet" and "set." When a prefix is added to describe what kind of FET it is, it is almost always pronounced as a two-syllable word. That is, JFET would be "jay-fet."

For a JFET, the side of the junction with one lead is called the "gate." The other side is the "channel" and the channel has two leads—the "drain" and the "source." Conduction through the channel is controlled by the gate. In this simple picture the drain and source look interchangeable, though in practice one will find that the device works much better one way than the other.

The device shown is an "*n*-channel JFET." The schematic symbol for the *n*–channel JFET is shown in Fig. 7.1b. Swapping the *p* and *n* regions will make a *p*-channel FET. The schematic for the *p*-channel JFET looks the same as for the *n*-channel except the arrow points in the opposite direction. The arrow is on the gate and points from "*p* towards *n*." If the arrow is drawn off center, it is closer to the

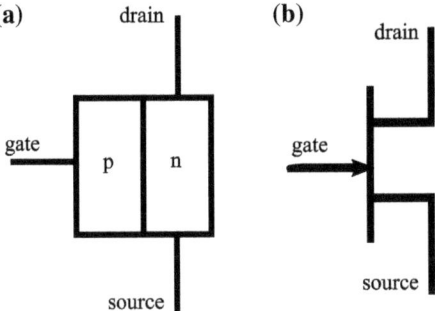

Fig. 7.1 (a) Basic construction of an *n*-channel junction field effect transistor (JFET) and (b) the corresponding schematic symbol

source. Most of the examples used here are based on the more common *n*-channel FET. For the characteristics of a *p*-channel FET, add a minus sign to all FET voltages and currents.

In normal JFET operation, the *p-n* junction is reverse biased. That means the current into/out of the gate will be very small (nano- or picoamps). If forward bias is applied to the *p-n* junction, the gate current becomes large and the device may be damaged.

With any three-lead device there are three potential differences (voltages) and three currents that can be specified. In this case the voltages are the drain to gate voltage (V_{dg}), the drain to source voltage (V_{ds}) and the gate to source voltage (V_{gs}). The three currents are the currents into each lead, I_d, I_g, and I_s. By convention, these currents are taken as positive for currents going *into* the device.

For FETs under normal operating conditions, the gate current is often negligible, so as an approximation, $I_g = 0$. This means that, using Kirchhoff's current law, $I_s = -I_d$. That is, in most circumstances, only one current need be specified. Also, it must be that $V_{dg} + V_{gs} = V_{ds}$ so only two voltages need to be specified. For a FET the two voltages V_{ds} and V_{gs} are usually the two specified. Under normal operating conditions an *n*-channel JFET will have $V_{gs} \leq 0$ and $V_{ds} > 0$.

A JFET works because of the changes in the size of the depletion region as a larger and larger reverse bias is applied to the *p-n* junction, as illustrated in Fig. 7.2. If a large enough negative bias is applied, the transistor will "pinch-off" and no current can flow. That value is known as the pinch-off voltage, V_p. When the applied bias is smaller than this, there will be current through the channel. If the current is too large, the voltage drop in the channel will result in pinch off—i.e., no current. Hence the current is self-restricting and cannot get too large. This internal feedback mechanism limits the current to an equilibrium value that is almost independent of the drain to source voltage. For a more detailed and more precise description of the physics behind JFET operation, the reader should refer to a more advanced text.

With the simplification that the gate current can be neglected, the FET "device rules" require the relationships between three values—two voltages and a current. Such a relationship can be described using a two-dimensional plot with a set of curves. The two common ways of presenting the FET characteristics are to plot I_d

Junction Field Effect Transistors

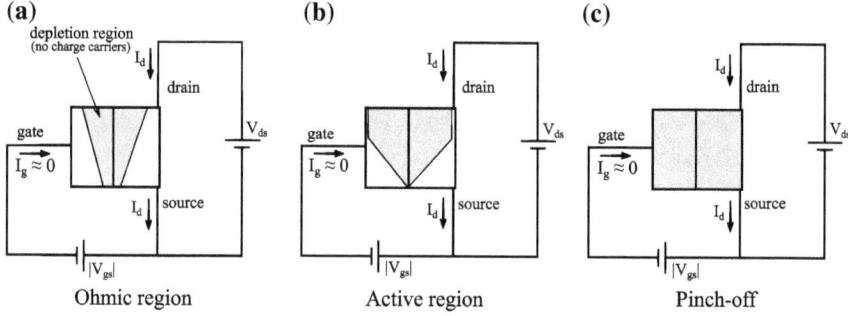

Fig. 7.2 Illustrating the effects on the depletion region as the reverse bias is increased from (a) to (c)

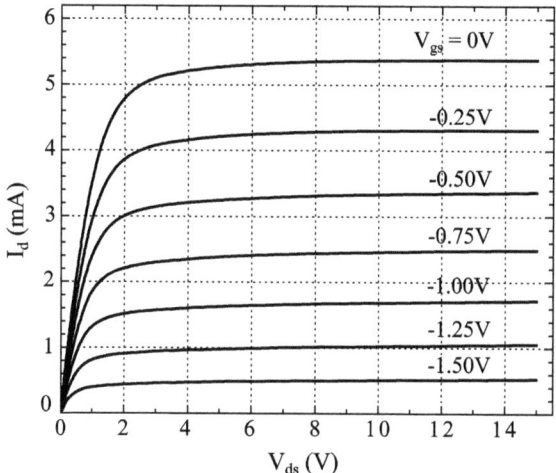

Fig. 7.3 Characteristic curves for a particular n-channel JFET

versus V_{ds} for different values of V_{gs} or to plot I_d versus V_{gs} for different values of V_{ds}. The latter is known as the "transfer characteristic."

A typical device characteristic for a low-power n-channel FET looks like the plot in Fig. 7.3. Note that for $V_{gs} > 0$ the gate is forward biased and those values are not used. By extrapolation, it can be seen that for this transistor the pinch-off voltage, the value of V_{gs} where the drain current, I_d, goes to zero, is about −1.8 V.

The so-called "active region" for this transistor, where the curves are roughly horizontal, corresponds to (roughly) $V_{ds} \gtrsim 2$ V. More generally it is usually the case that the active region corresponds to, very roughly, $|V_{ds}| > |V_p|$. Note that for an n-channel JFET in the active region, both V_{gs} and V_p will be negative; for a p-channel JFET, they are both positive.

An approximate formula, a mathematical model, can be used for the JFET in the active region away from the origin. It is the parabola,

$$I_d = I_{dss}\left(1 - \frac{V_{gs}}{V_p}\right)^2, \tag{7.1}$$

where I_{dss} and V_p are device parameters. The approximation is valid only for $0 \leq |V_{gs}| \leq |V_p|$, and where $|V_{ds}|$ is large enough to be in the active region. Do not try to use this formula for other types of FET's without verifying its validity.

The value of I_{dss} is the drain current when $V_{gs} = 0$, which is the maximum current one can expect. For the transistor characteristic shown in Fig. 7.3, $I_{dss} \approx 5.4$ mA. The value of V_p is the voltage when I_d first goes to 0. As mentioned above, for the curves shown in Fig. 7.3, $V_p \approx -1.8$ V. The values for I_{dss} and V_p depend somewhat on the particular operating conditions, and "typical values" are usually quoted. Due to device to device variations, those typical values are as accurate as one can expect to have in any case.

Circuit Analysis with a JFET

There are two types of analysis for FET (and other transistor) circuits: a large signal analysis, that is necessary when the non-linearities of the device are significant, and small signal analysis based on a linear model for small *changes* from a particular known starting solution. For a typical transistor amplifier both types of analysis are necessary. The first establishes an "operating point" and the second deals with small changes from that operating point. This same strategy is used in many areas of math and science—solve most of the problem one way then treat the rest as a perturbation.

The following two examples illustrate the analysis for larger signals. The first illustrates a design problem, where the components to achieve a given outcome are determined. The second illustrates an analysis problem for an existing circuit. That is, given a set of components, what is the outcome?

Example 1—Determine Circuit Components

Consider the circuit in Fig. 7.4, where the JFET characteristics are those of Fig. 7.3. If $V = 15$ V and it is desired to have $V_{gs} = -1$ V and $V_{ds} = 9$ V, what resistor values should be used?

Reading off the graph for the device characteristics in Fig. 7.3, if $V_{ds} = 9$ V and $V_{gs} = -1$ V, then $I_d = 1.6$ mA.

The gate current will be very small so the value of R_g is not too critical. There needs to be some path for that very small current, however. A value of $R_g = 1$ MΩ is usually more than sufficient to provide that path. Since the reverse current will be nano- or picoamps, the voltage drop across 1 MΩ is still small enough to be negligible. Thus, the voltage at the gate will be the same as the ground connection, or $V_g = 0$.

Circuit Analysis with a JFET

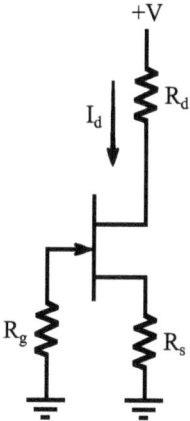

Fig. 7.4 Circuit for Examples 1 and 2

The voltage at the source, V_s, compared to ground, can be found from Ohm's law. The voltage drop from the source to ground is $I_d R_s$. Hence,

$$V_{gs} = V_g - V_s = 0\,\text{V} - 1.6\,\text{mA} \cdot R_s, \tag{7.2}$$

and so if $V_{gs} = -1$ V, $R_2 = 1$ V/1.6 mA = 625 Ω.

Using Kirchhoff's voltage law (KVL) from the power supply to ground,

$$15\,\text{V} - I_d R_d - V_{ds} - I_d R_s = 0, \tag{7.3}$$

and so

$$R_d = (15\text{V} - 9\text{V} - 1\text{V})/1.6\,\text{mA} = 3100\,\Omega. \tag{7.4}$$

If one or both of the computed resistor values had come out negative, then it would not have been possible to get the desired values using this circuit.

Example 2—Determine Operating Point

Consider the same circuit with $V = 15$ V, $R_d = 2400$ Ω, $R_s = 400$ Ω, and $R_g = 1$ MΩ. What are V_{gs}, V_{ds}, and I_d?

Once again, the current into the gate is negligible so $V_g = 0$ V.

There are two different approaches that can be used at this point: a graphical solution or the use of the JFET mathematical model. Since the graphical solution technique is more general and will always work, that will be considered first.

Kirchhoff's voltage law on the drain-source side of the transistor gives

$$15\,\text{V} - I_d R_d - V_{ds} - I_d R_s = 0, \tag{7.5}$$

or

$$I_d = (15\,\text{V} - V_{ds})/(R_d - R_s) = 5.4\,\text{mA} - V_{ds}/2800\,\Omega. \tag{7.6}$$

This is a straight-line relationship between I_d and V_{ds} and is known as the "load line." That line can be plotted along with the device characteristics. See Fig. 7.5a. Whatever the solution, it must be somewhere along that line. The intersections between the load line and the device characteristic curves, shown with solid circles, can be used to create a transfer characteristic. Those points are put on a plot of I_d versus V_{gs} and connected with a smooth curve as shown in Fig. 7.5b. The solution must be somewhere along that curve.

Now look at Kirchhoff's voltage law from the gate to source. This is called the bias line,

$$0\,\text{V} - V_{gs} - I_d R_s = 0 \text{ or } I_d = -V_{gs}/400\,\Omega, \tag{7.7}$$

which is a straight-line relationship between I_d and V_{gs}. Put that line on the second plot—the dashed line shown in Fig. 7.5b. The solution must be along that line.

Both conditions are satisfied at the point where the solid curve and the dashed line intersect, and so that point must correspond to the solution. In this case, close to $I_d = 2.2$ mA and $V_{gs} = -0.85$ V. Using those values in the load line equation,

$$V_{ds} = 15\,\text{V} - 2.2\,\text{mA} \cdot 2800\,\Omega = 8.8\,\text{V} \tag{7.8}$$

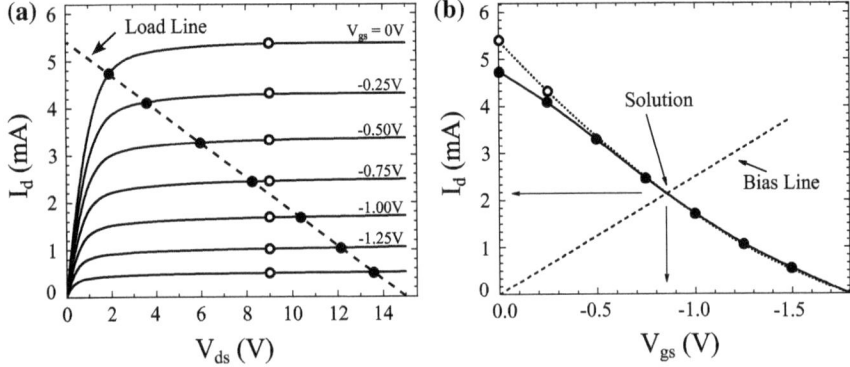

Fig. 7.5 To solve Example 2 graphically, the characteristic curves and load line in (**a**) are used to construct the transfer characteristic in (**b**). The bias line, from Kirchhoff's laws is added to (**b**) and the intersection is the solution

Circuit Analysis with a JFET

The process can be simplified a bit by assuming the solution is in the active region and noting that the curves in that region are reasonably flat. Then, rather than using the load line to extract data for the second plot, a simple vertical line will suffice. For example, simply look at the points corresponding to a fixed value of V_{ds} somewhere in the middle of the active region, say $V_{ds} = 9$ V. That is, the points in Fig. 7.5a that are indicated with open circles can be used instead of the solid circles. As long as the transistor is in the active region, the solution will be the same, at least within the accuracy obtainable by this method. The assumption that the solution is in the active region will need to be checked at the completion of the calculation.

A second solution method uses the JFET mathematical model. This method may not work well for other types of transistors. The first step is to extract the two parameters I_{dss} and V_p from the device characteristics, assuming the active region. For the transistor here, this was already discussed above with the results that $I_{dss} = 5.4$ mA and $V_p = -1.8$ V.

The solution should then simultaneously satisfy Kirchhoff's voltage law and the model. That is

$$I_d = -V_{gs}/400\,\Omega \tag{7.9}$$

and

$$I_d = I_{dss}\left(1 - \frac{V_{gs}}{V_p}\right)^2 = 5.4\left(1 - \frac{V_{gs}}{-1.8\,\text{V}}\right)^2 \text{ mA}. \tag{7.10}$$

Putting the first into the second,

$$I_d = 5.4\left(1 - \frac{I_d\,0.40\,\text{k}\Omega}{1.8\,\text{V}}\right)^2 \text{ mA} \tag{7.11}$$

and so with I_d in mA,

$$0.267 I_d^2 - 3.40 I_d + 5.4 = 0. \tag{7.12}$$

This is a quadratic that is solved using the quadratic equation. That is

$$I_d = \frac{3.40 \pm \sqrt{3.40^2 - 4 \times 0.267 \times 5.4}}{2 \times 0.267} = \frac{3.40 \pm 2.41}{0.533} = 1.86 \text{ or } 10.9\,\text{mA}. \tag{7.13}$$

Only the smaller solution is valid. The larger solution is out of the range where the model is valid and should be discarded.

The value obtained using the mathematical model and the value obtained directly from the data differ by about 15%. Such a disagreement should be expected. The model is not perfect. The value from the data is likely more accurate and should be used when possible.

The FET A.C. Model

A very simple model for the low frequency behavior of an *n*-channel FET transistor operating in its active region about some fixed operating point is shown in Fig. 7.6, where only linear circuit elements are used. Since there is only a negligible gate current, the model shows the gate lead with no connection—thus there will not be any gate current. The transistor includes a *dependent* current source where the current is proportional to the gate to source voltage, with proportionality constant "g_m." A model resistor, r_d, is also included, though in many applications its value is large enough that it can be omitted. A *p*-channel FET would look identical except the current source would point the other direction.

Note that the model parameters (here g_m and r_d) will depend on the operating point used—that is, on the large signal analysis—and also that *the model only applies to the changes* from the operating point, not to the total gate to source voltage, total drain current, etc. That is, V_{gs} in Fig. 7.6 is really only the change in V_{gs}, which might be written as ΔV_{gs}, however the "Δ" is typically omitted.

This linear model only works well when the changes from the operating point are small. The larger the input signals, the less accurate the model.

The "transconductance" (also called the "mutual conductance") g_m, is given by

$$g_m = y_{fs} = \left.\frac{dI_d}{dV_{gs}}\right|_{V_{ds} \text{ at op. pt.}} \tag{7.14}$$

and the model drain resistance is given by

$$r_d = 1/y_{os} = \left.\frac{dV_{ds}}{dI_d}\right|_{V_{gs} \text{ at op. pt.}} \tag{7.15}$$

That is, g_m is a measure of how much the drain current changes for a given change of V_{gs} (e.g., as one moves vertically on the characteristic curves) and $1/r_d$ is a

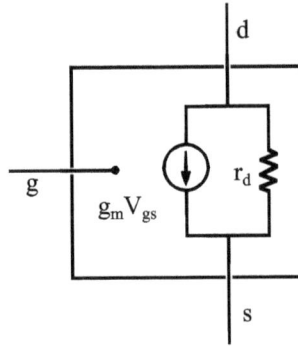

Fig. 7.6 A simple a.c. model for the JFET used for small changes from a fixed operating point

The FET A.C. Model

measure of the slope of the characteristic curves in the vicinity of the operating point. The alternate symbolic names shown (using y's with subscripts) are also common.

Values for g_m and r_d can be determined graphically from the characteristic curves. The value of g_m (but not r_d) can also be determined from the model equation (7.1) relating I_d and V_{gs}.

Example Consider the characteristic curves in Fig. 7.7 where the operating point (the result of the large signal analysis) has been determined (or chosen) to be at $V_{gs} = -1.5$ V, $V_{ds} = 8$ V. The appropriate model parameters for that operating point are determined using neighboring values read off the graph (also shown in Fig. 7.7):

$$g_m \approx \frac{7.7 - 3.5}{(-1) - (-2)} \frac{\text{mA}}{\text{V}} = 4.2 \, \text{mmho} = 4.2 \, \text{mS}$$
$$r_d \approx \frac{12 - 4}{5.4 - 4.9} \frac{\text{V}}{\text{mA}} = 16 \, \text{k}\Omega. \tag{7.16}$$

The units used for g_m are those of electrical conductivity: 1 mho = 1/(1 Ω) = 1 S, and 1 mmho = 1 milli–mho = 1 mS = 1/(1 kΩ). The "mho" (ohm spelled backwards, and pronounced like the name Moe) is an older unit for conductance (1/resistance) and has been replaced by siemens.[1] Note that the abbreviation for siemens is an uppercase "S" and must not be confused with seconds, which uses a lowercase "s." Units (and prefixes) are case sensitive.

There may be considerable variability in the transistor parameters, as much as a factor of two or more, from one transistor to the next even when they are of the same type. The curves given by manufacturers are "representative." Hence, there is usually no point in trying to get more than two or, in the best cases, three decimal places from any solution.

In Fig. 7.8, the simple linear model (dashed lines) for changes from the chosen operating point, derived from the graphically determined parameters, are compared with the actual transistor characteristics (solid lines). The model may be somewhat crude, however the fact that it contains only linear circuit elements makes it very useful. Note that the farther from the operating point, the poorer the agreement between the model and the actual characteristics.

Remember that the model only describes *changes* from the operating point.

At higher frequencies, capacitance within the transistor becomes important and will need to be included in the transistor model. Other additions to the model may be necessary in some other circumstances.

---

[1]Named for the German engineer, Werner von Siemens (1816–1892). Note that his name includes the final "s" so that the unit should also include the final "s." That is, the singular is "1 siemens" and not "1 siemen." After all, 1 °C is not "one degree celciu."

152   7 FETs

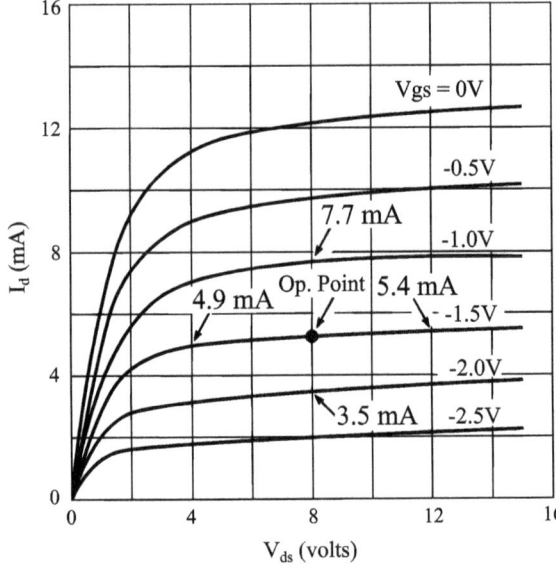

Fig. 7.7 The parameters for the a.c. model of Fig. 7.6 can be found graphically from the device characteristic curves

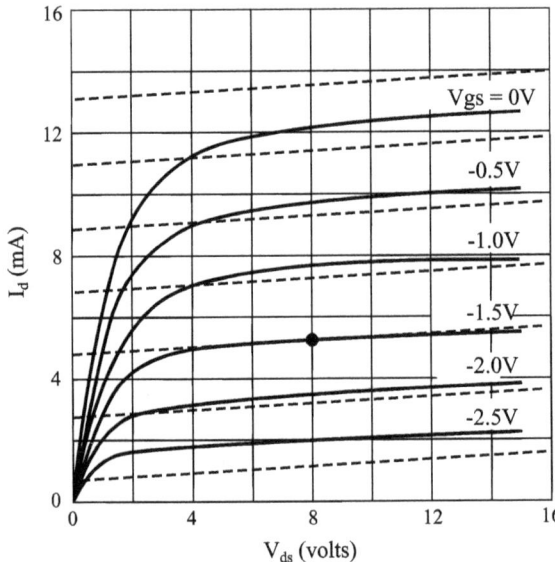

Fig. 7.8 A comparison of the measured characteristic curves (solid line) and the approximate behavior of the model of Fig. 7.6 using parameters derived graphically as in Fig. 7.7 (dashed line)

FET Amplifier Configurations

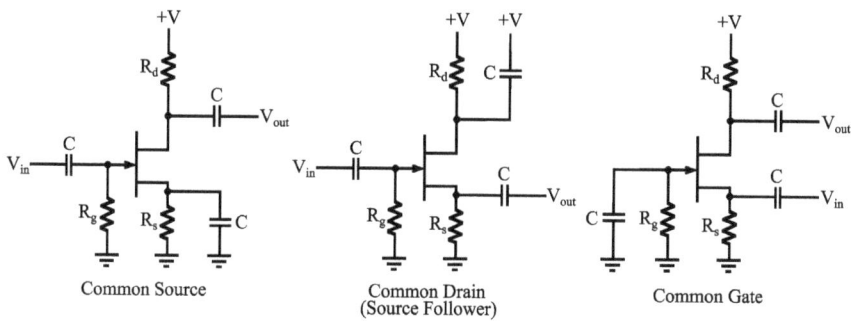

Fig. 7.9 The three basic configurations for single transistor JFET amplifiers

FET Amplifier Configurations

Figure 7.9 shows the three common configurations used for single-transistor FET circuits used to amplify time-dependent signals. They all are closely related, the only difference being where the input, output, and ground are connected. The d.c. (large signal) analysis, to find the operating point, for all of these is the same and is as was described in the previous section. For d.c. signals, the capacitors are open circuits and hence can be removed.

Once the d.c. analysis has been performed, the parameters g_m and r_d (if necessary) are used to look at small time-dependent changes from that point. That is, a linear FET model is used. This linear analysis works only if the transistor is in the active region and the input signals are small enough.

For the sake of simplicity, assume that the capacitors are large enough so that, at the frequencies of the signals to be amplified, they can be replaced by a short circuit (a wire). This is certainly not always the case, but in order to keep the amplifier analysis simple, that will be assumed here. Note that any constant (d.c.) voltage has no time-dependent part and thus is equivalent to every other d.c. voltage that has no time-dependent part, including ground (0 V). Hence *for the a.c. (time dependent) analysis all d.c. voltages look like ground*. For the a.c. analysis the three amplifier configurations will then reduce to the simplified circuits in Fig. 7.10.

It is then straightforward to compute the results shown in Table 7.1 from these models, where the "$\approx$" signs are valid for typical, but not all, applications.

Example Gain for common drain amplifier.

Compute the voltage gain for a common drain amplifier.

Consider the a.c. model for a common drain amplifier. The capacitors are assumed large enough so they can be replaced with wires and it is assumed that the transistor is in the active region where the simple FET model applies. In this case the desired quantity is the voltage gain, $A_v = V_{out}/V_{in}$. The circuit can be redrawn, as shown in Fig. 7.11 to aid in the analysis.

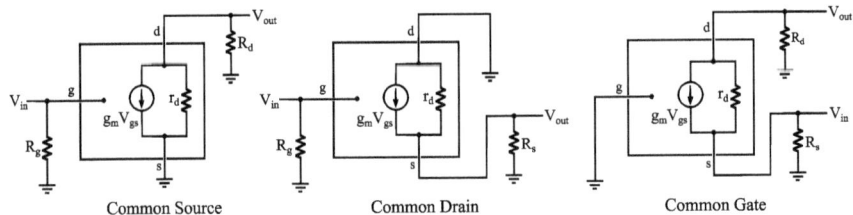

Fig. 7.10 Models for the circuits of Fig. 7.9 for small changes from a fixed operating point

Table 7.1 Voltage gain and input/output impedances for the three JFET amplifier configurations

Configuration	Voltage gain = V_{out}/V_{in}	Input impedance	Output impedance
Common source	$-g_m(R_d \| r_d)$	R_g	R_d
Common drain (source follower)	$\dfrac{g_m(R_s \| r_d)}{1+g_m(R_s \| r_d)} \approx 1$	R_g	$\dfrac{R_s \| r_d}{1+g_m(R_s \| r_d)} \approx \dfrac{1}{g_m}$
Common gate	$\dfrac{R_d}{R_s + (r_d + R_d)/(1+g_m r_d)}$	$R_s \left\| \left(\dfrac{r_d + R_d}{1+g_m r_d}\right) \approx R_s \right\| \dfrac{1}{g_m}$	$r_d + (1+g_m r_d)R_s$

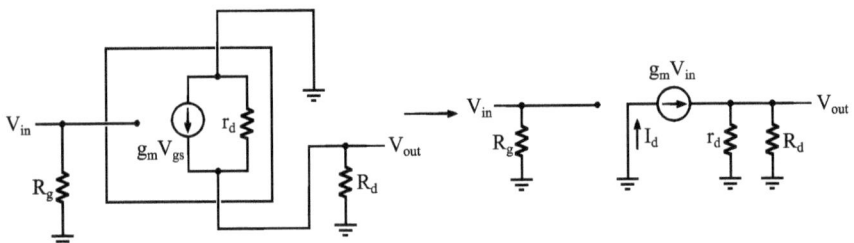

Fig. 7.11 The common drain circuit of Fig. 7.10 rearranged for analysis

Now

$$V_{gs} = V_g - V_s = V_{in} - V_{out}$$
$$V_{out} = I_d(R_s \| r_d) = g_m V_{gs}(R_s \| r_d) \qquad (7.17)$$
$$= g_m(V_{in} - V_{out})(R_d \| r_d),$$

so

$$V_{out}(1 + g_m(R_s \| r_d)) = g_m V_{in}(R_s \| r_d)$$
$$A_V = \dfrac{V_{out}}{V_{in}} = \dfrac{g_m(R_s \| r_d)}{1 + g_m(R_s \| r_d)}. \qquad (7.18)$$

FET Amplifier Configurations

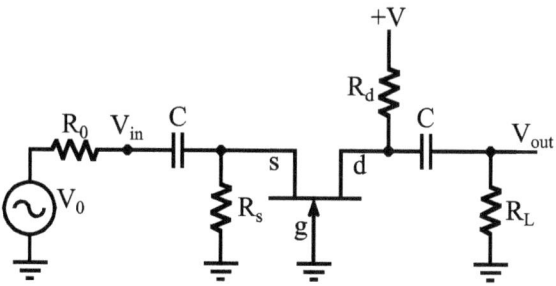

Fig. 7.12 The common gate amplifier with a source and load attached

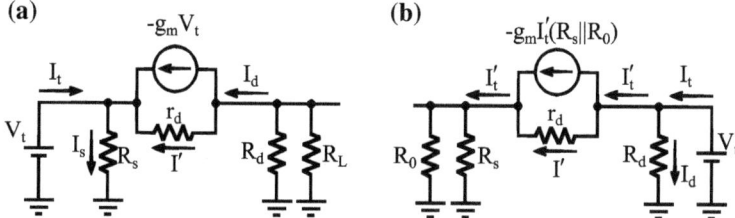

Fig. 7.13 The common gate amplifier with the JFET model set up to determine (**a**) the input impedance and (**b**) the output impedance of the amplifier

Example Impedances for common gate amplifier.

Compute the input and output impedances for a common gate amplifier.

Amplifiers will be connected to other circuitry at both the input and output. To determine how the amplifier affects this other circuitry, the amplifiers can be replaced by their Thevenin equivalent circuits as seen looking into either the input or output. The small signal models are used for these calculations. The problem can become a bit more complicated than previous Thevenin equivalent examples because the current source within the model is a dependent source—it cannot simply be turned off under all circumstances. The common gate amplifier is a good example where care must be taken.

The common gate amplifier with an input source, modeled by its Thevenin equivalent (source V_0, impedance R_0) and where the output is connected to a load, modeled by a simple resistor, looks like the circuit in Fig. 7.12. The same assumptions used to compute the amplifier voltage gain are used here. To compute the input and output impedances, a test voltage, V_t, is applied (on paper) to the input and output respectively and the resulting current, I_t, is computed. The ratio of the voltage to current is the impedance.

The models for the two impedance calculations are shown in Fig. 7.13. In both cases, the gate voltage is zero and, since it is not needed for the calculation, the gate connection is not shown.

Applying straightforward linear analysis (KVL and KCL),

$$Z_{in} = R_s \left\| \left(\frac{r_d + R_d \| R_L}{1 + g_m r_d} \right) \right.$$ (7.19)

and

$$Z_{out} = R_d \| [(1 + g_m r_d)(R_s \| R_0) + r_d].$$ (7.20)

The Ohmic Region

The discussion above concentrated on the transistor in the active region. Sometimes the ohmic region, at lower values of V_{ds}, is of use. In that region, the transistor can be used as a voltage variable resistor. As V_{gs} is changed, the (average) slope of the line changes, and hence the effective resistance changes. See Fig. 7.14. Of course, the FET will act like a resistor only for smaller changes in V_{ds}, where the line is relatively straight, and care is required to avoid values that are negative.

MOSFETs

MOSFETs use a very thin metal oxide barrier between the gate and the rest of transistor. This oxide layer is an insulator and hence the gate current is even smaller than the gate current for a JFET. In addition to an extremely small gate current, the

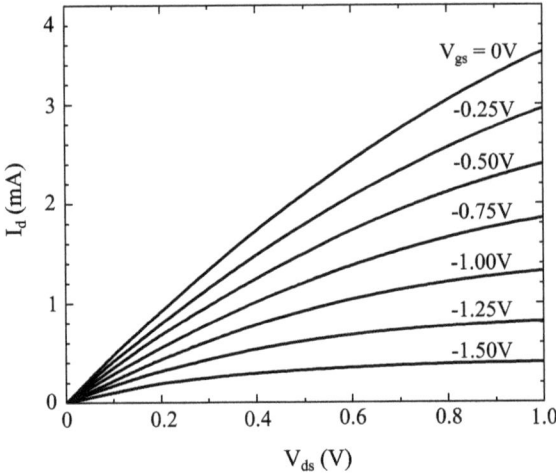

Fig. 7.14 A close-up of the ohmic region of the characteristic curves of Fig. 7.3

gate voltage may be both positive and negative. In a MOSFET the conducting channel is controlled by the electric field created by a gate to source voltage. "MOSFET" is almost always pronounced as a two-syllable word (Moss–fet). MOSFETs are also referred to as IGFETs—insulated-gate FETs. IGFET is a more general term.

There are two basic types of MOSFETs: the enhancement and the depletion MOSFET.

In the enhancement MOSFET, impurities are introduced so that there is no channel when the gate to source voltage is zero. When a voltage is applied, a channel is created by the minority carriers. For example, a *p*-channel MOSFET will be made from an *n*-doped substrate. When a negative gate to source voltage is applied, a net positive carrier concentration is induced near the gate, thus creating a *p*-channel in the *n*-doped material. This is illustrated in Fig. 7.15.

In the depletion mode MOSFET a conducting channel in the absence of a gate voltage is created by impurities. The gate voltage then modifies the width of the channel. Figure 7.16 illustrates an *n*-channel depletion mode MOSFET with both a negative (a) and positive (b) gate voltage. When the negative voltage is applied, the channel size is reduced, thus reducing the current. When a positive voltage is applied, the size of the channel is increased, increasing the current.

In addition to low-power applications, MOSFETs can be made for high voltages (hundreds of volts) and large currents (>100 A) and hence they are good for high power applications. Enhancement mode devices are commonly used in on/off switching applications. Depletion mode devices are nice for amplification of a.c. signals since the gate to source voltage can be both positive and negative.

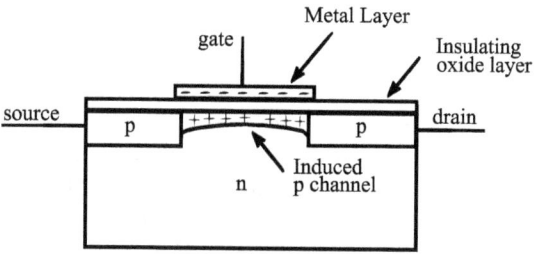

Fig. 7.15 Basic construction of an enhancement mode MOSFET

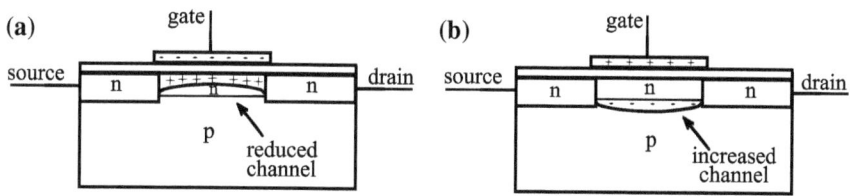

Fig. 7.16 The construction of a depletion mode MOSFET shown at (a) with $V_{gs} > 0$ and in (b) with $V_{gs} < 0$

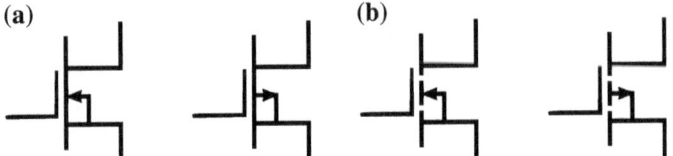

Fig. 7.17 Schematic representations of *n*- and *p*-channel MOSFETs. Depletion mode devices are shown in (**a**) and enhancement mode in (**b**) with *n*-channel devices on the left and *p*-channel devices on the right in both cases

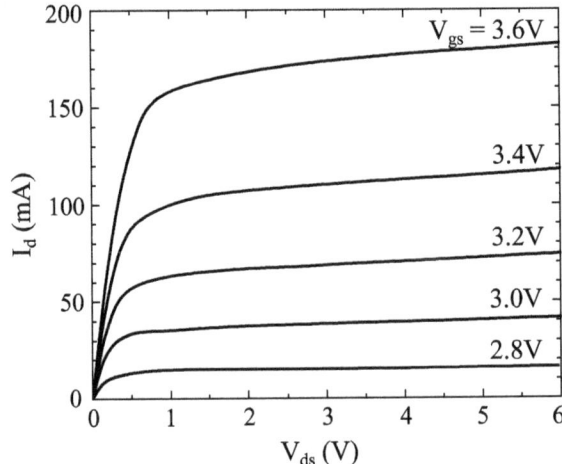

Fig. 7.18 Measured device characteristics for a particular enhancement mode MOSFET

Schematic representations of some MOSFETs are shown in Fig. 7.17. The very thin oxide layer of the MOSFET is easily damaged by static electricity and many devices include a built-in protection diode to reduce the possibility of damage. That protection diode may appear within the schematic for the MOSFET.

Characteristic curves for two sample MOSFETs are shown in Figs. 7.18 (enhancement mode) and 7.19 (depletion mode). These curves can be used to obtain graphical solutions in a manner identical to the curves for the JFET. There is nothing particularly special about these two examples—they were chosen to illustrate some of the variety in behavior that can be expected from different transistors.

Additional Application—Dynamic Memory

A capacitor connected to a MOSFET gate can hold its charge for a relatively long time. Figure 7.20 shows how this fact can be used to create a simple 2-state memory element from an enhancement mode MOSFET. To store a value, the

Additional Application—Dynamic Memory

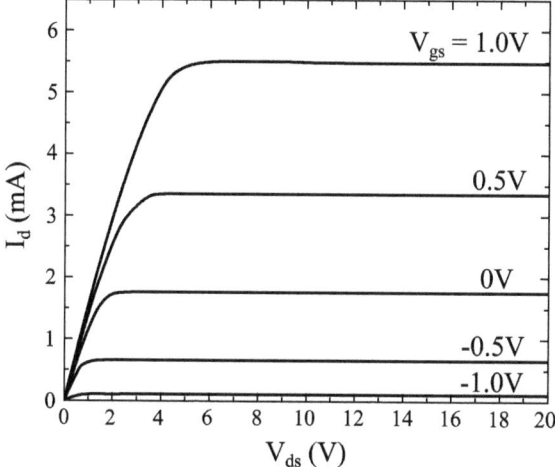

Fig. 7.19 Measured device characteristics for a particular depletion mode MOSFET

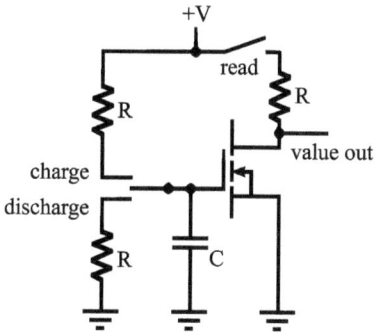

Fig. 7.20 The long RC time constant for a capacitor connected to the gate of a MOSFET can be used to make a two-state memory element

capacitor is momentarily connected to either +V (to charge it) or ground (to discharge it). When disconnected from both, the capacitor will hold its state of charge for a time governed by the RC time constant. With the extremely large resistance at the MOSFET gate, R_{gate}, even a very small capacitance (a few pF or even less) can lead to an appreciable time constant. In fact, the inherent capacitance of the gate connection may be large enough to be useful without an additional capacitor. For the example of Fig. 7.20 the state of the capacitor (charged or discharged) can be read at a later time by connecting the drain to +V through a resistor. A low output value, near 0 V, indicates the capacitor is charged, and a higher value, near +V, that the capacitor is discharged.

This type of memory will need to be refreshed from time to time and is referred to as dynamic memory. The output is simply read often enough, compared to the

RC time constant, and the result is used to either recharge or re-discharge the capacitor. The simplicity of the memory unit makes up for the need to periodically refresh the memory when a large amount of memory is being used.

Problems

1. In the circuit of Fig. 7.P1, the gate to source voltage is $V_{gs} = -1.0$ V. What is the drain to source voltage, V_{ds}?

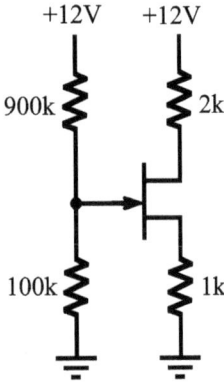

Fig. 7.P1 Problem 1

2. An *n*-channel JFET is characterized by the model parameters $I_{dss} = 5$ mA and $V_p = -3$ V. Sketch, based on this information, the characteristic curves for $V_{gs} = 0, -1$ V, and -2 V.
3. The *n*-channel JFET shown in Fig. 7.P3 is characterized by model parameters $I_{dss} = 5$ mA and $V_p = -3$ V. What are the gate to source voltage, V_{gs}, the drain to source voltage, V_{ds}, and the drain current, I_d for this circuit?

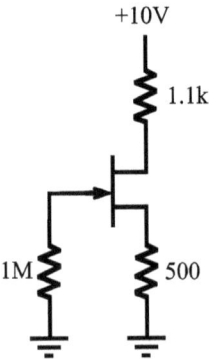

Fig. 7.P3 Problem 3

Problems

4. If the MOSFET characterized in Fig. 7.18 is used near an operating point of $V_{ds} = 12$ V and $I_d = 2$ mA, what is the appropriate value for the transconductance, g_m?

5. If the MOSFET characterized in Fig. 7.17 is used in the ohmic region, $V_{ds} < 0.5$ V, and with $V_{gs} = 3.6$ V, what is the effective drain to source resistance of the device?

6. (a) Show that the JFET mathematical model can be written in the alternate form

$$I_d = K(V_{gs} - V_t)^2.$$

   (b) If $I_{dss} = 10$ mA and $V_p = -3$ V, what are the corresponding parameters, K and V_t?

   (c) Show, based on this model and at an operating point with drain current I_d, that $g_m = 2\sqrt{KI_d}$.

7. Though they might not be identified as such, introductory mechanics courses include several results based on small changes. In hindsight, can you identify one (or more) of these?

Chapter 8
Bipolar Junction Transistors

The bipolar junction transistor, or BJT, is a three-lead device made by stacking three layers of doped semiconductor material with alternating *p*- and *n*-doping. Hence, there are NPN or PNP type transistors. While the details of the electronic behavior of these transistors differs from that seen for the FET, the general approach for circuit analysis remains the same. Solutions are found for an operating point based on device characteristics, that is, the "large signal behavior," then, if necessary, linearized models are used to look at small changes from that operating point. While the FET can be considered a voltage-controlled device, the BJT is considered current-controlled.

For the BJT, the three leads are called the collector, base, and emitter. The basic structure of an NPN transistor looks something like Fig. 8.1. While that figure gives the appearance that the emitter and collector could be switched, in real transistors the geometry is much less symmetrical and the performance with these two leads switched is usually quite poor.

Schematically, BJT transistors are drawn as in Fig. 8.2, where the arrow is on the emitter and points from "P" towards "N".

A BJT in normal operation will have the base-emitter junction forward biased and the collector-base junction reverse biased. To a good approximation, the basic behavior of the BJT can be specified in many circumstances by the relationships between I_b, I_c (or I_e), and V_{ce}, without worrying about V_{be} too much. The forward biased base-emitter junction will result in a significant base current, I_b, and a small base to emitter voltage, V_{be}.

Despite the fact that the collector-base junction will be reverse biased, the collector current can be quite large due to the extra charge carriers that will be present in the base region when the base-emitter junction is conducting. The motion of these extra "injected" charge carriers is dominated by thermal diffusion. If the base region is physically small enough, once those charges get into the base region, there is a much better chance that they will diffuse to the collector rather than out through the base. When there are no extra injected carriers the collector current will be a reverse saturation current—essentially zero for most practical purposes.

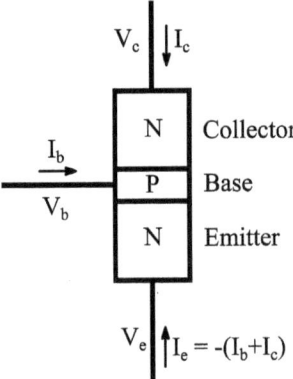

Fig. 8.1 The three-layer structure of an NPN bipolar junction transistor

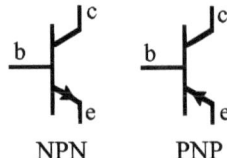

Fig. 8.2 Schematic symbols for NPN and PNP transistors

For the purposes of basic circuit analysis, however, only the electrical characteristics need to be known, and it is less important why the characteristics are what they are. Readers interested in learning more about semiconductor theory, and how it applies to transistors, are referred to more advanced texts on the subject.

As was the case for the FET, the BJT has multiple voltages and multiple currents that are relevant. As already mentioned, of typical concern will be the collector current (similar to the drain current for the FET), the collector to emitter voltage (similar to the drain to source voltage for the FET) and the current into the base. For the FET, the gate (to source) *voltage* acted to control the *drain* current. For the BJT, the base *current* acts to control the *collector* current.

The large signal characteristics for an NPN BJT can be measured using a circuit such as shown in Fig. 8.3, where V_{be}, I_c, and I_b are measured, for various values of V_{ce}, resulting in a family of curves. The measured results for a particular transistor are shown in Fig. 8.4. The curves on the left (8.4a) are the characteristic curves analogous to those for the FET. When V_{ce} is larger than about 1 V, the transistor is in the active region. When V_{ce} is smaller than about 1 V, the transistor is in the saturation region. As was the case for the JFET, the active region is used for amplifiers. In the saturation region, both junctions may be forward biased causing the transistor to behave like a wire. The saturation region is used for on/off applications including basic switching and some digital electronics.

8 Bipolar Junction Transistors

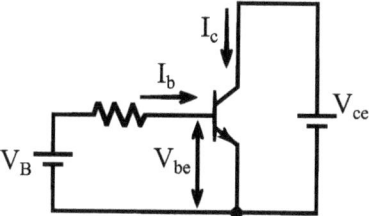

Fig. 8.3 Basic circuit used to define transistor characteristics. Shown for NPN. For PNP, all currents and voltage sources are reversed

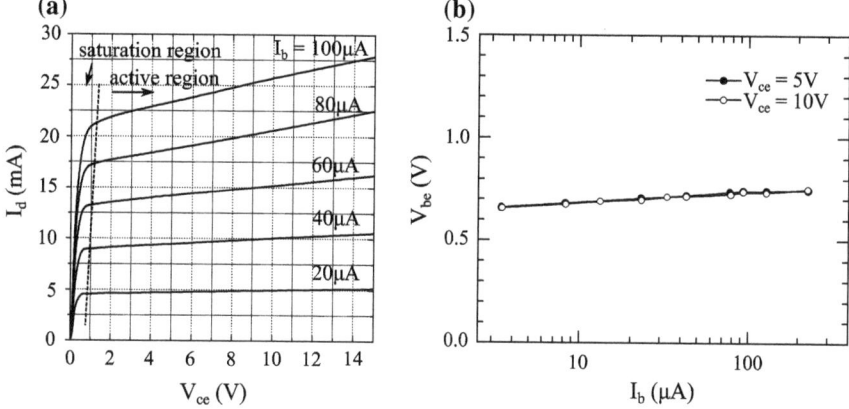

Fig. 8.4 Measured transistor characteristics for a small-signal NPN transistor

The measured curves on the right (8.4b) show that in the active region V_{be} does indeed appear to be a "turn-on voltage of about 0.5 V" (in this case, close to 0.7 V) over a wide range of conditions. While these curves are for a particular NPN transistor, similar behavior can be expected for other NPN transistors. For a PNP BJT, add a minus sign to all voltages and currents in these curves.

BJT D.C. Model

The "transfer characteristic" refers to I_c as a function of I_b, for different values of V_{ce}. That is, for selected fixed values of V_{ce} the data is read off the previous graph and those points are connected with a smooth curve. Figure 8.5a shows the transfer characteristics corresponding to the BJT characteristics shown in Fig. 8.4. Notice that to a reasonable approximation, and over a wide range of values for V_{ce} (but always in the active region) the transfer characteristic is described approximately by a straight line and it is almost the same line over a wide range of values for V_{ce}.

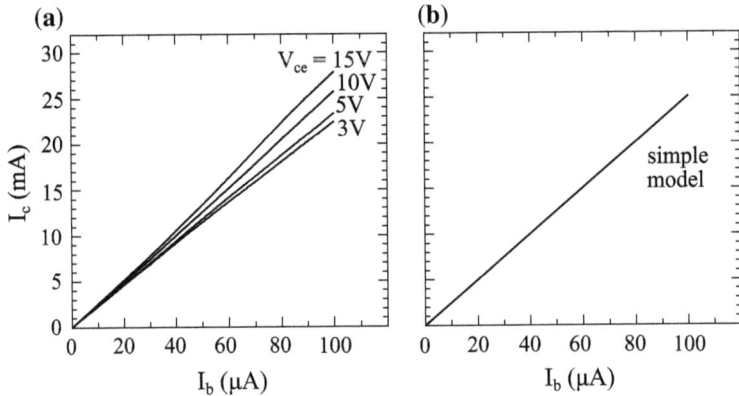

Fig. 8.5 Collector current as a function of base current for (**a**) measured values for several different collector-emitter voltages and (**b**) for a simple model that approximates the measured values

Hence, as a model, it is possible to write $I_c = \beta I_b$ for some constant β. Such a line is shown in Fig. 8.5b for this transistor, with $\beta \approx 250$. This value for β would be referred to as the "DC current gain."

Since β is the ratio of two currents, it does not need any units. Typical values for β are 30 to 300. Hence, approximations that result when $\beta \gg 1$ can be used routinely. Note also that β can vary by as much as a factor of 2 from one transistor to another of the same type, so it is rare that the computation of highly accurate values will be necessary.

Measurements of V_{be} for this same transistor as a function of base current, I_b, for different values of V_{ce}, shown in Fig. 8.4b, are also simple to model. For the active region, there is very little dependence on V_{ce}. Hence, the extremely simple model of Fig. 8.6 might be appropriate—that is a constant voltage (~ 0.7 V in this case) such as from a battery. The data shows a slight slope upwards for higher currents. If that trend is important, then the model can also be given a slight upwards tilt (i.e., a small series resistance).

Combining these ideas, a simple model for a BJT made from linear components, valid *in the active region*, would look like those of Fig. 8.7, where r_b is a small-valued resistor and r_c a large-valued resistor. The battery provides the few tenths of a volt appropriate for V_{be}, but of course it is not a real battery and cannot supply power. Here the battery is shown as 0.5 V, a convenient value to remember and use. The dependent current source supplies a collector current proportional to the base current. As is the case for the battery, it is not a real current source and cannot supply power on its own.

The small-valued resistor, r_b, is often optional, and provides a small upward slope on the line as seen in the measured values of V_{be} versus I_b. The larger resistor, r_c,

BJT D.C. Model

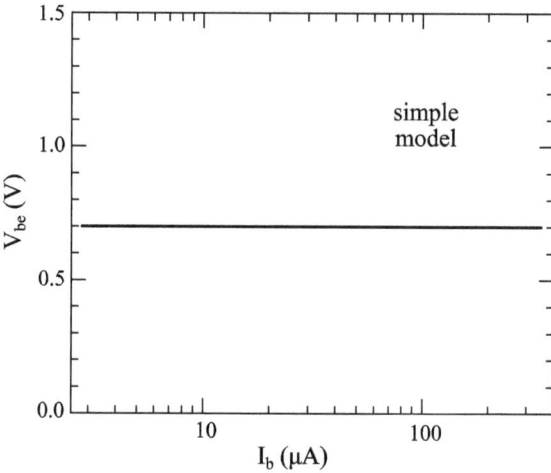

Fig. 8.6 A simple model that can be used to approximate the data of Fig. 8.4b

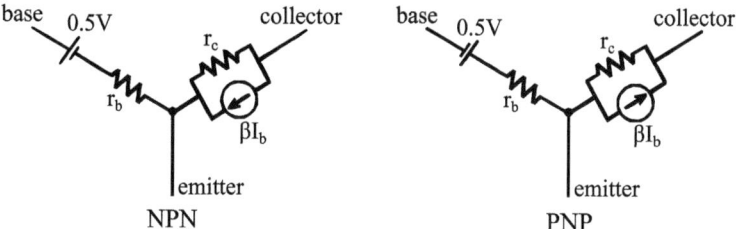

Fig. 8.7 Simple, linearized models for BJTs when used in the active region

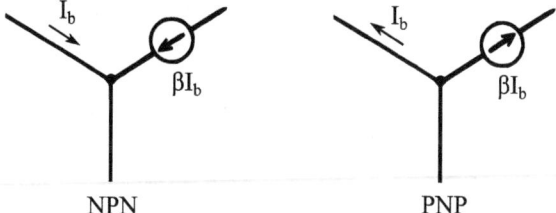

Fig. 8.8 Extremely simple models for BJTs. These models can be used if the resistors and sources shown in Fig. 8.7 are negligibly small compared to other components in the circuit

plays the same role as did r_d for the JFET, and allows for the fact that the characteristic curves are not ideally flat. In many applications it can be omitted. The convention being followed here is that lower case r's are used for resistors within the model, reserving upper case R's for real resistors.

In many cases, a circuit might be understood with an even simpler model for the transistor. The battery voltage is taken to be zero, $r_b = 0$ (a wire), and $r_c \to \infty$ (an open circuit) resulting in simplest BJT models shown in Fig. 8.8.

A first look at some of the following transistor problems might lead one to think that neglecting components is routine and is done in a somewhat cavalier manner—that is, using a technique sometimes referred to as hand waving. Whether or not any components of any model can be neglected depends, of course, on the values of *other* components in the circuit. Hence the very simple models should always be used with caution. The examples below will illustrate how such simplifications can be justified based on the relative size of component values. Whether or not such simplifications are valid must be considered on a case-by-case basis and should always be justified, or at least, justifiable.

Note that when the simple models work, the FET and BJT can be described similarly except that the FET is voltage controlled (by V_{gs}, with negligible gate current) and the BJT is current controlled (by I_b, with small/negligible base-emitter voltage). In both cases an "output current" (I_d for the FET, I_c for the BJT) is what is being controlled.

The appropriate model parameters may depend on how the transistor is being used. The models above are motivated by the large signal behavior. For the BJT those models can be used to find the operating point—that is the results of the large signal analysis in the active region. Of course, the graphical method, as was used for the FET, can be used for BJTs as well.

BJT A.C. Model

When used for an amplifier, the same basic BJT model can be used, though with parameters appropriate for small changes in the voltages and/or currents. To consider small changes from an operating point, proceed as was done for the JFET. The result is that the BJT transistor model will look the same as above, but the values of the parameters may need some adjustment.

When considering small changes from an operating point, the slope of the transfer characteristic near the operating point is considered. That is,

$$\beta = \left.\frac{dI_c}{dI_b}\right|_{V_{ce} \text{ constant}} \approx \left.\frac{\Delta I_c}{\Delta I_b}\right|_{V_{ce} \text{ constant}}, \tag{8.1}$$

which can depend on the operating conditions. If the transfer characteristic is a straight line, this value is the same as the dc current gain. It is often the case that the dc and small signal current gains can be taken to be the same—they are close enough in value—and at the risk of some confusion, the same symbol, β, is used for both. In some cases, the change in the slope of the transfer characteristic may be important. For the transfer characteristic shown above near $I_b = 100$ μA, β (for

BJT A.C. Model

small changes) will be somewhat larger for $V_{ce} = 15$ V than for $V_{ce} = 3$ V. More complicated models for these transistors refer to "h-parameters" (hybrid parameters), of which there are many, and the h-parameter corresponding to β is "h_{fe}."

For an idealized BJT, the characteristic curves in the active region would be flat. In cases where the slope is important, a parameter can be used to describe the slope of those lines,

$$\frac{1}{r_c} = h_{oe} = \left.\frac{dI_c}{dV_{ce}}\right|_{I_b \text{ constant}} \approx \left.\frac{\Delta I_c}{\Delta V_{ce}}\right|_{I_b \text{ constant}}, \tag{8.2}$$

evaluated at a point on the curves near where the transistor is being used. Typically, $r_c \approx$ 1k to 10k. Aside from some name changes, this is identical to what was done to find r_d for the JFET model.

Another parameter that is sometimes important, and which may not be evident in the characteristic curves, relates the voltage across the base to emitter junction to the resulting current into that junction, I_b. That junction looks somewhat like a diode, so an appropriate diode model can be used. It is convenient for simple calculations to use a piece-wise linear model such as

$$V_{be} = V_{be0} + I_b R_b, \tag{8.3}$$

where V_{be0} is the "turn on voltage" for the junction, which is about 0.5 V, and

$$r_b = h_{ie} = \left.\frac{dV_{be}}{dI_b}\right|_{V_{ce} \text{ constant}}, \tag{8.4}$$

evaluated near the operating point. Typically, $r_b \approx$ 10 to 1000 Ω, and the value depends strongly on the operating point. The constant d.c. value, V_{be0}, will not be important for a.c. signals.

BJT Large Signal Example

Graphical Solutions

For the following example, the transistor has the characteristics shown in Fig. 8.9a. The open circles indicate data points that are read off of the characteristic for $V_{ce} = 7$ V to create the transfer characteristic, Fig. 8.9b. That value of V_{ce} was chosen only because it is near the middle of the active region shown. For reference, several other such curves for different choices of V_{ce} are shown with the dashed lines. The dotted line is discussed below.

Now consider such a transistor in the circuit of Fig. 8.10. What is the collector current I_c, and the collector to emitter voltage, V_{ce}? Remember that all grounds are connected together and are considered to be the zero-volt reference.

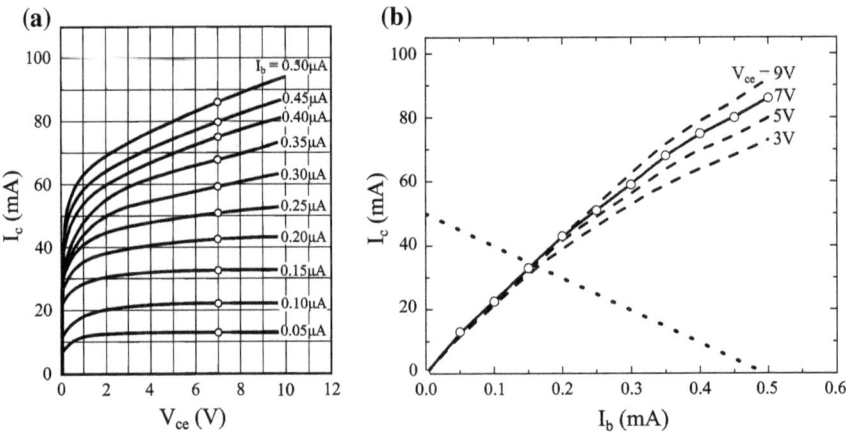

Fig. 8.9 The transistor characteristics shown in (**a**) are used to create the transfer characteristics in (**b**) as part of a graphical solution to a BJT circuit

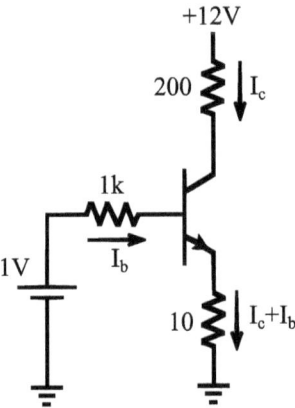

Fig. 8.10 An example BJT circuit to be solved

Applying Kirchhoff's voltage law from the 12 V supply, through the transistor and out the emitter generates the so-called "load line" for this configuration.

$$12\,\text{V} - 200I_c - V_{ce} - 10(I_c + I_b) = 0. \tag{8.5}$$

Similarly, going through the 1 V supply and the out the emitter, generates the so-called "bias line,"

$$1\,\text{V} - 1\text{k}I_b - V_{be} - 0.01\text{k}(I_c + I_b) = 0. \tag{8.6}$$

There are too many unknowns to be able to solve these equations by themselves. The transistor properties need to be included in some way, at least approximately.

BJT A.C. Model

As a very simple approximation, as mentioned above, take $V_{be} \approx 0.5$ V to get

$$0.5 \text{ V} - 1.01\text{k}\, I_b = 0.01\text{k}\, I_c, \text{ or } I_c = 50 \text{ mA} - 101 I_c. \tag{8.7}$$

This is a straight-line relationship between I_b and I_c and is plotted on the transfer characteristic (see the dotted line in Fig. 8.9b). While a value for V_{ce} is still unknown at this point, it is clear from the graphical solution that, at least if $3 \text{ V} < V_{ce} < 9 \text{ V}$, the current I_c must be very close to 34 mA and I_b will be close to 0.16 mA. Putting these values back into the first equation gives

$$V_{ce} = 12 \text{ V} - 0.2\text{k} \cdot 34\text{mA} - 0.01\text{k} \cdot 34.16 \text{ mA} \approx 12 \text{ V} - 0.21\text{k} \cdot 34 \text{ mA} = 4.9 \text{ V}. \tag{8.8}$$

This value of V_{ce} is indeed within the active range and this solution should be reasonable.

The two important properties of the BJT that made this solution possible are that within the active region V_{be} is almost constant and that the transfer characteristic does not depend on V_{ce} very much.

Single Supply Operation

In the example above, two sources of power are shown—one a 12 V supply and the other a 1 V supply. In practice this is inconvenient and rarely necessary. Instead, the single larger supply is used, along with Thevenin's theorem, to create an equivalent circuit.

Consider the circuit shown in Fig. 8.11. The two "biasing resistors," R_1 and R_2, are to be chosen so that the circuit is equivalent to that of Fig. 8.10. A Thevenin equivalent (refer to Chap. 2) is created for the biasing circuit, as shown on the right, where

$$V_b = \frac{R_2}{R_1 + R_2} 12 \text{ V}$$
$$R_b = R_1 \| R_2. \tag{8.9}$$

Solving to match the values in the example above, that is $V_b = 1$ V and $R_b = 1\text{k}$, yields $R_1 = 12\text{k}$, $R_2 = 1.1\text{k}$. When solving single-supply circuits, a first step is often the reverse process—to create the Thevenin equivalent two-supply circuit from a pair of biasing resistors.

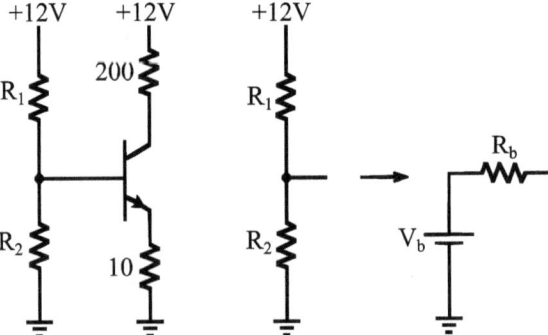

Fig. 8.11 Thevenin's theorem is used to create an equivalent circuit to that shown in Fig. 8.10 that requires only one voltage source

Solutions from Parameters

It is often the case that full transistor characteristics are not available and solutions must be obtained using limited parameters that have been supplied. For the transistor in the example above, the transfer characteristic shows that $\beta \approx 200$. Consider that same transistor in the circuit in Fig. 8.12, assuming the value of β is all that is known about the transistor.

As a first step in this analysis, replace the biasing circuit (the resistors on the left plus the power supply) with its Thevenin equivalent (Fig. 8.13). Now use the transistor model, which assumes the active region, to get the linear circuit of Fig. 8.14. Using Kirchhoff's voltage law around the left and right loops,

$$1.1\,\text{V} - I_b(0.91\text{k} + r_b) - 0.5\,\text{V} - (I_c + I_b)(0.1\text{k}) = 0$$
$$12\,\text{V} - I_x(0.5\text{k}) - V_{ce} - (I_c + I_b)(0.1\text{k}) = 0.$$
(8.10)

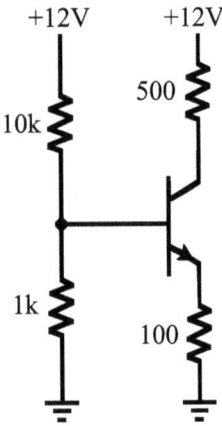

Fig. 8.12 A second example BJT circuit to be solved

BJT A.C. Model

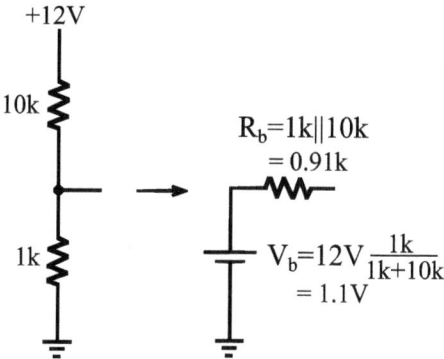

Fig. 8.13 The first step to solve the circuit of Fig. 8.12 is to use Thevenin's theorem for the biasing resistors

Here V_{ce} equals the voltage across the current source. Since r_c will be significantly larger than 500 Ω, r_c will have a much smaller current through it and can be neglected in comparison (see Problem 6 of Chap. 1). With that approximation, $I_c = \beta I_b$. The equations become

$$(1.1 - 0.5)\text{V} - I_b(0.91\text{k} + r_b + (1+\beta)(0.1\text{k})) = 0$$
$$12\,\text{V} - V_{ce} - I_b(\beta(0.5\text{k}) + (1+\beta)(0.1\text{k})) = 0. \quad (8.11)$$

In the first equation, with $\beta = 200$, the 100 Ω resistor acts like it is close to 20k, and since $r_b \ll 10\text{k}$, r_b can be neglected in comparison. Then the first equation can be solved for I_b (and hence also $I_c = \beta I_b$) and then that result is put into the second equation to find V_{ce}. For this example, $I_b = 29$ μA, $I_c = 5.8$ mA, and $V_{ce} = 8.5$ V. Since $V_{ce} \gtrsim 1$ V, this is in the active region, as was assumed above, and so the solution should be reasonable.

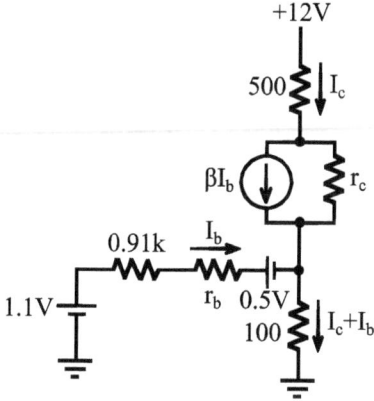

Fig. 8.14 The second step is to replace the BJT with the linearized model

Table 8.1 The β dependence for the operating point of the example calculation

β	I_b (μA)	I_c (mA)	V_{ce} (V)
50	100	5.0	9.0
100	54	5.5	8.8
200	29	5.8	8.5

Using this example, it is possible to estimate some of the effects of the inevitable variability from transistor to transistor. Solving the equations above with $\beta = 50$, 100, and 200 yields the results shown in Table 8.1. Notice that I_c and V_{ce} do not change very much. If the equations are solved leaving β as a variable, then

$$I_c = \frac{\beta(0.6\,\text{V})}{(1+\beta)(100\,\Omega)+910\,\Omega}, \tag{8.12}$$

and it can be seen that I_c will be roughly independent of β provided $\beta \gg 1$, which is typical, and also if $\beta(100\,\Omega) \gg 910\,\Omega$, which is certainly valid for this example.

One good reason to include the $\sim 100\,\Omega$ emitter resistor in such a circuit is to minimize some of the effects of the significant variations in β that occur from transistor to transistor. As will be seen later, the emitter resistor also can have a significant impact on amplifier gain.

BJT Amplifiers

There are three basic configurations for single-transistor BJT amplifiers, shown for NPN transistors in Fig. 8.15. The common base configuration is usually drawn so that the input is on the opposite side of the drawing from the output. Here it is drawn to facilitate comparison between the three amplifiers. Note that the "large signal analysis" or "d.c. analysis" to find the operating point is identical for all of these. That large signal analysis was shown above.

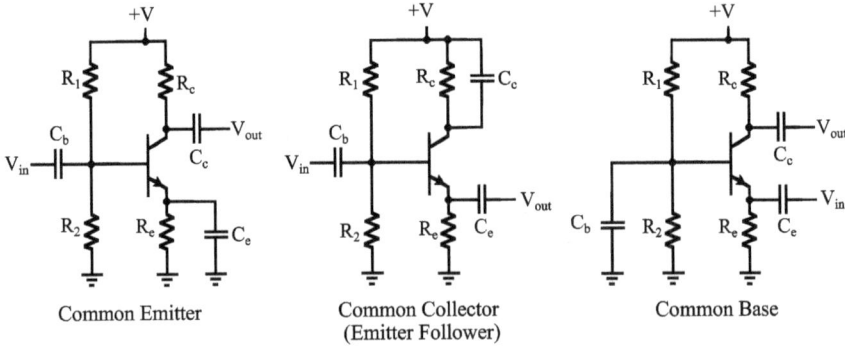

Fig. 8.15 The three single-transistor amplifier configurations used with BJT's. Shown for NPN transistors

BJT Amplifiers

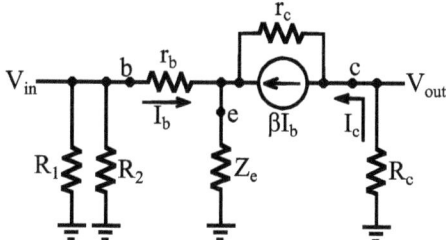

Fig. 8.16 The BJT model is used to determine the voltage gain for small a.c. signals. Note that it is assumed the transistor is in the active region. Note also that all of the d.c. voltages are 0 V a.c

Consider the Common Emitter (CE) amplifier in some more detail. The operating point is found as before and it is assumed that this has been done and the transistor is indeed in the active region. The next step is to look at the a.c. small signal analysis. The overall procedure here is identical to what was done for the FET transistor amplifiers, however some important details have changed.

To keep matters simple, consider time dependent input signals, and hence also output signals, that have a frequency large enough so that the impedance of the input and output capacitors is negligible (or alternatively, that "the capacitors are large enough"). Such an assumption is certainly not necessary and is not always appropriate. As was the case for the FET amplifiers, it is done here to keep the analysis simple. With that assumption, those capacitors can be replaced with wires for this part of the analysis. In the case of the CE amplifier, the emitter capacitor may or may not be present and so the more general situation is treated here—the combined emitter resistor and capacitor are treated as an impedance Z_e.

Now the transistor model is used with all d.c. voltage sources "turned to 0 V" (i.e., replaced with wires to ground). Putting all the 0 V connections (the grounds) at the bottom of the figure, the common emitter amplifier now looks, for the a.c. signals, as shown in Fig. 8.16. Remember that the voltages and currents here refer *only* to changes from the operating point and when one is considering changes, *any* constant voltage is the same as zero—after all, there is zero change for a constant voltage. Kirchhoff's law from the input and down through the emitter gives

$$V_{in} - I_b r_b - (I_b + I_c) Z_e = 0. \tag{8.13}$$

Kirchhoff's law around the right loop gives

$$-I_c R_c - (I_c - \beta I_b) r_c - (I_b + I_c) Z_e = 0, \tag{8.14}$$

which can be solved to get

$$I_c = I_b \frac{\beta r_c - Z_e}{r_c + R_c + Z_e} \approx I_b \frac{\beta r_c}{r_c + R_c + Z_e} \approx \beta I_b, \tag{8.15}$$

where the first approximation will be very good in virtually all circumstances and the last approximation is valid if r_c is large enough compared to the collector and emitter resistors, which is often, though not always, true. Substituting back into the first equation, and noting that $V_{out} = -I_c R_c$, yields

$$V_{out} = -I_c R_c = -V_{in} \frac{\beta R_c r_c}{r_c(r_b + (1+\beta)Z_e) + R_c(r_b + Z_e)} \approx -V_{in} \frac{\beta R_c}{r_b + (1+\beta)Z_e}, \quad (8.16)$$

where the approximation is certainly valid if r_c is large compared to R_c. Consider also the case where there is no emitter capacitor, and $\beta R_e \gg r_b$. Then $V_{out}/V_{in} \approx -R_c/R_e$. On the other hand, if an emitter capacitor with a large valued capacitor is included, so the total emitter impedance is very small at the frequencies of interest, then $\beta |Z_e| \ll r_b$, so $V_{out}/V_{in} \approx -\beta R_c/r_b$. The advantage of the former (no emitter capacitor) is that the voltage gain is independent of β, and hence is much more under the designer's control. The advantage of the latter is that the gain is much larger. Do recall however, that the emitter resistor, which is part of Z_e, is very important for the dc analysis, the "biasing," and so that resistor cannot be simply omitted to get a large voltage gain—hence the use of a bypass capacitor to short the a.c. signals to ground without affecting the d.c. behavior.

Also of interest for an amplifier is knowledge of its input and output impedances. That is, what does the amplifier look like to the other circuits that are connected to it? For the amplifier above, if $|Z_e|$ is small enough so that it can be replaced by a wire, the input impedance is obviously $(R_1 \| R_2) \| r_b$ which will be relatively small since r_b is generally small. On the other hand, if there is no emitter capacitor, the calculation is somewhat more complicated due to the fact that the current source is now a dependent current source. The superposition principle, where each source is considered independently, does not apply for dependent sources.

To do this calculation, use the fact that the input impedance, Z_i, is given by $Z_i = V_{in}/I_{in}$ and

$$I_{in} = \frac{V_{in}}{R_1 \| R_2} + I_b = V_{in} \left(\frac{1}{R_1 \| R_2} + \frac{r_c + R_c + Z_e}{r_c(r_b + (1+\beta)Z_e) + r_b(R_c + Z_e)} \right), \quad (8.17)$$

to get the very complicated and non-intuitive result,

$$Z_i = \left(\frac{1}{R_1 \| R_2} + \frac{r_c + R_c + Z_e}{r_c(r_b + (1+\beta)Z_e) + r_b(R_c + Z_e)} \right)^{-1}, \quad (8.18)$$

and so if $|Z_e|$ is very small, the expected result is obtained. If that is not the case, the situation is more complicated. Looking at the case where both r_c and β are large, then

$$Z_i = (R_1 \| R_2) \| (r_b + (1+\beta)Z_e). \quad (8.19)$$

In the case where there is no emitter capacitor it is often true that $(1+\beta)R_e \gg r_b$, due to the large values of β, so that r_b can be neglected.

The output impedance, Z_o, is the Thevenin equivalent impedance at the output of the amplifier. If the dependent current source depends only on the input voltage, but not the output voltage, it is possible to use the superposition principle and "look back into the circuit with the sources set to 0" to find the Thevenin equivalent resistance. For the common emitter amplifier with a large emitter capacitor, so $|Z_e|$ is small, the latter method works and the output impedance is then $R_c \| r_c \approx R_c$ if r_c is large. If $|Z_e|$ is not negligibly small, however, more care is necessary. In that case, the current source depends on the voltage at the output even if the input voltage is "turned off."

Recall that the Thevenin impedance is the ratio of the open circuit voltage and the short circuit current. The open circuit output voltage, V_{out}, was computed above (see Eq. 8.16). The a.c. model with the output connected to a short circuit looks like Fig. 8.16 but with $R_c = 0$, and so the short circuit output current, I_{os}, is I_c, as computed above, but with $R_c = 0$. Then

$$Z_o = \frac{V_{out}}{I_{os}} = \frac{R_c r_c (r_b + (1+\beta) Z_e)}{r_c(r_b + (1+\beta) Z_e) + R_c(r_b + Z_e)}, \qquad (8.20)$$

and it can be seen that if r_c and/or β is large enough so that the second term in the denominator is negligible, then $Z_o \approx R_c$. If the output capacitor, C_o, is not negligible, its impedance will need to be added to the above result.

Now for some more concrete examples.

Common Emitter Amplifier Example

Estimate the a.c. voltage gain for the circuit in Fig. 8.17 that uses an NPN transistor with $\beta = 36$.

First, assume the transistor is operating in the active region—that assumption will be checked along the way.

Now do the DC analysis with the input signal off. At DC all the capacitors look like open circuits and can be removed from the analysis. Simplify the biasing on the

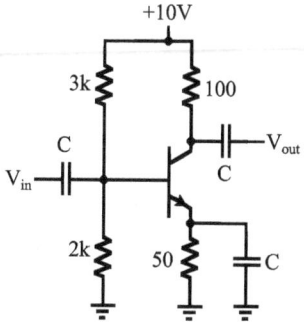

Fig. 8.17 A common emitter amplifier to be analyzed as an example

left using the Thevenin equivalent as shown in Fig. 8.18, and put this into the transistor model to get the linear circuit in Fig. 8.19. Now one can hope to neglect r_c (remove it) since typically $r_c > 1$k and the 100 and 50 Ω resistors are both much smaller than 1 kΩ.

Using Kirchhoff's voltage law,

$$4\,\text{V} - (1.2\text{k} + r_b)I_b - 0.5\,\text{V} - (1+\beta)I_b \cdot 50\,\Omega = 0, \tag{8.21}$$

so with $\beta = 36$ and assuming r_b is small enough to be negligible (here that means $r_b \ll 1.2\text{k}$),

$$I_b = 3.5\,\text{V}/3\text{k} = 1.17\,\text{mA}, \tag{8.22}$$

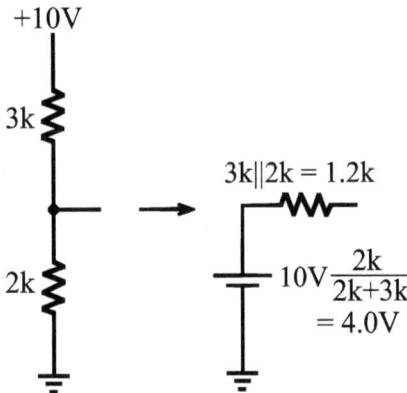

Fig. 8.18 The first step to analyze the amplifier of Fig. 8.17 is to use Thevenin's theorem to simplify the biasing circuitry

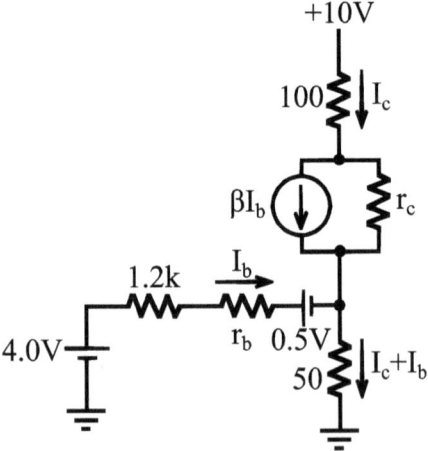

Fig. 8.19 The second step to analyze the amplifier of Fig. 8.17 is to assume the active region and put in the (d.c.) transistor model. Here the capacitors are treated as open circuits

BJT Amplifiers

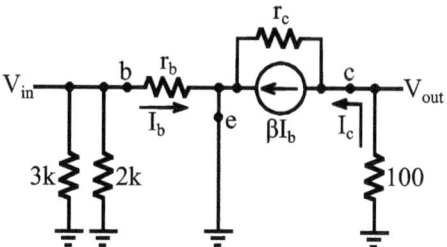

Fig. 8.20 The last step to analyze the circuit for small a.c. signals. Here the capacitors are approximated as short circuits. If the frequency of the signal is not high enough, then the impedance of the capacitors must be included

so $I_c = 42$ mA giving $V_{ce} = 10$ V $-$ (42 mA)(100 Ω) + (43 mA)(50 Ω) = 3.6 V. That is in the active region (i.e., $V_{ce} > 1$ V), validating the initial assumption.

Now look at the a.c. (i.e., time-dependent) model. For simplicity, assume the capacitors (with unknown values) are large enough so that they can be treated as short circuits. Then, realizing that the d.c. voltages are "0 volts a.c." the equivalent circuit for the time-dependent signals is as shown in Fig. 8.20. Then $I_b = V_i/r_b$ and neglecting r_c (it is large compared to 100 Ω on one side and shorted to ground at the emitter), $I_c = 36 I_b$ and $V_{out} = 100 I_c$. This gives

$$V_{out}/V_{in} = (36/r_b) \cdot 100 \, \Omega, \tag{8.23}$$

and to get a value, r_b needs to be known. Note that if r_b had been neglected here, the result would have been unusable—in this case r_b is not "small enough compared to other resistors" since there are no others between the input and ground.

Using the exponential model for the base to emitter p-n junction, the resistance is given (at room temperature) roughly by[1]

$$r_b = 0.026 \, \text{V}/I_b = 0.026 \, \text{V}/1.17 \, \text{mA} = 22 \, \Omega. \tag{8.24}$$

where I_b is the value at the operating point (from the d.c. analysis). Hence a voltage gain of about 160 should be expected.

Common Collector Amplifier Example

Estimate the a.c. voltage gain for the circuit in Fig. 8.21 *that uses an NPN transistor with* $\beta = 36$.

---

[1]The exponential model for diodes is in Chap. 6. The resistance for small changes is found from the derivative, $dV_d/dI_d = 1/(dI_d/dV_d)$. Such an estimate is accurate only to factors of order unity. Many manufacturers supply graphs from which one can find $r_b = h_{ie}$ for specific transistors under a variety of operating conditions.

This is the same as Example 1, except using the common collector configuration. Hence, the large-signal d.c. analysis is identical to Example 1 and need not be repeated.

Again, for simplicity assume the capacitors act as shorts for the time dependent signals. At lower frequencies, the impedance of the capacitors will need to be considered.

Putting in the a.c. model for the BJT and again noting that the constant 10 V source is 0 V a.c., the a.c. model for the circuit is as shown in Fig. 8.22, and so

$$V_{in} - I_b r_b - (I_b + I_c) \cdot 50 = 0$$

$$I_c \approx \beta I_b = \frac{\beta V_i}{r_b + (1+\beta) \cdot 50}, \qquad (8.25)$$

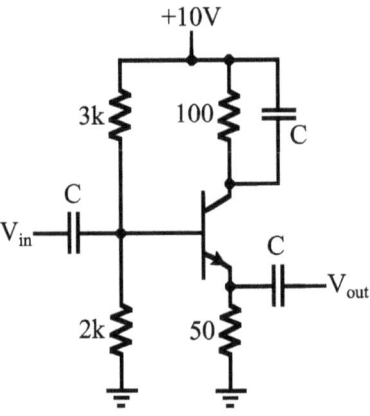

Fig. 8.21 A common collector amplifier analyzed as an example. The biasing for this example is the same as for the example circuit of Fig. 8.17

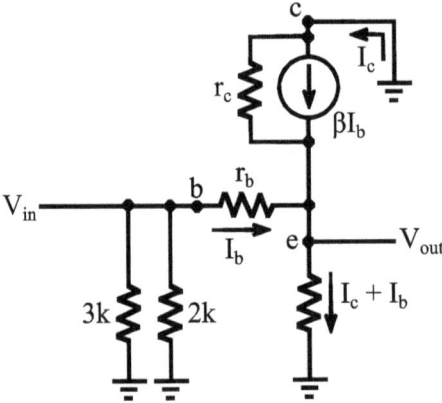

Fig. 8.22 The transistor model is placed into the circuit of Fig. 8.21 to find the a.c. small signal gain

BJT Amplifiers

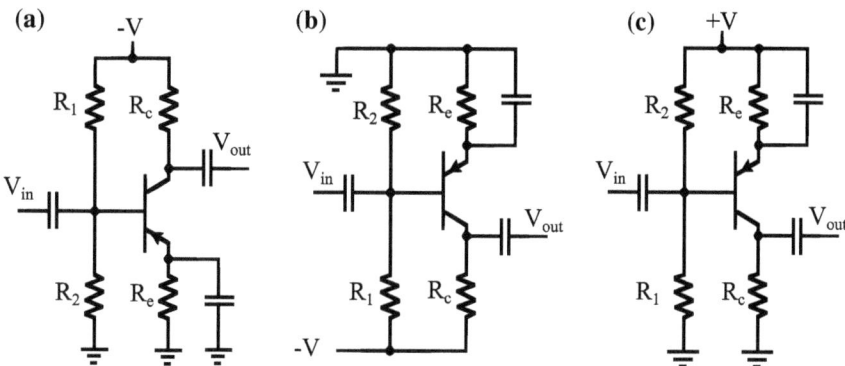

Fig. 8.23 The PNP common emitter amplifier at (**a**) is derived from the NPN amplifier by replacing the transistor and reversing the power supply. The circuit is more commonly shown with the higher voltage at the top as in (**b**), for a negative supply, or (**c**), for a positive supply

and if $(1 + \beta) \cdot 50 \gg r_b$, which is certainly the case here, then

$$V_{out} = (I_c + I_b) \cdot 50 = V_{in}. \tag{8.26}$$

The voltage "gain" is very close to 1. That is, the output voltage "follows" the input voltage.

This amplifier has a low output impedance (here, 50 Ω) whereas the input impedance is about β times larger. Hence there is a large current gain, and hence a large power gain, even though there is little change in the voltage.

In summary, the same basic procedure is used for all transistor amplifiers. First, the large signal analysis with no input is used to find the "operating point." Then parameters for a suitable model are found to describe small changes about that operating point. The model is kept simple and linear. The model is placed in the circuit to look at the changes that occur (e.g., for an a.c. signal). Along the way, various approximations might be possible to simplify the analysis. At each step, of course, those approximations need to be justified.

The examples above use NPN transistors. If a PNP transistor is used, the voltages and currents will be reversed, as shown for the common emitter amplifier in Fig. 8.23a. It is common practice to place larger voltages toward the top of the page, so the common emitter amplifier with a PNP transistor will appear as in Fig. 8.23b if a negative power supply is used and as in Fig. 8.23c if a positive supply is used. Remember that only the difference in voltage matters.

Using the Saturation Region

Transistors are also used as on/off devices, both for switching and for digital logic. When a BJT is used in this way, *the transistor model does not apply*. However, that model can be used to estimate the results if the model is used carefully.

Example 1 Consider the circuit of Fig. 8.24 where V_{in} is either 0 V or 5 V (relative to ground). Assume that for this transistor, $\beta \approx 250$. Going around the input loop Kirchhoff's voltage law yields

$$V_{in} - 10k I_b - 0.5\,\text{V} = 0. \tag{8.27}$$

so $I_b = 0$ or 0.45 mA depending on V_{in}.

When $I_b = 0$, then $I_c = 0$ and $V_{out} = 5\,\text{V} - 1\text{k}\,I_c = 5\,\text{V}$. In this state the transistor is "off" and the output sees only the 5 V power source through the 1k resistor.

When $I_b = 0.45$ mA, the model is used to compute $I_c = 250$ and $I_b = 112.5$ mA. Then $V_{out} = 5\,\text{V} - 1\text{k}\,I_c = -107.5\,\text{V}$. Since the transistor cannot supply power, it merely reroutes power, it must be the case that $0\,\text{V} < V_{out} < 5\,\text{V}$ and so this solution is clearly invalid. The output will, in some sense, get as close to this solution as it can while still being within the allowed range. Hence the solution is very close to 0 V.

What happens can be seen graphically by plotting the load line ($V_{ce} = 5\,\text{V} - 1\text{k}\,I_c$) on the characteristic curves (Fig. 8.25). The solution based on the model is negative, but *the model does not apply* on the negative side of the V_{ce} axis. The real transistor characteristic gives a solution in the saturation region with a very small voltage for V_{ce}. A small voltage with a large current looks like a wire. In that state the transistor is said to be "on".

This particular circuit takes an input voltage of 0 V or 5 V and produces a corresponding 5 V or 0 V as the output. In a sense, the circuit provides "the opposite of the input."

Example 2 The saturation region is also used to switch higher power devices using a low power control signal. As an example, consider a 12 V, 55 W light bulb that is to be switched using an NPN transistor and a 5 V signal (Fig. 8.26).

If the bulb is specified as a "12 V" bulb and a 12 V power supply is used, that means to achieve full brightness, virtually all of the 12 V available must be across the bulb, and not the transistor. Hence, to turn on the bulb (to full brightness), V_{ce} must be close to 0 V. If it is a 55 W bulb, then the current can be computed:

$$P = VI \rightarrow I = P/V = 55\,\text{W}/12\,\text{V} = 4.6\,\text{A}. \tag{8.28}$$

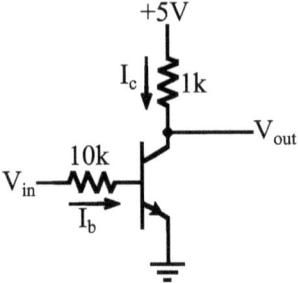

Fig. 8.24 A very simple single transistor circuit for on/off applications

Using the Saturation Region

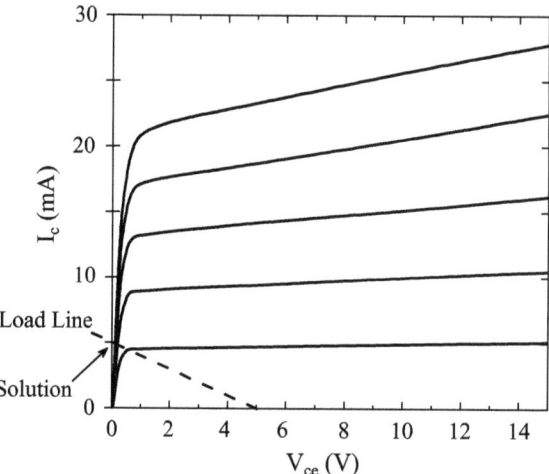

Fig. 8.25 A graphical solution for the circuit of Fig. 8.24

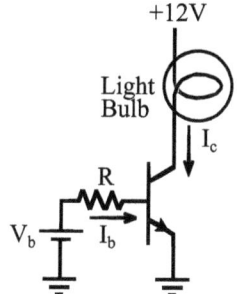

Fig. 8.26 An example of a transistor being used as an on/off switch for a light bulb

A transistor must be chosen which can handle that much current (or more). Such information is available on the manufacturer's transistor data sheet. Also on that data sheet will be the expected range for the dc current gain, β. To be safe, the minimum expected value should be used for calculation. For use as an example, assume that, for the transistor chosen, that minimum is $\beta = 50$.

Then to turn on the light, you need $I_b > 4.6\text{A}/50 = 92$ mA. If the signal, $V_b = 5$ V, then solving for R,

$$5\text{ V} - I_b R - 0.5\text{ V} = 0, \tag{8.29}$$

so $R < 49\ \Omega$.

The strategy here was to use β, a parameter for the active region, to estimate the values necessary to end up in the saturation region.

Problems

1. For the transistor characteristics shown in Fig. 8.4a, estimate the a.c. current gain and the value of the model resistor r_c near an operating point with $V_{ce} = 7$ V and $I_c = 80$ mA.
2. Derive an expression for the voltage gain of the common base configuration using the transistor model and appropriate approximations. Be sure to justify any approximations. Assume the biasing has already been done and that the transistor is in the active region.
3. The circuit in Fig. 8.P3 can provide a current through a load, R_L, which is almost constant as the applied voltage, +V, changes, provided that voltage is within a certain range. If the transistor has $\beta = 150$ and the desired applied voltage range is 5 V < +V < 15 V, what current, I, is expected and what is the maximum value that can be used for R_L?

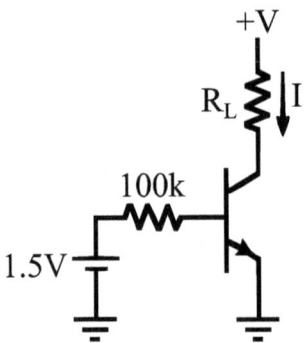

Fig. 8.P3 Problem 3

4. Electric Eye: For the circuit in Fig. 8.P4, the NPN transistor is specified as having $h_{fe} = \beta = 75$. The resistor R_L is a model for a 12 V, 1 W light bulb (i.e., the light bulb is designed to be used with 12 V across it—significantly more and it will burn out, significantly less and it will not be at full brightness. Significant here is approx. ±10%). The variable resistor, R_s, is a photoresistor[2] where the resistance when exposed to a light beam is 100 Ω and when the beam is blocked, it is 100 kΩ. Estimate an appropriate value, or range of values, for R_x so that the light bulb is (nearly) fully on when the light beam is blocked, and is (essentially) off when the beam is unblocked.

---

[2]A photoresistor is a bit of semiconducting material that can be exposed to light. If the photons of light have an energy larger than the semiconductor band gap, electrons and holes can be created by the light. As the concentration of these charge carriers increases, the resistance of the device decreases.

Problems

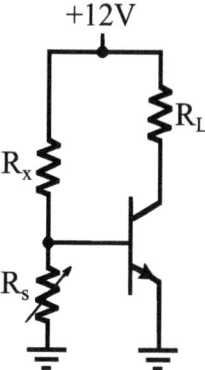

Fig. 8.P4 Problem 4

5. Solve for all the currents and for V_{ce} for the PNP transistor circuit shown in Fig. 8.P5 assuming the transistor has $\beta = 150$.

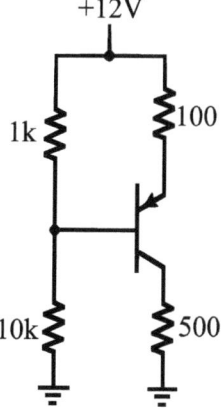

Fig. 8.P5 Problem 5

Chapter 9
More on Amplifiers

The previous chapters showed, in some detail, how to make amplifiers using a single transistor. Further detailed analysis for any particular amplifier is left to those who find it necessary. Here some additional ideas applicable to amplifiers are presented more generally. To start, yet another general theorem for linear circuits is considered. After that, some useful configurations that rely on multiple transistors are presented, as well as some more general concepts that apply to any linear amplifier.

Miller's Theorem

While the inner workings of an amplifier may involve non-linear devices, the amplifiers shown can be treated as a device that is linear—at least as long as the input and output signals do not become too large or go out of range. The basic device rule for a linear amplifier is that there is an output voltage (or current) that is proportional to the input voltage (or current).

Miller's theorem is useful for linear amplifiers, in particular for understanding the origin of the upper frequency limits of many amplifiers. Miller's theorem can also be used for some circuit analysis when feedback is employed. Feedback will play a very important role for the op-amp circuits discussed in the next chapter.

To start, consider a box (containing electronics) that has three connections where one of them is the zero-volt reference (i.e., ground). The other two are labeled "1" and "2." The current out of these connections is I_1 and I_2 respectively, and the voltages measured relative to ground are V_1 and V_2. Miller's theorem applies if the box has the property that $V_2 = kV_1$, where k is some constant. Now consider such a box with an impedance Z connected between 1 and 2 as shown in Fig. 9.1a.

It is a simple matter to compute the relationships between I_1 and V_1, and I_2 and V_2,

© Springer Nature Switzerland AG 2020
B. H. Suits, *Electronics for Physicists*, Undergraduate Lecture Notes in Physics,
https://doi.org/10.1007/978-3-030-39088-4_9

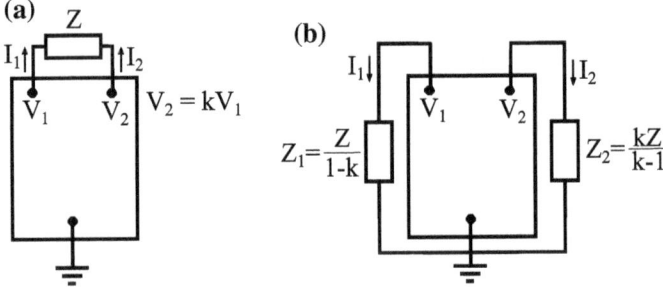

Fig. 9.1 Miller's theorem shows that the circuitry inside the box will be unchanged if the circuit in (**a**) is converted to the circuit in (**b**)

$$I_1 = \frac{V_1 - V_2}{Z} = \frac{V_1(1-k)}{Z} = \frac{V_1}{(Z/(1-k))}$$
$$I_2 = \frac{V_2 - V_1}{Z} = \frac{V_2(1-1/k)}{Z} = \frac{V_2}{(kZ/(k-1))},$$
(9.1)

and, so far as the electronics within the box is concerned, the original configuration is equivalent to attaching an impedance Z_1 from connection 1 to ground and an impedance Z_2 from connection 2 to ground, provided you use the values

$$Z_1 = \frac{Z}{1-k}; \quad Z_2 = \frac{kZ}{k-1},$$
(9.2)

as shown in Fig. 9.1b.

To see how this applies to understanding the frequency dependence of a transistor amplifier, consider the "hybrid pi" BJT transistor model shown in Fig. 9.2a and a similar model for a JFET in Fig. 9.2b. These models are a bit more sophisticated than what was used previously and include some of the capacitance associated with the *p-n* junctions that had been previously ignored.

Figure 9.2 includes typical ballpark values for a low-power transistors. Capacitance values for higher power transistors can be much larger. If the BJT transistor shown is used in a common emitter amplifier, which has a large voltage gain, the small collector to base capacitance of roughly 3 pF is directly connected between the input (the base) and the output (the collector). If the voltage gain of this amplifier is $-A$, Miller's theorem shows this is equivalent to connecting a capacitance from the input to ground with a value $(1 + A)$ larger, illustrated in Fig. 9.3. With the common emitter configuration, it is easy to achieve values such as $A \approx 100$. Hence there may will be a significantly larger *effective* capacitance connected between the input and ground. That capacitance will short out higher frequency input signals before they can be amplified.

The larger the amplifier gain, the larger is this effective capacitance between the input and ground, and hence the lower the maximum frequency that will be amplified. Similar behavior occurs in all linear amplifiers. A result is that the

Miller's Theorem

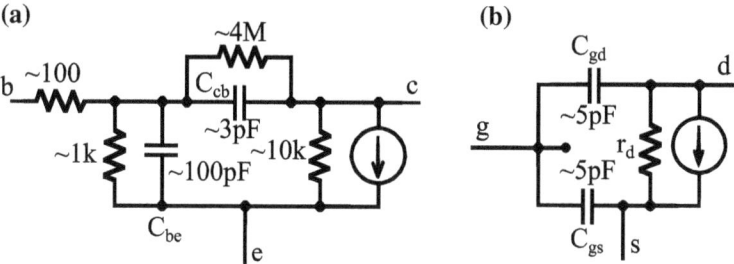

Fig. 9.2 A more detailed model for (**a**) a BJT and (**b**) an FET transistor that includes some of the internal capacitance

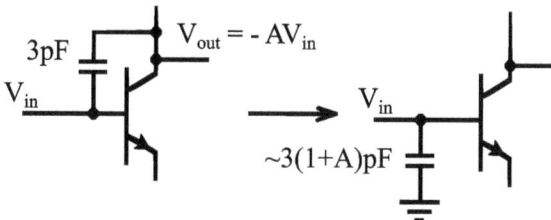

Fig. 9.3 Miller's theorem applied to the BJT in Fig. 9.2 when the transistor is used in a common emitter amplifier

product of the gain and the bandwidth (in this case the bandwidth is the upper frequency) is roughly constant. If you try to vary the bandwidth or the gain, the other changes as well, keeping the product constant. In fact, the frequency response of many amplifiers will be specified using the "gain-bandwidth product" and/or the "unity gain bandwidth" for this reason. For example, if an amplifier circuit has a unity gain bandwidth of 100 kHz and it is used with a gain of 4, the bandwidth would be expected to be about 25 kHz.

Two-Transistor Configurations

There are several interesting configurations involving two (or more) transistors that can accomplish a task which a single transistor cannot. Several of these are illustrated in this section.

The Cascode Configuration

To make a (transistor) amplifier that works over a wider range of frequencies, the capacitance of the device and the results of Miller's theorem (above) must be taken into account. One way to improve the high frequency response is to use a pair of transistors configured as a cascode pair. For BJT transistors, this is essentially a common emitter amplifier directly connected to a common base amplifier, as shown in Fig. 9.4a. For the JFET, it would be a common source connected to a common gate amplifier, as shown in Fig. 9.4b.

The common base and common gate amplifiers have a low input impedance and hence tend to short out the voltage gain (A) from the first transistor. Hence, the first transistor has a small voltage gain, and thus a small Miller effect with a large current gain. The capacitance between the output of the second transistor and the input is very small and hence the Miller effect is greatly reduced.

The Darlington Pair

One way to increase the current gain beyond that of a single BJT is to create a "Darlington pair," as shown in Fig. 9.5. Ignoring 1 in comparison to β_1, and β_2 (e.g., $(1 + \beta_i) \approx \beta_i$), the effective current gain is now $\beta_1 \times \beta_2$ which can be very large indeed. Most often the reason to do this is not to get large currents at the output, but to have small currents at the input—that is, a large input impedance.

Of course, the use of β to describe the transistor(s) assumes they are in their active region. Maintaining that condition is much more difficult with the Darlington pair and virtually impossible if one extends this idea to three or more transistors.

The Darlington pair works because the output of the first transistor, a current, is an appropriate input for the second transistor. With FETs, which use a voltage to control a current, such a pairing is problematic.

Complementary Symmetry Amplifier ("Push-Pull")

One problem with the single transistor linear amplifiers is that they use a lot of power even when there is no signal. Another problem is that they do not work well for constant (d.c.) signals. Both of these effects are a result of the need for biasing—the need to stay in the active region away from the origin of the device

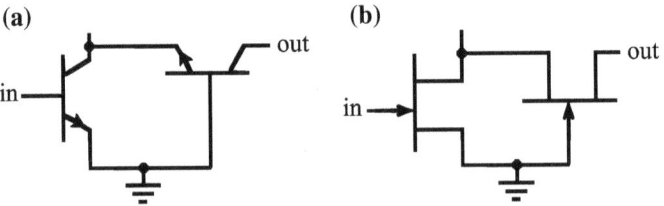

Fig. 9.4 The cascode configuration uses two transistors in place of one to reduce the effects shown in Fig. 9.3

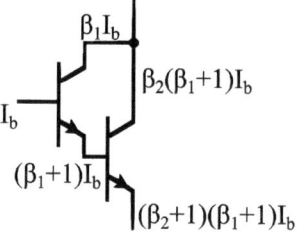

Fig. 9.5 A Darlington pair of transistors can be used to get a very large current gain

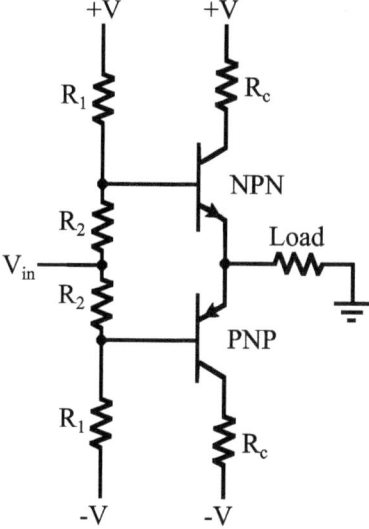

Fig. 9.6 The complementary, or "push-pull," amplifier uses one transistor for a positive input, and a second for a negative input

characteristics. One way around this is a two-transistor amplifier known as the complementary symmetry amplifier or "push-pull" amplifier. An example of such an amplifier made using BJT transistors is shown in Fig. 9.6. Similar circuits can be produced using FETs—in particular MOSFETs.

The biasing resistors are such that when there is no signal, both transistors are just barely "off." If a positive signal is present, the upper transistor conducts and acts as an amplifier. The lower transistor is still off and does not do anything. On the other hand, if a negative signal is present, the situation is reversed. Note that there is a problem with this amplifier if the load is disconnected. During operation only one transistor is on and is trying to send current to the load, however, if the load is absent, the only place for the current to go is through the other transistor, which is off. If no precautions are taken, this can, in some circumstances, cause the ratings of one or the other transistor to be exceeded and possibly in the destruction of one or the other transistor.

Differential Amplifier

The differential amplifier (or difference amplifier) has two inputs and the output is proportional to the difference between the two inputs. For schematics, it is common to use a triangle to indicate an amplifier if the internal contents are unknown or not of immediate importance. Using that symbol, the differential amplifier is as shown in Fig. 9.7.

The basic two-transistor differential amplifier configuration for BJT's and for FET's is shown in Fig. 9.8, where all biasing resistors have been omitted from the drawing for convenience. During normal operation the transistors should be in their active region. For the best results, identical pairs of transistors are used and they will have identical biasing components.

The BJT amplifier can be analyzed using the simple transistor model to get

$$V_{out} = \frac{-\beta R_s}{r_b}(V_1 - V_2). \tag{9.3}$$

A real differential amplifier will not have components that are exactly matched and will have an output that can be written

$$V_{out} = A_d(V_2 - V_1) + A_c \frac{V_2 + V_1}{2}, \tag{9.4}$$

where A_d is the "differential voltage gain" and A_c is the "common mode voltage gain." An ideal differential amplifier would have $A_c = 0$. One measure of the quality of a differential amplifier is known as the "Common Mode Rejection Ratio" or CMRR, ρ, where

$$\rho = \left|\frac{A_d}{A_c}\right|, \tag{9.5}$$

and for a good differential amplifier ρ should be a large value ($\gg 1$). The CMRR is usually expressed in decibels (dB)

$$\text{CMMR in dB} = 20\log_{10}(|A_d/A_c|). \tag{9.6}$$

To achieve a large CMRR for the BJT example above, R_e should be large. This can be accomplished in practice by replacing R_e with a constant current source (e.g., another transistor to produce a fixed current). The constant current source provides the correct biasing, but looks like an infinite resistance for *changes*.

Ideally, the output of a differential amplifier, or any amplifier for that matter, should not depend on (small) changes or transients from the power supply. A measure of this is known as the power supply rejection ratio, or PSRR. It is defined in a manner similar to the CMRR and is also usually expressed in dB. The appropriate ratio is the change in the power supply voltage divided by the change in the output voltage, for a fixed input.

Two-Transistor Configurations

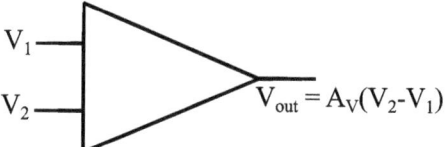

Fig. 9.7 A general schematic for a differential, or difference, amplifier

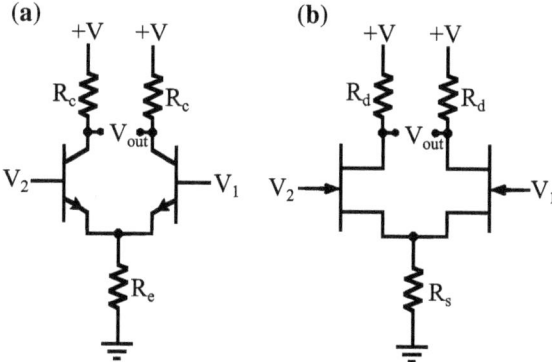

Fig. 9.8 The two-transistor configurations for a difference amplifier based on (a) BJTs and (b) JFETs

Current Mirror

Another interesting application using two transistors is known as the "current mirror." The basic current mirror made with NPN transistors looks like the circuit of Fig. 9.9. The circuit takes as an input the current I_1 and produces a copy $I_2 = I_1$ provided the transistors are in their active regions and are closely matched in their properties. When they are matched, the base currents through the two transistors will be identical, and hence so will be the collector currents. Using identical transistors, any changes in temperature, etc., will be the same for both and so the balance is maintained.

One use for such a circuit would be to create a controlled current source. For example, the current I_1 is controlled using a known circuit, and that forces the current I_2 to follow no matter what load is attached to it (provided the transistor stays in the active region). Such circuits are also often found as part of the biasing circuitry for more advanced integrated circuit amplifiers.

Silicon Controlled Rectifiers (SCR) and Triacs

Another useful way to connect two transistors is shown in Fig. 9.10a. While sometimes used as a two-transistor device, this is accomplished in a single device known as a silicon controlled rectifier (SCR). An SCR consists of four layers with alternate types of doping as shown in Fig. 9.10b. SCR's are sometimes also referred

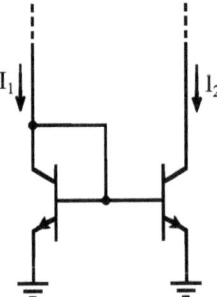

Fig. 9.9 The current mirror uses two matched transistors to produce the same current in two separate paths

to as "thyristors."[1] The schematic for an SCR is shown in Fig. 9.10c and the current-voltage characteristics are illustrated in Fig. 9.10d.

The SCR has a double valued response,[2] and so the solution depends on the history of the device. If you start at zero (or negative) voltage, the current will stay small until you reach a threshold voltage, at which point the current jumps to a large value. Once you are at the larger current, the device looks like a resistor (with a low resistance) until you get below the minimum holding voltage (or corresponding current), at which point the device turns off.

The addition of a "gate" (or sometimes "trigger") connection is used to control the threshold voltage, and thus determines when the device turns on. To be completely "on" typically requires a few tens of milliamps between the gate and the cathode. That is, more or less, sufficient to "turn-on" the gate-to-cathode *p-n* junction.

The threshold voltage with no gate current can be 100's of volts, and the minimum holding voltage about 1 V. Usually the minimum holding current, rather than a voltage, is specified and will be in the range of 1 to 100 mA for typical SCRs.

Note that the gate cannot be used to turn the device off. The device only turns off when the anode to cathode voltage drops below the holding voltage (or current). As a switch, this works for AC since a sinusoidal voltage will eventually go through zero volts. DC power can be used if a separate switch is available to interrupt the current and hence turn off the device—for example a "reset" button such as is found on an alarm circuit.

Two SCR's connected in parallel, but in reversed directions, make a device known as a triac. The schematic symbol for a triac is shown in Fig. 9.11. A triac produces a symmetric output and is useful for AC power control. Triacs are used, for example, in common household light dimmers and some motor speed controls. In those applications, the gate of the triac is used to turn on the signal somewhere in the middle of the AC input cycle, then when the AC goes through zero the traic

---

[1] If the gate is omitted, the device may be referred to as a "four-layer diode" or a "Shockley diode."
[2] The function is actually triple valued, however one of the solutions is always unstable and so the observed response is double valued.

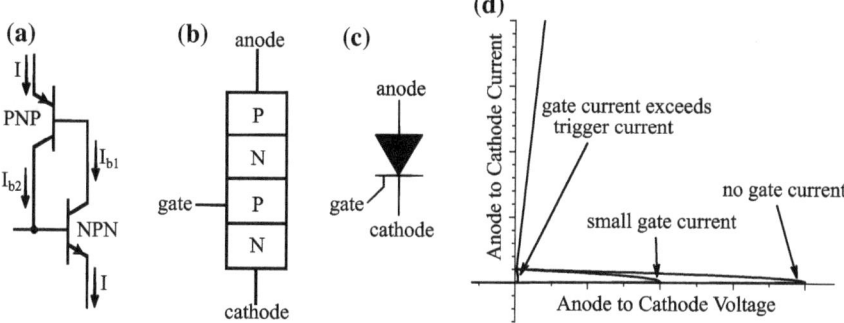

Fig. 9.10 The two-transistor configuration in (**a**) can be constructed as a four-layer single device (**b**), a silicon controlled rectifier or SCR, that has the schematic shown in (**c**) and general characteristics illustrated in (**d**)

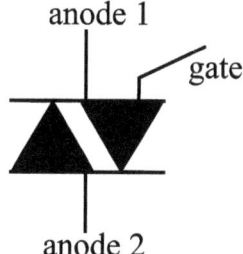

Fig. 9.11 Two back-to-back SCR's as a single device is known as a triac, and is used for switching a.c. signals

turns off. Hence, the sinusoidal input power will be on for only a fraction of the total time each cycle. If the gate turns on the signal just after each zero crossing, the AC will be on for almost all of the cycle. On the other hand, if the gate turns on the signal just before each gate crossing, the signal will turn off almost immediately, and little power gets through.

Connecting Amplifiers

In general, linear amplifiers such as those discussed so far (which are referred to as "class A amplifiers"), no matter how constructed, can be modeled using Thevenin equivalents. This is illustrated in Fig. 9.12. In that figure, A_v is the "open circuit voltage gain," and Z_i and Z_o are the input and output impedances respectively. All of the values may, of course, be frequency dependent. Also shown is a signal source, with its equivalent output impedance, and the output of the string of amplifiers is connected and a resistive load, R_L. It is not hard to find the following relationships

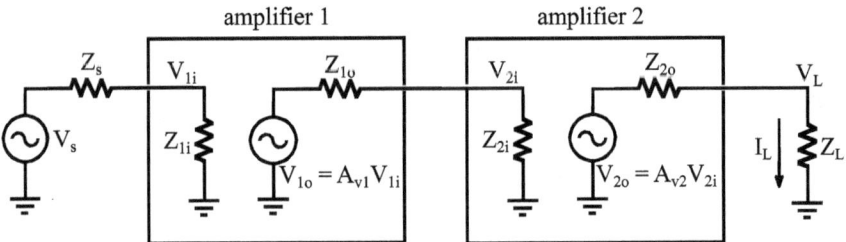

Fig. 9.12 Thevenin's theorem can be used to model the connection from a signal source, through the amplifiers, and then to a final load, R_L

$$V_{1i} = V_s \frac{Z_{1i}}{Z_s + Z_{1i}}$$
$$V_{2i} = A_{v1} V_{1o} \frac{Z_{2i}}{Z_{1o} + Z_{2i}} \qquad (9.7)$$
$$V_L = A_{v2} V_{2i} \frac{R_L}{Z_{2o} + R_L},$$

which describe the voltage amplification. Current and power amplification are treated in a similar manner.

Impedance Matching

Sometimes it is appropriate to maximize the power gain. For the example in Fig. 9.12, the power supplied to the load,

$$P_L = V_L I_L = \left(\frac{A_{v2} V_{2i}}{Z_{2o} + R_L}\right) R_L, \qquad (9.8)$$

for fixed gain A_{v2} will be a maximum when $Z_{2o} = R_L$, a condition that is known as an "impedance match." To show this, find Z_{2o} that satisfies

$$\frac{dP_L}{dZ_{2o}} = 0. \qquad (9.9)$$

Likewise, the maximum power delivered to amplifier 2 occurs when $Z_{1o} = Z_{2i}$, and the maximum power is delivered to amplifier 1 when $Z_s = Z_{1i}$.

Impedance matching is also useful when the interconnecting cable lengths become long. With longer cables, signals can reflect back to the source causing a length-dependent impedance transformation (see Chap. 4). To avoid these reflections, the characteristic impedance of the cable should match the input impedance of the following amplifier. If maximum power transfer is also desired, then both of those impedances should also match the output impedance of the original source.

Problems

1. Consider the circuit shown in Fig. 9.P1. Assume the transistors are in their active region and have the same current gain β, and assume that all capacitors have a small enough impedance for a.c. signals so that they can be considered short circuits for those a.c. signals. Draw a carefully labeled circuit diagram appropriate for analyzing the small-signal a.c. behavior of this circuit. Such a diagram includes replacing the transistors with the appropriate transistor model for the active region, simplifying the diagram where possible, labeling currents, etc. It is not necessary to do the analysis.

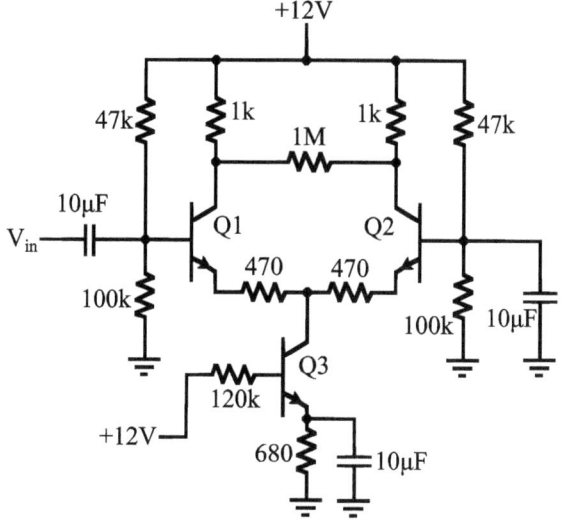

Fig. 9.P1 Problem 1

2. For the circuit of Fig. 9.P2, and assuming an ideal voltage source, prove that the maximum power is delivered to R_L when $R_L = R_o$. In contrast, what value of R_0 gives the maximum current through, and what value gives the maximum voltage across, the load, R_L?

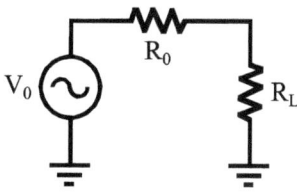

Fig. 9.P2 Problem 2

3. Derive the differential voltage gain for the two-transistor BJT differential amplifier (i.e., derive Eq. 9.3). Assume the biasing has already been done, that the transistors are in their active region, and that the transistors are exactly matched.
4. Use the FET transistor model to derive the differential voltage gain for the FET differential amplifier. Assume any necessary biasing has already been done, that the transistors are in their active region, and that the transistors are exactly matched.
5. A differential amplifier with a differential voltage gain of 100 has an output of 0 V when the two inputs are both 0 V. If the amplifier has a CMRR of 60 dB, what is the magnitude of the output when both inputs are 1 V?

Chapter 10
The Ideal Op-Amp

An operational amplifier (op-amp) is a special type of differential amplifier that has some very useful properties. Op-amps are available as inexpensive self-contained integrated circuits and the detail of what goes on inside the device is rarely of concern. In this chapter, the ideal op-amp, treated as a device, is introduced and many useful circuits based on the ideal op-amp with feedback are presented. After that, some discussion of properties of a real op-amp are presented.

Ideal Op-Amp Properties

An ideal op-amp is a perfect differential amplifier with the following additional properties:

- Input impedance infinite (no current enters the inputs)
- Output impedance zero (any current at output, as needed)
- Differential voltage gain is infinite
- CMRR infinite (i.e., common mode gain is zero)
- Frequency range infinite
- No offset voltages
- Characteristics are independent of power supply (PSRR infinite).

An op-amp as a device is shown symbolically using a triangle, as shown in Fig. 10.1, in the same way as is any differential amplifier. One input signal is amplified with a minus sign, the "inverting input," the other with a plus sign, the so-called "non-inverting input." The power supply connections to the op-amp are shown in Fig. 10.1, however they are often omitted from a schematic diagram, or shown separately from the op-amp, unless there is a good reason to include them. Of course, the op-amp will not work without power, so those connections must be included during circuit construction. Op-amps will usually have two power connections. For many amplifier applications, one supply connection is usually a

© Springer Nature Switzerland AG 2020
B. H. Suits, *Electronics for Physicists*, Undergraduate Lecture Notes in Physics,
https://doi.org/10.1007/978-3-030-39088-4_10

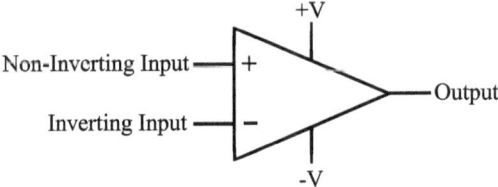

Fig. 10.1 The op-amp schematic, including power connections

positive voltage and the other negative.[1,2] Often those two supplies are equal in magnitude, but there certainly are exceptions. Since the device requires external power, an op-amp is referred to as an "active device."

Note that op-amps rarely have a ground connection—only the potential difference between the two power connections is important for op-amp operation. Many of the circuits that follow include a ground connection external to the op-amp. The relationship between that ground and the two op-amp power sources is important for proper circuit operation. Some op-amps have additional connections, not discussed here, to allow for adjustments to make the behavior closer to ideal.

Op-amps are often used with feedback. Feedback is an external connection between the output and the input. If the output of an amplifier is connected back to the input, directly or indirectly, with a net negative sign, this is referred to as "negative feedback." Linear applications of op-amps use negative feedback of one sort or another. Without that feedback, the large (ideally infinite) gain would result in theoretically unrealistic, and practically unusable, output signals.

Negative feedback results in a correction signal that maintains the output at a certain level. Imagine two cars on a highway. Car A is traveling at speed, V. Car B is supposed to match the speed of car A. If car B is going too fast, the error in its speed, ΔV, is positive. To correct for this, the speed needs to be changed by $-\Delta V$. On the other hand, if car B is going too slowly, the error in its speed is negative, $-\Delta V$. Now to correct for that, car B needs to change its speed by $-(-\Delta V) = +\Delta V$. In each case, the error is multiplied by a minus sign to get the correction. That minus sign is what is referred to as negative feedback. With positive feedback, if car B is going too fast, it will speed up even more, and it will never match speeds with car A.

When an ideal op-amp is in a circuit and has negative feedback, the output of the op-amp will adjust, as best as it can, so that the two inputs to the op-amp are equal. Many op-amp circuits can be solved using this special "device property." Why this technique works is demonstrated at the end of this chapter. First, a number of useful

---

[1] For general purpose use, ±3 V to ±15 V would be typical. The appropriate range of values for a specific op-amp will be found on the manufacturer's datasheet.

[2] A common beginner's mistake during construction is to confuse the non-inverting (+) and inverting (−) inputs with the positive (+) and negative (−) power supply connections.

Ideal Op-Amp Properties

example circuits are solved using this simplified device property. The second important property used for these solutions, and important to remember, is that no current enters (or leaves) either ideal op-amp input.

Linear Op-Amp Circuits

Op-amps have utility both as linear amplifiers as well as for some non-linear applications, including digital circuitry. This section deals with linear applications. When the op-amp is used with negative feedback, the circuit analysis can usually be simplified by assuming an ideal op-amp and, as mentioned above, that due to the negative feedback, the output will adjust (if it can) to force the inverting and non-inverting inputs to be equal. This method is shown by example in what follows.

Example 1—Buffer

Assuming an ideal op-amp, the relationship between the input and output voltages for the circuit in Fig. 10.2 is easy to determine. First, write down the voltages at the inverting ("−") and non-inverting ("+") inputs in terms of the voltages and any other components shown. Use the fact that no current enters the op-amp inputs (an ideal device rule). In this simplest case these input voltages are easily found to be

$$V_+ = V_{in} \quad V_- = V_{out}. \tag{10.1}$$

There is negative feedback (due to the direct connection between the output and V_-) so we set $V_+ = V_-$ (the simplified "device rule" for the ideal op-amp *when there is negative feedback*) and so clearly,

$$V_{in} = V_{out}. \tag{10.2}$$

This circuit has a voltage gain of 1, but an extremely large current gain. The current into the non-inverting input is extremely small and the output current can be many orders of magnitude larger. The circuit is referred to as a buffer. It separates the source of the input voltage from any load that may follow so that virtually no power from the source is required. All the power is supplied by the op-amp power supplies (not shown in the diagram).

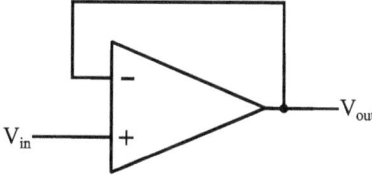

Fig. 10.2 Example 1, op-amp buffer

Example 2—Inverting Amplifier

In some circuits, it is beneficial to look at the current through the feedback component. Consider the circuit of Fig. 10.3. The voltage at the non-inverting input is obviously 0 V (ground). Since there is negative feedback, the non-inverting (+) and inverting (−) inputs are taken to be equal (i.e., $V_+ = V_- = 0$ V). Then it is easy to see that $I_{in} = V_{in}/R_1$. Since that current does not enter the op-amp input, it must go around (as shown) through R_2. Hence, V_{out} can be determined using the voltage drop from the input,

$$V_{out} = V_- - I_{in}R_2 = 0V - V_{in}\frac{R_2}{R_1} = -\frac{R_2}{R_1}V_{in}. \qquad (10.3)$$

The voltage gain of the circuit is $-R_2/R_1$. Note that for this configuration, the gain is completely under the control of the circuit designer. The input impedance of the circuit, that is, the impedance "seen" by the input source, will be R_1. Be careful not to confuse the input impedance of the op-amp (ideally infinite) with the input impedance of the circuit, which includes the resistors.

Op-amp circuits with negative feedback and where the non-inverting input is connected to ground (i.e., 0 V) will have a "virtual ground" at the inverting input. The virtual ground will be 0 V, however unlike a real ground that can source and sink current, the current here needs somewhere else to go.

In summary, the two basic strategies that are used to solve ideal op-amp problems with negative feedback are:

- Compute the inverting and non-inverting input voltages in terms of the input and output voltages and any additional components, then set the inverting and non-inverting inputs voltages equal to each other and solve.
- Find the non-inverting input voltage, then set the inverting and non-inverting input voltages equal. Now look at the voltage drop due to the current that goes through the negative feedback component(s) and where that current came from and where it can go.

Now look at some more examples that use these strategies.

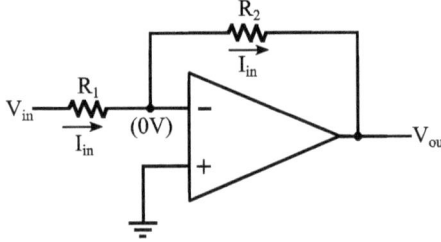

Fig. 10.3 Example 2, inverting amplifier

Linear Op-Amp Circuits

Example 3—Non-inverting Amplifier

Figure 10.4 shows a non-inverting amplifier. It is obvious that $V_+ = V_{in}$. Using the voltage divider equation, valid since no current enters the op-amp inputs,

$$V_- = V_{out} \frac{R_1}{R_1 + R_2}. \tag{10.4}$$

There is negative feedback due to the connection between the output and V_-, so $V_+ = V_-$, and

$$V_+ = \frac{R_1}{R_1 + R_2} V_{out} \rightarrow V_{out} = \frac{R_1 + R_2}{R_1} V_{in}. \tag{10.5}$$

The input impedance for the non-inverting configuration is determined by the particular op-amp used and will be very large.

Example 4—Difference Amplifier

Figure 10.5 combines the circuits of Examples 2 and 3 and uses two input voltages. The voltage divider equation can be used for each of the op-amp inputs to get

$$V_+ = V_B \frac{R_4}{R_3 + R_4} \quad V_- = V_{out} \frac{R_1}{R_1 + R_2} - V_A \frac{R_2}{R_1 + R_2} \tag{10.6}$$

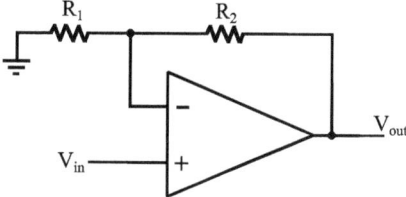

Fig. 10.4 Example 3, non-inverting amplifier

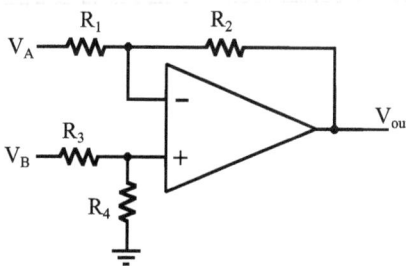

Fig. 10.5 Example 4, difference amplifier

Since there is feedback from the output back to V_-, set $V_+ = V_-$, and so

$$V_{out} = \frac{R_1 + R_2}{R_1}\left(V_B \frac{R_4}{R_3 + R_4} - V_A \frac{R_2}{R_1 + R_2}\right). \tag{10.7}$$

For the special case where $R_1 = R_3$ and $R_2 = R_4$,

$$V_{out} = \frac{R_2}{R_1}(V_B - V_A), \tag{10.8}$$

and this is a simple difference amplifier.

Example 5—Summing Amplifier

The circuit of Fig. 10.6 also has two inputs. Since the voltage at the non-inverting input is fixed at zero (ground) and there is negative feedback through R_3, there is a virtual ground at the inverting input. Hence, the two inputs provide a total current of

$$I = I_1 + I_2 = \frac{V_A}{R_1} + \frac{V_B}{R_2}, \tag{10.9}$$

which has nowhere to go except around the op-amp. Hence, starting at zero volts at the inverting input and computing the drop across the feedback resistor, R_3,

$$V_{out} = -R_3\left(\frac{V_A}{R_1} + \frac{V_B}{R_2}\right). \tag{10.10}$$

For the special case where $R_1 = R_2$,

$$V_{out} = -\frac{R_3}{R_1}(V_A + V_B). \tag{10.11}$$

This configuration adds two signals together. If the two input resistors are not equal, the circuit provides a weighted sum. This example was shown for two inputs, but this idea can be extended to any number of inputs, each with its own weighting

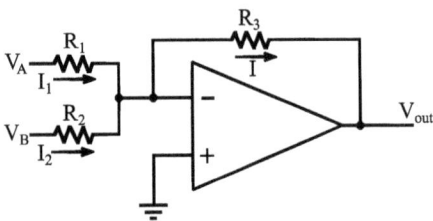

Fig. 10.6 Example 5, summing amplifier

resistor. If the minus sign on the output is of concern, the addition of an inverting amplifier (Example 2) can be used to remove it.

Example 6—Integrator

Consider the circuit in Fig. 10.7. Since the non-inverting input is connected to ground, there is a virtual ground at the inverting input provided that the capacitor gives negative feedback—that is, since $V_+ = 0$, the feedback will keep $V_- = 0$ as well. Hence the current through R is $I = V_{in}/R$. The solution here will use the same strategy as Example 2 above, but with the additional complications due to the capacitor.

The current, I, does not enter the op-amp and so it must go to the capacitor. The voltage across the capacitor will then grow with time as the charge builds up. The voltage across the capacitor, V_c, is given by

$$V_c(t) = V_c(0) + \frac{Q}{C} = V_c(0) + \int_0^t \frac{I}{C} dt, \tag{10.12}$$

and since there is a virtual ground (0 V) on the left side of the capacitor, and the output on the other, it must be that $V_{out} = -V_c$.

Putting in the expression for the current, the output is related to the input by

$$V_{out} = -V_c(0) - \frac{1}{RC} \int_0^t V_{in}(t) dt, \tag{10.13}$$

and so this circuit provides an output linearly related to the integral of the input. It is an "integrator." If a provision for discharging the capacitor at $t = 0$ is included (for example, a push-button switch across the capacitor) then the output is proportional to the definite integral of the input starting at $t = 0$.

If the input is a sinusoid with amplitude V_i at angular frequency ω, the circuit could instead be analyzed using complex impedances. The type of analysis that is appropriate will depend on the application.

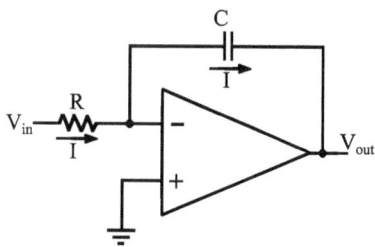

Fig. 10.7 Example 6, integrating amplifier

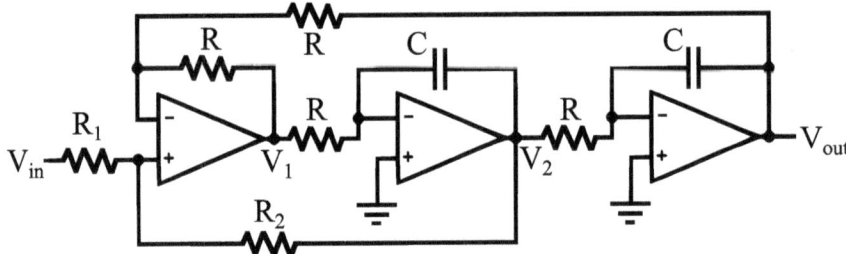

Fig. 10.8 A use of integrator circuits to model the driven harmonic oscillator

Using complex impedances (and the fact no current enters the ideal op-amp) proceed as in Example 2. The a.c. input current through R is V_{in}/R and hence the voltage drop from the inverting input to the output gives

$$V_{out} = -\frac{V_{in}}{R}\frac{1}{i\omega C} = \frac{i}{\omega RC}V_{in}, \qquad (10.14)$$

a type of "low-pass filter" that has a large gain at small frequencies and a small gain at larger frequencies. Remember that the "i" in the numerator is a 90° phase shift (a sine becomes a cosine, a cosine becomes minus sine), so aside from a minus sign, this is exactly what happens when you integrate a sinusoidal signal.

Any offset voltages on the inputs to the integrator are particularly annoying, since the integral of a constant will simply grow linearly with time until the op-amp is out of range. Such offsets may need to be addressed so that their effects are negligible.

One interesting application of the integrator is the circuit of Fig. 10.8. Four of the resistors are taken to have the same resistance for simplicity, though that is not necessary. The output, V_{out}, is proportional to the integral of V_2, which in turn is proportional to the integral of V_1. Hence, setting V_1 to

$$V_1 = (RC)^2 \frac{d^2V}{dt^2}, \qquad (10.15)$$

then

$$V_2 = -(RC)\frac{dV}{dt} \text{ and } V_{out} = V. \qquad (10.16)$$

The inputs to the summing op-amp on the left will be

$$V_+ = V_{in}\frac{R_2}{R_1+R_2} + V_2\frac{R_1}{R_1+R_2} \qquad (10.17)$$
$$V_- = (V_{out}+V_2)/2.$$

Linear Op-Amp Circuits

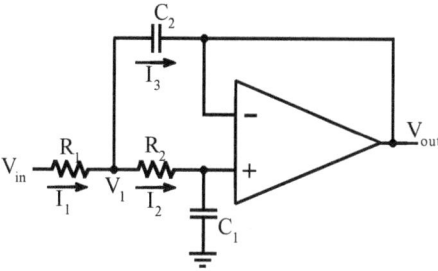

Fig. 10.9 Example 7, low-pass filter

Since there is negative feedback, these inputs are equal. The resulting equation can be written

$$\frac{d^2V}{dt^2} + \frac{1}{RC}\left(\frac{2R_1}{R_1+R_2}\right)\frac{dV}{dt} + \frac{1}{(RC)^2}V = \frac{1}{(RC)^2}\left(\frac{2R_2}{R_1+R_2}\right)V_{in}, \quad (10.18)$$

which is same as the differential equation for a driven, damped harmonic oscillator with natural frequency $\omega = 1/(RC)$. In electronics applications, this circuit is often referred to as a "state variable filter." [3]

It is straightforward to show that if the resistor and capacitor in Fig. 10.7 are swapped, the output is, aside from some scale factors and a minus sign, the derivative of the input.

Example 7—Low-Pass Filter

Figure 10.9 illustrates one of several circuits that can be used as a low-pass filter. That is, lower frequencies are passed through while higher frequencies are blocked. When circuits start to get more complicated, it is often useful to label some intermediate voltages or currents to facilitate a solution. In this case, the node voltage V_1 has been added to the diagram.

There is a direct connection from the output to the inverting input and so there is negative feedback and $V_{out} = V_- = V_+$. Since no current enters the non-inverting input, the voltage divider equation can be used to get

$$V_+ = V_1 \frac{1/i\omega C}{R_1 + 1/i\omega C} = \frac{V_1}{1+i\omega R_2 C_1} = V_{out}, \quad (10.19)$$

or

---

[3] Taking outputs from V_1, V_2, and V_{out} results in a high-pass, band-pass, and low-pass filters respectively. The center frequency, gain, and damping terms can be adjusted independently.

$$V_1 = V_{out}(1 + i\omega R_2 C_1), \text{ so}$$
$$V_1 - V_{out} = i\omega R_2 C_1 V_{out}. \quad (10.20)$$

Using KCL and Ohm's law,

$$I_1 = I_2 + I_3$$
$$I_2 = \frac{V_1 - V_+}{R_2} = \frac{V_1 - V_{out}}{R_2} = i\omega C_1 V_{out} \quad (10.21)$$
$$I_3 = \frac{V_1 - V_{out}}{1/i\omega C_2} = \frac{i\omega R_2 C_1 V_{out}}{1/i\omega C_2} = -\omega^2 R_2 C_1 C_2 V_{out},$$

and so

$$V_{in} = I_1 R_1 + V_1 = (I_2 + I_3)R_1 + V_{out}(1 + i\omega R_2 C_1)$$
$$= V_{out}\left(1 + i\omega(R_1 + R_2)C_1 - \omega^2 R_1 R_2 C_1 C_2\right). \quad (10.22)$$

Then

$$V_{out} = V_{in} \frac{1}{1 - \omega^2 R_1 R_2 C_1 C_2 + i\omega C_1 (R_1 + R_2)} \quad (10.23)$$

The low-pass behavior can be seen by considering limits. In the limit that the frequency is very low ($\omega \to 0$) the denominator becomes 1 and $V_{out} = V_{in}$. When the frequency is very high ($\omega \to \infty$) the denominator becomes very large in magnitude and so $V_{out} \to 0$.

For comparison, solve the passive circuit of Fig. 10.10 using the voltage divider equation to get

$$V_{out} = V_{in} \frac{1}{1 - \omega^2 LC + i\omega RC}. \quad (10.24)$$

This is of the same form as Eq. 10.23 and will have the same behavior if

$$LC = R_1 R_2 C_1 C_2 \text{ and } RC = (R_1 + R_2)C_1. \quad (10.25)$$

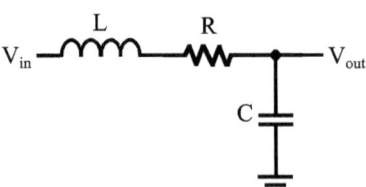

Fig. 10.10 Equivalent circuit for Fig. 10.8

Hence, Figs. 10.9 and 10.10 are, in a sense, equivalent circuits.[4] Inductors are notoriously non-ideal and tend to be large, heavy, and expensive compared to resistors and capacitors, especially at lower frequencies. The circuit in Fig. 10.9 achieves the same result without an inductor, and can, in fact, act more ideally than would real components used with the circuit of Fig. 10.10. The so-called "cut-off" frequency for the filter is where the output is reduced by 3 dB, or in this case, the output voltage is reduced by $1/\sqrt{2}$ from its low frequency value.

Swapping the capacitors and resistors in this circuit changes the response to that of a high-pass filter. A slightly different arrangement can produce a band-pass filter. These filters are generally referred to as "second-order active filters" and there are several different configurations that can be used for each of them.

Example 8—Instrumentation Amplifier

It is often necessary to determine a small voltage difference, perhaps from a sensor or the output from a bridge circuit (see Chap. 2). The difference amplifier above (Example 4) can sometimes be useful, however such an amplifier does not treat the two input signals identically and may have an input impedance that is too small to allow for an ideal voltage measurement. A much more symmetric and sensitive treatment can be achieved using a so-called instrumentation amplifier, such as is shown in Fig. 10.11. Note that the input impedance seen by both sources is R_{in}, which is under the designer's control and is often chosen to be relatively large (~ 1 MΩ).

This is a three-op-amp design. Labels for the intermediate voltages V'_A and V'_B are added to the diagram for convenience. The final op-amp (U_3) is configured identically to that of the difference amplifier of Example 4, and so[5]

$$V_{out} = \frac{R_3}{R_2}(V'_B - V'_A). \qquad (10.26)$$

Both of the input op-amps (U_1 and U_2) clearly have negative feedback and thus the voltage across R_g is $V_A - V_B$. No current goes into the inverting input of the op-amps so the current that goes through both of the resistors R_1 and the resistor R_g must be the same. Using Ohm's law and equating the currents

$$I = \frac{V'_A - V'_B}{2R_1 + R_g} = \frac{V_A - V_B}{R_g}, \qquad (10.27)$$

---

[4]To make the equivalence complete, a buffer (e.g., see Example 1) should be added to the output of the circuit in Fig. 10.10 so that any load present will not draw current from the filter.

[5]It is common to label integrated circuits, such as op-amps, using a "U." Sometimes the two-letter non-standard label "IC" is used instead of "U."

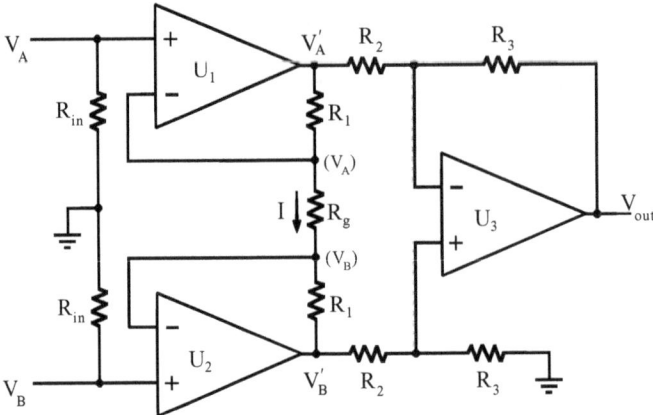

Fig. 10.11 Example 8, instrumentation amplifier

so

$$V'_A - V'_B = \frac{2R_1 - R_g}{R_g}(V_A - V_B). \quad (10.28)$$

Putting 10.28 into 10.26,

$$V_{out} = \frac{R_3}{R_2}\left(1 + \frac{2R_1}{R_g}\right)(V_B - V_A). \quad (10.29)$$

High-performance instrumentation amplifiers can be purchased relatively inexpensively as a single pre-packaged device with laser-trimmed matched resistors. Often there will be provisions to adjust the gain resistor (R_g), that need not be matched to any others, and possibly some other (usually optional) connections to make the behavior even more ideal.

Example 9—A Capacitive Sensor for Smaller Values of Capacitance

An op-amp circuit using a fixed known value of capacitance, C, can be used to measure an unknown capacitance C_x, using the circuit in Fig. 10.12a, where $V(t)$ is a known sinusoidal input voltage at angular frequency ω.

Linear Op-Amp Circuits

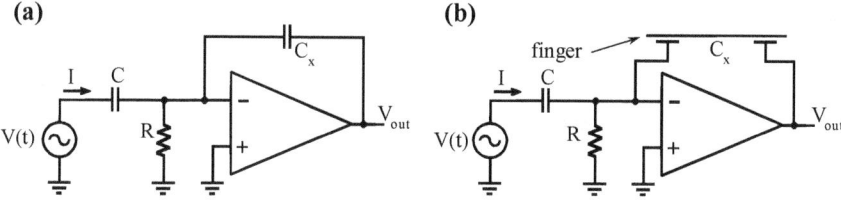

Fig. 10.12 Example 9, a circuit to measure a small capacitance, C_x. In (**b**) the circuit is used to detect a nearby conducting object, in this case a finger

The resistance R is chosen to be a large value compared to $1/\omega C$ and is only necessary to provide a DC path to ground on the input, often necessary for a real op-amp. It is ignored in the following analysis, which assumes an ideal op-amp.

Due to feedback, there is a virtual ground at the inverting input. The current into the inverting input is

$$I_{in} = i\omega CV, \tag{10.30}$$

which, due to the high input impedance of the amplifier, must go through C_x. Hence,

$$V_{out} = -\frac{I_{in}}{i\omega C} = -\frac{C}{C_x}V. \tag{10.31}$$

Since $V(t)$ can be made very small, C_x can be significantly smaller than C and the output can still be easily measured. That measured value is used to determine C_x.

This idea has been used to make a sensitive distance detector[6] where the distance between the plates of C_x is measured. Alternatively, the capacitor configuration in Fig. 10.12b has been used to detect fingers.[7] In that case, the third capacitor "plate" is the portion of the tip of a finger just above the sensor. The size of the such sensing plates can be very small. The distance to the nearby conducting skin determines C_x. Many touch-screen and touch-pad devices use an array of capacitive sensors. The fixed capacitor plates can be made from a layer of metal so thin that it is transparent. Since such devices rely on the electrical conductivity of fingers to work, if a pen or pencil is used instead of a finger, or even if the finger used is particularly dry, the capacitive touch devices will not respond.

---

[6]See, for example, Pigage et al. (1968).

[7]For a similar application to detect fingerprints, see Tartagni and Guerrieri (1998). who use a charge-based method rather than the a.c. impedance method described here.

Example 10—Negative Resistor

Consider the circuit in Fig. 10.13. This circuit differs from the previous examples in that it has an input but no output. This circuit demonstrates "impedance transformation" using an op-amp.

Assuming negative feedback is present so the inputs to the op-amp must be equal, it is straightforward to write down

$$V' = V - IR_1$$
$$V = V''$$
$$V'' = V' \frac{R_2}{R_1 + R_2} = (V - IR_1) \frac{R_2}{R_1 + R_2} \quad (10.32)$$
$$V\left(1 - \frac{R_2}{R_1 + R_2}\right) = -I \frac{R_1 R_2}{R_1 + R_2},$$

which is easily solved to get $V = -IR_2$. That is, this circuit acts just like a resistor to ground but with a negative resistance. The value of R_1 does not show up in the final equation and is determined by the current and voltage limitations of the device. That is, R_1 needs to be a value that keeps the device in a region of operation that provides negative feedback, but otherwise its value is not important.

Note that in the circuit and analysis above, any impedance Z_2 could have been used in place of R_2 without changing the basic result. As a specific example, if a capacitor were used in place of R_2, the circuit adds a minus sign to the capacitive impedance, making the circuit act like an inductor to ground.

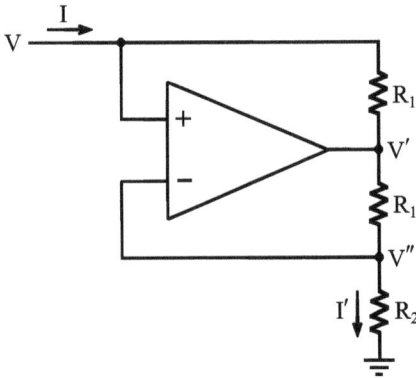

Fig. 10.13 Example 10, a negative resistor to ground

Example 11—Constant Current Source

There are several configurations that can be used for a constant current source. The so-called Howland configuration uses a combination of positive and negative feedback. Such a circuit could also be called a voltage-to-current converter. The circuit diagram in Fig. 10.14 is referred to as the "improved Howland configuration."

Assuming an ideal op-amp where the negative feedback is strong enough to win out over the positive feedback, the output current is found using KCL, KVL, and the properties of the ideal op-amp with feedback. With feedback, $V_+ = V_- \equiv V$, and so

$$V_{ref} - I_1 R_1 = V$$
$$V = I_2(R_2 + R_3) \tag{10.33}$$
$$V_{ref} = 2I_1 R_1 = V'.$$

Combining these,

$$V' = -V_{ref} + 2I_2(R_2 + R_3). \tag{10.34}$$

Now,

$$V'' = I_2(R_2 + 2R_3) = 2I_2(R_2 + R_3) - I_2 R_2, \tag{10.35}$$

so

$$V'' - V' = V_{ref} - I_2 R_2$$
$$\frac{V'' - V'}{R_2} = -I' = \frac{V_{ref}}{R_2} - I_2 = -I_{out} - I_2 \tag{10.36}$$
$$I_{out} = -\frac{V_{ref}}{R_2}.$$

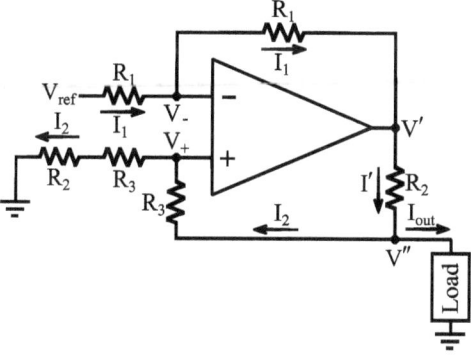

Fig. 10.14 Example 11, constant current source

Since the output current is independent of the load, the same output would be expected for any load, at least to the extent the op-amp stays within its usable range. The values of R_1 and R_3, which do not directly affect the output current, are chosen to ensure the op-amp stays within that range. Most real op-amps can provide a maximum current up to no more than about 10 or 20 mA. If needed, the output current can be amplified (e.g., using a BJT) to get much higher values.

Other Op-Amp Circuits

There are useful circuits that use the op-amp as a linear amplifier where the circuit itself is not linear. Several such circuits are illustrated here as examples.

Example 12—Non-linear Element in Feedback

Consider the circuit of Fig. 10.15 where the FET characteristics are shown in Fig. 10.16. Assuming the current I is positive, there is a connection, i.e., feedback, from the output to the input through the (n-channel) FET. This circuit can be solved with a graphical method.

Since the non-inverting input is connected to ground (0 V), the voltage at the inverting input should also be 0 V and so $I = 7$ V/1.5 kΩ = 4.67 mA. The input impedance of the op-amp is very large so this current must go through the FET. Since the FET gate and source are connected together, $V_{gs} = 0$, and the solution must be on the $V_{gs} = 0$ curve of the transistor characteristics. Follow the curve corresponding to $V_{gs} = 0$ V until $I_{ds} = 4.67$ mA. Now read off V_{ds}. Thus, for this circuit $V_{out} = -V_{ds} = -1.8$ V.

Note that if the input voltage were 15 V then $I = 9.33$ mA if *negative feedback is assumed*. However, no solution exists for $I = 9.33$ mA. This, indirectly, shows

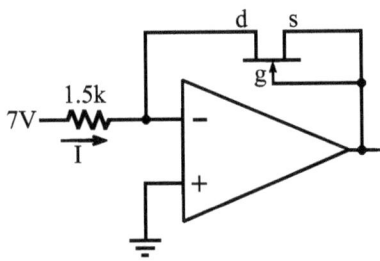

Fig. 10.15 Example 12, an op-amp with a non-linear feedback component, in this case an FET

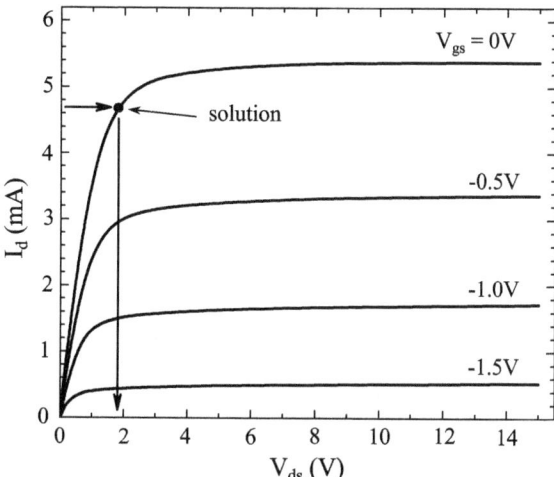

Fig. 10.16 FET characteristics used to solve the circuit of Fig. 10.15

that the feedback cannot make the two inputs equal and the assumption of negative feedback is invalid for this case.[8] It is necessary to go back and rethink the problem.

Example 13—Ideal Diode

As far as the voltage is concerned, the circuit in Fig. 10.17 acts almost like an ideal diode. There will be a connection between the output and inverting input only when the diode is forward biased. When the diode is reverse biased, the feedback is broken and the output of the amplifier will tend toward minus infinity. In practice the output will end up at the maximum negative value the op-amp can provide (which is usually close to the negative power supply voltage).

Assuming that there is negative feedback, the two inputs to the op-amp will be equal, and hence $V_{out} = V_{in}$. It is clear, however, that the diode is only conducting when $V_{out} > 0$. The assumption of negative feedback is only valid when $V_{in} > 0$. When $V_{in} < 0$, the output of the op-amp goes negative and the feedback is broken. With the diode reversed, no current is supplied to the load (R) and so the output voltage at the load, V_{out}, is zero.

Hence, for the ideal op-amp, the output voltage is what you expect for an ideal diode:

$$V_{in} > 0 \rightarrow V_{out} = V_{in}$$
$$V_{in} < 0 \rightarrow V_{out} = 0. \tag{10.37}$$

---

[8]It is always good to check assumptions, even if a solution does exist.

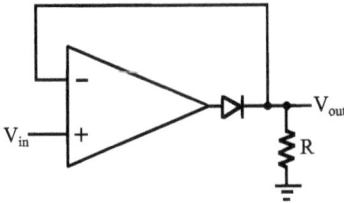

Fig. 10.17 Example 13, an ideal diode circuit

A more careful analysis shows that the effective diode turn-on voltage is not quite reduced to zero, but is reduced by a factor comparable to the gain of the op-amp (e.g., a factor of 10^5 or more). Hence, instead of having a turn-on voltage of about 0.5 V, the effective turn-on voltage is about 5 μV for a typical op-amp. That is negligible compared to typical op-amp offset voltages (~ 1 mV) present for real op-amps. Note also that this circuit only acts as an ideal diode for the input voltage. Any current from the input source is blocked by the op-amp.

During the negative part of the input signal, the op-amp output goes as low as it can. Since it can take a while for the op-amp to recover from that, this circuit may only work well at lower frequencies. How low will depend on the particular op-amp. Should that frequency dependence be an issue, there are similar circuits that use strategies that provide an alternate connection for the output when the diode is off, and that alternate path keeps the output from the going to the extremes. One of these is schemes is included in the next example.

Example 14—Peak Follower

Consider the circuit of Fig. 10.18. The second op-amp, U_2, is a buffer (see Example 1) and hence the output will be equal to the voltage across the capacitor, V_C. The output from the first (U_1) is labeled V_1.

If V_1 is greater than V_C then diode D_1 is conducting and diode D_2, which sees V_C via the output of U_2, is "off." In that case, U_1 gets negative feedback from the output of U_2, equivalent to the feedback in the previous example. The output of U_1 follows the input and the capacitor follows along.

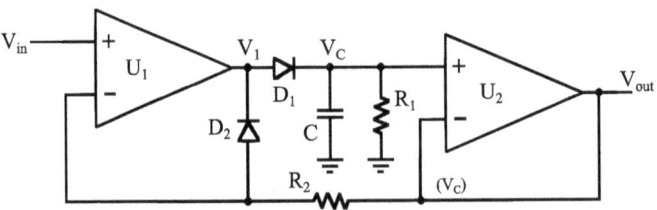

Fig. 10.18 Example 14, a peak follower

Other Op-Amp Circuits

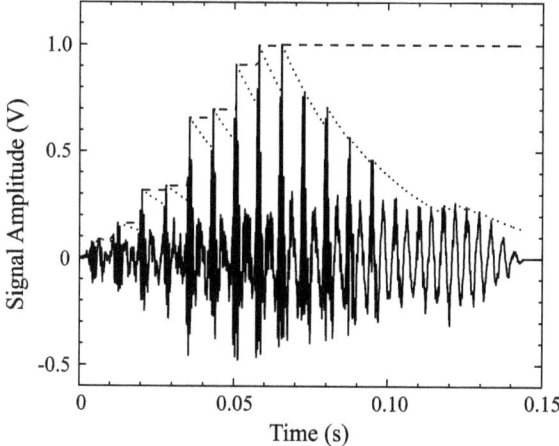

Fig. 10.19 A short audio signal (solid line) sent to the peak follower saves the peak if a large time constant is used (dashed) or follows the "envelope" of the signal if a moderate time constant is used (dotted)

If V_1 is less than V_C, then diode D_1 is "off" and diode D_2 conducts. With diode D_2 conducting, U_1 has negative feedback directly from its own output and it forms a buffer, so that $V_1 = V_{in}$. Since that signal is blocked by D_1, it is of no real consequence except that U_1 is prevented from going to its extremes. So if V_{in} is smaller than V_C, the capacitor is disconnected from the input. When disconnected, the capacitor will eventually discharge through R_1, there being no other place for the current to go.

With no resistor R_1, the capacitor will charge up to the maximum value of V_{in} and just sit there, storing the peak value of the signal. Including that resistor allows the capacitor to "reset" with time. Alternatively, a (transistor) switch could be used to discharge the capacitor once the stored peak value is no longer required.

How rapidly the circuit will recover from a peak in the input depends on the $R_1 C$ time constant. Figure 10.19 illustrates the behavior for a very large and a moderate time constant. The latter might be used to try to follow the envelope of the input. If the time constant is made very short, for example if the capacitor is simply removed, the output is the same as that of the ideal diode circuit (Example 13).

Example 15—Log Amplifier

If the input voltage, V_{in}, shown in Fig. 10.20, is positive, then the diode will be conducting. With the diode "on" there is feedback—the diode acts like a wire for changes in the voltage or current—and hence a virtual ground at the inverting input. The input current is then V_{in}/R. The output voltage is then determined from the voltage drop across the diode, V_d. That is, $V_{out} = -V_d$. Using the analytic model for the diode (Eq. 6.1), the input and output voltages are related by

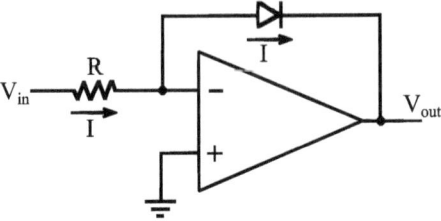

Fig. 10.20 Example 15, a log amplifier

$$I_d = \frac{V_{in}}{R} = I_0 \left(\exp\left(-\frac{V_{out}}{\eta V_T} \right) - 1 \right) \qquad (10.38)$$

Recall that for a semiconductor diode, I_0 is very small, so that if I_d is any appreciable value (say, greater than about 1 μA), then the exponential must be very large compared to 1 and the "−1" can be neglected in comparison. The output in terms of the input can be found by multiplying both sides by R, taking the log of both sides, and then using properties of the logarithm. That is,

$$\log(V_{in}) = \log(RI_0 \exp(-V_{out}/\eta V_T)) = -V_{out}/\eta V_T + \log(RI_0)$$
$$V_{out} = -\eta V_T \log(V_{in}/RI_0). \qquad (10.39)$$

Aside from some scale factors and a minus sign, the output is the logarithm of the input.

Note that if the input voltage is negative, the diode does not conduct and the feedback is broken. The op-amp will go to its largest possible output voltage. Since there are no (real) logarithms for negative numbers, such behavior should not be surprising for negative inputs.

A logarithm circuit is useful for indicating signal levels in dB, since, after all, dB's are a logarithm with a scale factor. Common signals with an amplitude that is convenient to express in dB include those that arise from the detection or transmission of sound, light, and radio signals.[9]

If the positions of the diode and resistor are swapped, the circuit becomes an "anti-log" or exponential circuit. The combination of log and anti-log circuits, along with sum and/or difference circuits, can, in principle, be used to multiply or divide two values. It is difficult to get a precision multiplication from such circuits due to imperfections in the diode model and the temperature dependence, particularly of V_T.

---

[9] One place to observe such measurements is the "VU meter" that may be on the output of an audio amplifier or on an audio sound board.

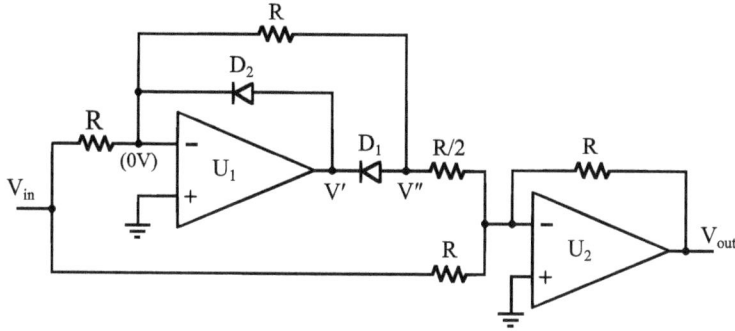

Fig. 10.21 Example 16, an absolute value circuit

Example 16—Absolute Value Circuit

Figure 10.21 illustrates a cousin to the ideal diode circuit. Here op-amp U_2 is configured as a summing amplifier (see Example 5) and so

$$V_{out} = -(V_{in} + 2V''). \tag{10.40}$$

If $V' > V_{in}$, then diode D_2 conducts and there is negative feedback and a virtual ground at op-amp U_1. With a virtual ground at both op-amps and D_1 off, it must be that $V'' = 0$. If $V' < V_{in}$, then D_2 is off and U_1 gets negative feedback through a resistance R, forcing $V'' = -V_{in}$. There is feedback in either case. The former occurs when $V_{in} < 0$ and the latter when $V_{in} > 0$. Thus,

$$\begin{aligned} &\text{If } V_{in} < 0, V_{out} = -(V_{in}) \\ &\text{If } V_{in} > 0, V_{out} = -(V_{in} - 2V_{in}) = V_{in} \end{aligned} \tag{10.41}$$

or, equivalently, $V_{out} = |V_{in}|$. Because of this, this circuit is sometimes referred to as a full-wave rectifier, however, as for the ideal diode (Example 13), no significant current from the input passes to the output. To get good results with this circuit, the resistors should be well-matched. To help achieve a good match with the other resistors, the resistor with value "R/2" can be obtained using two equal valued resistors, R, in parallel, so that all the resistors used have the same value and can be matched to great precision, for example by measuring each in turn using the Wheatstone bridge seen in Chap. 2.

More Power

Op-amps are low power devices used primarily to process or create signals. You should not expect to get a significant amount of power from an op-amp. Op-amps will not run motors or light up bright lights by themselves. In simple cases where more power is required transistors can be added in a manner similar to the circuits in Fig. 10.22.

The first circuit (10.22a) might be used to control a positive voltage across a load such as a small motor or a modest heating element. The feedback will cause the voltage across the load to match the input voltage, as long as the values stay within the ranges allowed by the devices.

This second circuit (10.22b) might be used for audio signals, or other signals that are both positive and negative. Again, the feedback causes the output across the load to follow the input voltage. The resistors shown, other than the load, help to keep the current within acceptable ranges and will depend on the particular application.

For any such circuits, where feedback is taken from the output of the power transistors, the tendency is that any attempts to achieve more power at the output will likely reduce the maximum usable frequency.

Less Than Ideal Difference Amplifiers

Finite Input Resistance and Gain

The solution for a general differential amplifier is a bit more complicated to obtain than it is for the ideal op-amp. The main purpose of this section is to illustrate how

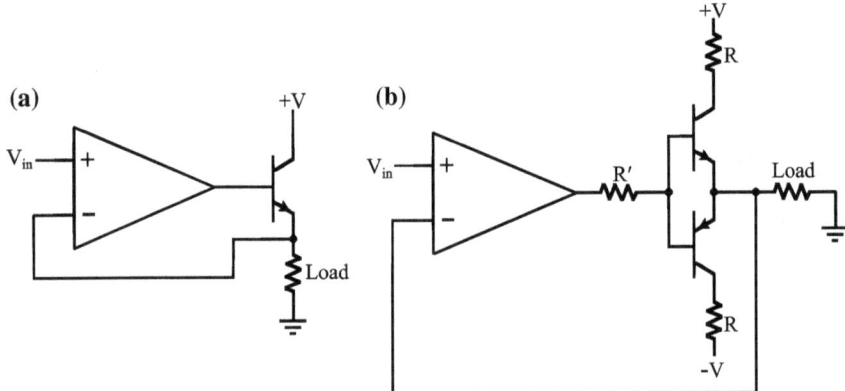

Fig. 10.22 Two circuits showing the use of transistors to boost the output power from an op-amp

Less Than Ideal Difference Amplifiers

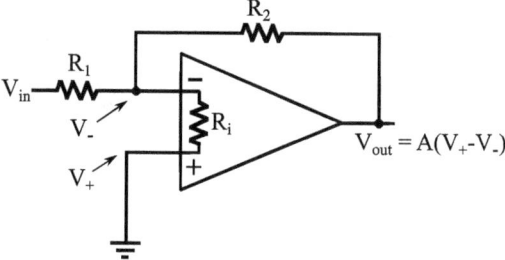

Fig. 10.23 A differential amplifier including an input impedance between the two inputs

you might do that and then to look at the limit as the amplifier becomes ideal. The results found in that limit justify the simplified solution method used for all the previous op-amp examples.

Consider the simple circuit shown in Fig. 10.23 that is constructed using a differential amplifier with finite differential voltage gain, A, and an (effective) internal resistance between the inputs, R_i. In terms of the input voltage, V_{in}, and unknown output voltage, V_{out}, the inputs to the differential amplifier can be found using simple superposition.

$$V_+ = 0$$
$$V_- = V_{in} \frac{R_i \| R_2}{R_i \| R_2 + R_1} + V_{out} \frac{R_i \| R_1}{R_i \| R_1 + R_2}, \tag{10.42}$$

and these can be substituted back into the expression for V_{out}:

$$V_{out} = A(V_+ - V_-) = -A\left(V_{in} \frac{R_i \| R_2}{R_i \| R_2 + R_1} + V_{out} \frac{R_i \| R_1}{R_i \| R_1 + R_2}\right). \tag{10.43}$$

Now solve this for the output voltage to get

$$V_{out} = -\frac{\frac{R_i \| R_2}{R_i \| R_2 + R_1}}{1 + A \frac{R_i \| R_1}{R_i \| R_1 + R_2}} V_{in}. \tag{10.44}$$

Now if the input resistance, R_i, is very large compared to both R_1 and R_2, this reduces to

$$V_{out} = \frac{-A \frac{R_2}{R_2 + R_1}}{1 + A \frac{R_1}{R_1 + R_2}}, \tag{10.45}$$

and in the limit that A becomes very large this is

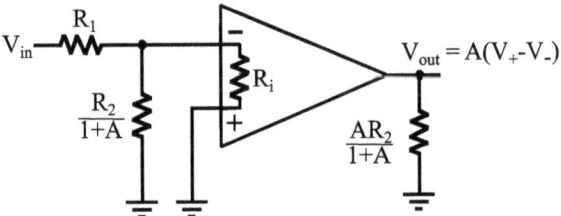

Fig. 10.24 An equivalent to the circuit of Fig. 10.23 obtained from Miller's theorem

$$V_{out} = -\frac{R_2}{R_1}V_{in}, \tag{10.46}$$

as expected (e.g., see the inverting amplifier, Example 2 above). Now to justify the simpler solution method used for ideal op-amps, put this solution back into the expression for V_- (with R_i large),

$$V_- = V_{in}\frac{R_2}{R_1+R_2} + V_{out}\frac{R_1}{R_1+R_2} = V_{in}\frac{R_2}{R_1+R_2} + \left(-\frac{R_2}{R_1}V_{in}\right)\frac{R_1}{R_1+R_2} = 0, \tag{10.47}$$

and it can be seen that the feedback has forced $V_- = V_+$ in this limit.

This problem can be addressed in a completely different way using Miller's Theorem (see Chap. 9). Using Miller's theorem, the circuit of Fig. 10.23 is equivalent to the circuit in Fig. 10.24.

If $R_2 \ll (1 + A)R_i$, a condition easy to meet if both A and R_i are large, then

$$V_+ = 0; \quad V_- = V_{in}\frac{R_2/(1+A)}{R_2+R_2(1+A)}$$
$$V_{out} = -AV_- = -A\left(V_{in}\frac{R_2/(1+A)}{R_1+R_2(1+A)}\right) = -\left(\frac{R_2A/(1+A)}{R_1+R_2(1+A)}\right)V_{in} \tag{10.48}$$

and the previous result is obtained in the limit that A becomes very large. Referring back to Fig. 10.24, note also that once again as A becomes large, V_- approaches V_+ due to the feedback.

Thus, a real differential amplifier with feedback will act like an ideal op-amp if the amplifier gain, A, is very large compared to 1 and the amplifier input impedance is large compared to the feedback impedance.

Available integrated circuit op-amps typically have an "open loop gain," i.e., with no feedback, between about 10^5 and 10^6, which is certainly quite large compared to 1. Op-amp voltage gain is so large it is often expressed in units of "volts per millivolt," or V/mV, so 10^5 would be 100 V/mV.

Op-amp input impedances can range from about 10^6 to 10^{12} ohms. The latter is comparable to the resistance of typical wire insulation, circuit board material, and many other insulators—hence, if such high impedance values are of importance, some care must be taken to avoid reducing that resistance through nearby insulating materials. With op-amps, one crude, but simple method to do this is "dead bug construction." The op-amp case is glued to a mounting board with the connecting leads pointed up in the air, thus avoiding any contact with mounting sockets, circuit board material, or other insulators.

Finite Frequency Range

A general purpose, inexpensive integrated circuit op-amp can be expected to have a unity gain bandwidth of about 1 MHz. That is, the buffer of Example 1 can be used up to 1 MHz, the inverting or non-inverting amplifiers (Examples 2 and 3) with a gain of 10 can be used to 100 kHz, and so on. Such op-amps are usually adequate for signals in the audio range (0–20 kHz). With slightly more expense, op-amps with a unity gain bandwidth into the 100's of MHz are available.

While it may be tempting to use an op-amp with a very high frequency response for lower frequency operation, after all that would be more "ideal," such use can lead to trouble. Remember that the amplifier is still working at its highest frequencies even if it is not being used at those high frequencies. At 100 MHz and above, even relatively short wire lengths become very important (see Chap. 4) and can act as impedance transformers. With a high frequency op-amp, the circuit design and layout need to take the high-frequency response into account even if the input signals are much lower in frequency. It is very easy make such circuits behave very poorly due to a bad layout or to have these high-frequency amplifiers turn into high-frequency oscillators[10] due to unintended inappropriate feedback.

Small Signals and Drift

An amplifier's offset voltages can be a nuisance when trying to amplify very small signals (~ 1 mV). While the offset can be compensated at any given time, offsets tend to drift over time, especially with changes in device temperature. There are various schemes to deal with this issue and there is not time and space to go into all of them here. One particularly simple scheme is to turn a relatively slowly varying input signal into a time dependent one. Since the drift is slow and appears as a d.c. offset, that offset can be removed with a capacitor. Such a "chopping" scheme is illustrated Fig. 10.25, where the peak-to-peak amplitude of the output is proportional to the low-level signal input. The electronically controlled switch would be implemented with a timing circuit and transistor switches (available as integrated

---

[10]Such unintended oscillation is referred to as "parasitic oscillation."

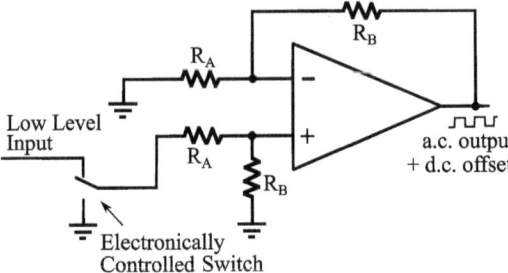

Fig. 10.25 One scheme to reduce problems with low-frequency drift is to convert the input signal to a higher frequency using an on/off switch

circuits). Raising the frequency of the desired signal before the amplifier also has some advantages for noise reduction since amplifiers tend to produce more electrical noise at lower frequencies than at higher frequencies.

Oscillations

Any amplifier with positive feedback can oscillate if conditions are right. Consider the non-inverting circuit of Fig. 10.4 (Example 3). Then

$$V_- = \frac{R_1}{R_1 + R_2} V_{out}; \quad V_+ = V_{in}, \text{ and so}$$
$$V_{out} = A(V_+ - V_-) = A V_{in} - A \frac{R_1}{R_1 + R_2} V_{out}, \quad (10.49)$$

or defining the fraction of the output that is used for feedback, β, as

$$\beta = -\frac{R_1}{R_1 + R_2}, \quad (10.50)$$

where the minus sign indicates negative feedback is being used. Then

$$V_{out} = \frac{A}{1 - \beta A} V_{in}. \quad (10.51)$$

In Example 3, $\beta < 0$ (so βA is negative), so all is as before.

Now consider what happens for the circuit of Fig. 10.4 if the two inputs are swapped. That has the same effect as changing A into $-A$. Then if components are chosen so that $\beta A = 1$, the denominator goes to zero in which case an output is expected even in the limit that the input becomes negligibly small. This is known as the Barkhausen criterion. That criterion is much more interesting when capacitors and/ or inductors are involved in the feedback. Then β can become a frequency dependent

Oscillations

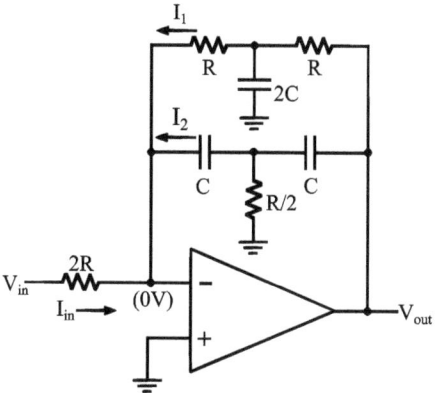

Fig. 10.26 The twin-T oscillator made using an op-amp

complex value. If $A = 1$ at some non-zero frequency, an output signal at that frequency can be expected even in the absence of an input. That is, the circuit oscillates.

One example of an op-amp oscillator is the twin-T oscillator, shown in Fig. 10.26. The circuit can be solved using KCL, the virtual ground, and the currents shown. The result is

$$V_{out} = \frac{1 - i\omega RC}{1 - (\omega RC)^2} V_{in}. \qquad (10.52)$$

The denominator goes to zero, and oscillation should be expected, when $\omega RC = 1$.[11] The combined phase shifts from the capacitors have turned what looks to be negative feedback into positive feedback.

The Transconductance Amplifier

A cousin to the op-amps discussed above is the so-called operational transconductance amplifier (OTA). Like the op-amp, there are two high impedance differential inputs. However, the output is a current rather than a voltage. That is,

$$I_{out} = g_m(V_+ - V_-). \qquad (10.53)$$

The output voltage will depend on the load. There is often an additional separate current input that can be used to control g_m. Applications often take advantage of that extra control to make voltage-controlled volume controls, voltage-controlled filters, and similar circuits.

---

[11] In the language of Chap. 5, there are poles at $s = \pm i/(RC)$.

Problems

1. (a) Show that for sinusoidal input signals, the circuit in Fig. 10.P1 gives an output that has the same magnitude as the input, but a different phase, and that the resistor R can be adjusted to get different phase shifts. (b) What is the maximum range of the phase shifts that can be expected? Assume an ideal op-amp. This circuit is sometimes called an "all-pass filter" since all frequency components pass through with no change in amplitude.

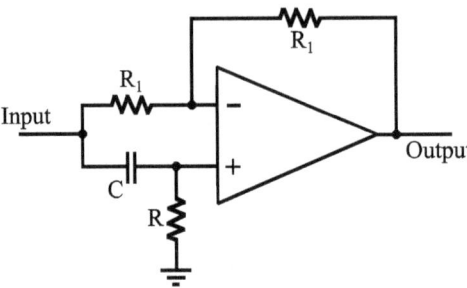

Fig. 10.P1 Problem 1

2. For the circuit in Fig. 10.P2, how is the output related to the input for the case where the switch is open and when it is closed?

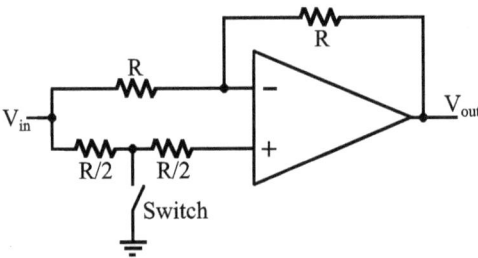

Fig. 10.P2 Problem 2

Problems

3. Compute the voltage gain (V_{out}/V_{in}) for the circuit in Fig. 10.P3, assuming an ideal op-amp.

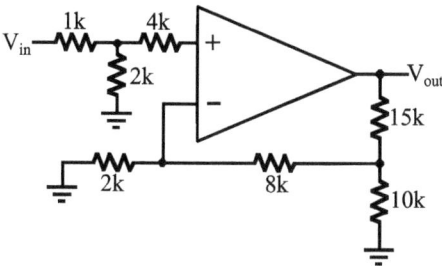

Fig. 10.P3 Problem 3

4. In the circuit of Fig. 10.P4, the non-linear circuit element "x" has the current-voltage relationship $V_x = bI_x^3$, where $b = 1.5\text{ V}/(\text{mA})^2$ is a constant. What is the output voltage, V_{out}?

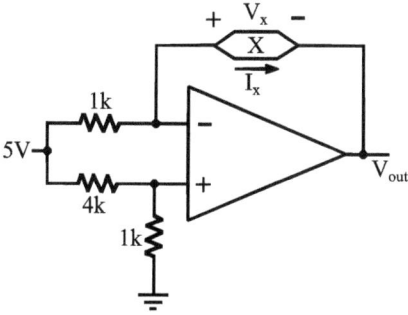

Fig. 10.P4 Problem 4

5. A solar cell (or photovoltaic cell) is a two-lead device used to detect and extract electric power from light. A small solar cell can be used as the detector for a light meter. Such a cell connected to a short circuit will emit a current roughly proportional to the number of incident photons that are within the cell's detection range. Design a circuit with a solar cell, an op-amp, and resistors (as required) that provides an output *voltage* proportional to the number of incident photons.

6. Assume ideal op-amps and derive the relationship between V_{out} and the two inputs, V_A and V_B for the circuit in Fig. 10.P6.

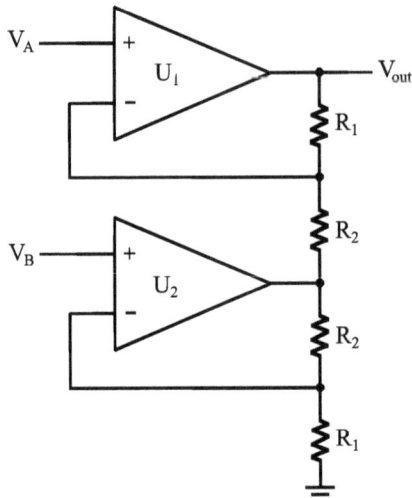

Fig. 10.P6 Problem 6

7. Show that the op-amp circuit in Fig. 10.P7a behaves the same as the circuit in Fig. 10.P7b. These circuits act as a band pass filter, allowing signals to pass through only if their frequency is near to the filter's center frequency. If the center frequency for the circuit in Fig. 10.P7b is $f_0 = (2\pi\sqrt{LC})^{-1}$, what is the center frequency for the op-amp circuit?

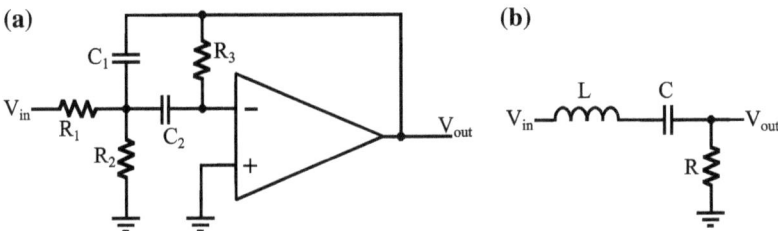

Fig. 10.P7 Problem 7

8. Derive V_{out}/V_{in} for the op-amp circuit in Fig. 10.P8. Assume an ideal op-amp.
9. Assuming an ideal diode and op-amp, show that the output voltage of the circuit in Fig. 10.P9 is the absolute value of the input voltage. How does the behavior change for a real semiconductor diode?

Problems

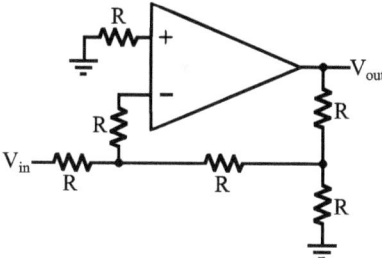

Fig. 10.P8 Problem 8

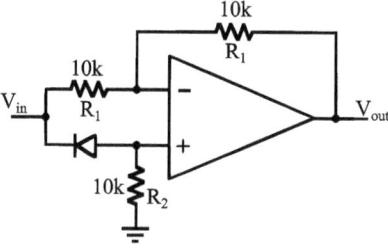

Fig. 10.P9 Problem 9

10. For the circuit of Fig. 10.P10 what is V_{out}? Assume ideal op-amps and that for any input one of the diodes is conducting. Consider both positive and negative V_{in}. Why will it never be the case that both diodes are conducting?

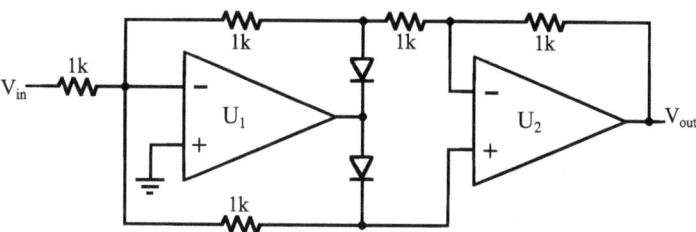

Fig. 10.P10 Problem 10

11. (Challenge Problem) Assume ideal op-amps and compute the input impedance $R_{in} = V_{in}/I_{in}$ and the current through R_5 for the circuit shown in Fig. 10.P11. (Hint: the answers are relatively simple mathematical relations).

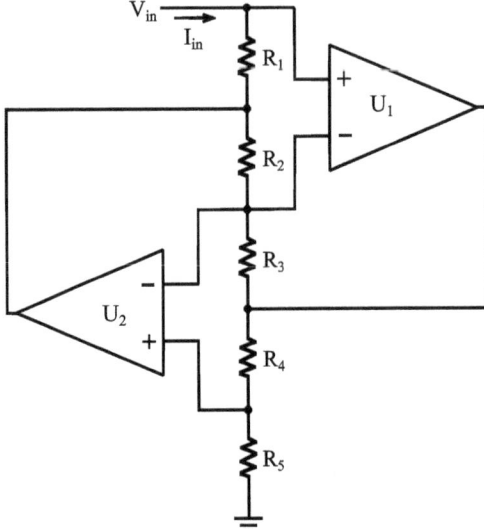

Fig. 10.P11 Problem 11

12. (Challenge Problem) Analyze the circuit shown in Fig. 10.P12 assuming ideal op-amps with negative feedback to determine V_o in terms of V_1 and V_2. Why is it ok to assume negative feedback is being used in this circuit even though the upper op-amp has feedback applied only to the non-inverting input?.

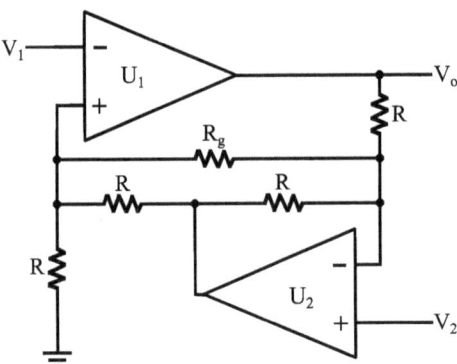

Fig. 10.P12 Problem 12

References

R.F. Pigage, A.B. Ryan, R.E. Smith, *Method for Inspecting and Adjusting Cutter Blades*, US Patent 3,641,431 (1968)

M. Tartagni, R. Guerrieri, A fingerprint sensor based on the feedback capacitive sensing scheme. IEEE J. Solid-State Circ. **33**, 133–142 (1998)

Chapter 11
Non-linear Uses of Op-Amps

In the previous chapter, the op-amp was used as a linear amplifier. A real op-amp has limitations that can be exploited to make useful circuits in which the op-amp is no longer linear. These applications use the very high gain of the op-amp along with the limited output range to create new circuits. Of particular interest is use as a comparator.

Limited Output Range

The output range of integrated circuit op-amps is limited, or "clipped," near the level of the power supplies. That is, if V^+ is the positive supply and V^- the negative,[1] then the output voltage, V_{out}, is restricted to the range $(V^- + \delta) < V_{out} < (V^+ - \delta)$, where, for general purpose op-amps, $\delta \approx 1$ V. There are slightly more expensive op-amps, the so-called "rail-to-rail" op-amps, where δ can be as small as a few millivolts.

The inputs for most op-amps are also limited by the supply voltages. For general purpose op-amps, the inputs are bounded by the supply voltages—they should be no smaller than V^- and no larger than V^+. For some more expensive op-amps, which includes many of the "rail-to-rail" op-amps, the inputs can exceed the supplies by as much as a volt without causing damage. The manufacturer's datasheet for the op-amp should be consulted for details.

Provided the op-amp extremes are not exceeded, the limiting behavior can be of use for some interesting applications. Consider, for example, the circuit of

---

[1] Power supply connections are often labeled using $\pm V_{cc}$, V_{cc} and V_{ee}, and/or with V_{dd} and V_{ss}. V_{cc} and V_{dd} being the higher voltage of their pair. The letters in the subscripts originally referred to the collector, emitter, drain, and source of the transistor(s) in the amplifier.

© Springer Nature Switzerland AG 2020
B. H. Suits, *Electronics for Physicists*, Undergraduate Lecture Notes in Physics,
https://doi.org/10.1007/978-3-030-39088-4_11

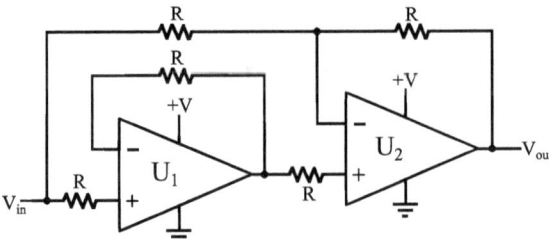

Fig. 11.1 Absolute value circuit using rail-to-rail op-amps

Fig. 11.1, assumed to be constructed using rail-to-rail op-amps.[2] The "negative" supply is set at ground (0 V). The first op-amp (U_1) is wired as a buffer, however the output cannot go negative. Hence U_1 only amplifies when the input is positive. The output of the first op-amp will look like that of a half-wave rectifier. When the input is positive, the second op-amp (U_2) will see V_{in} from U_1 at its non-inverting input. The negative feedback will adjust so the inverting input is also equal to V_{in}, and it does that by adjusting the output voltage to V_{in}. When the input is negative, U_1 outputs 0 V, since it is limited by the power supply, and that creates a virtual ground at the inverting input of the second op-amp. The negative feedback on U_2 adjusts the output to $-V_{in}$ (a positive value) to achieve this. Thus, this combined circuit outputs the absolute value of the input, without using any diodes.

There may seem to be several more resistors in Fig. 11.1 than are really necessary. For example, the two connected to the inputs of U_1 and the resistor to the non-inverting input of U_2 could be wires and the circuit analysis would look the same. These resistors will ensure that the voltage at the inputs to the (real) op-amps do not go too far outside the supply range. Those "extra" resistors limit the possible current that can occur. The values of those resistors are less critical. The device datasheet will provide values for the maximum allowable voltages and currents for the particular device used.

The Op-Amp Comparator

The simple circuit in Fig. 11.2, an op-amp without feedback, creates a comparator.[3] For this circuit, the difference between the two input voltages is greatly amplified. For the ideal op-amp, the output would approach ± infinity. As mentioned above, for a real op-amp the output is limited to values between the two supply voltages. Thus, for the real op-amp, the two input voltages are compared and the output has one of two

---

[2]For example, using the LMC6482 dual op-amp, $R = 10k$ and $+V = 5$ V.
[3]A "digital comparator" will compare two digitally stored numeric values. Be aware that the word "digital" is sometimes omitted. Digital comparators are functionally quite different from the comparators discussed here.

The Op-Amp Comparator

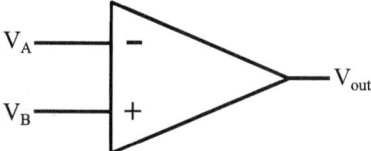

Fig. 11.2 Schematic for a comparator—An op-amp without feedback

values, the maximum and minimum that the op-amp can produce, depending on which input is larger. For the real op-amp, this will be true except for a very small range of values where the two inputs are extremely close together—closer than about 10 μV or even less. The comparator is a highly non-linear application for the op-amp.

In summary, V_{out} is near the positive supply voltage, V^+, if $V_B > V_A$ or the negative supply voltage, V^-, if $V_B < V_A$. In practice, making $V_A = V_B$ to get an output ½-way in between is virtually impossible for a real op-amp and that possibility is simply ignored here.

The comparator is useful for many applications and several examples are shown below. It should not be surprising that comparators are often part of the interface between analog electronics and digital (on/off) electronics.

There are devices available that are designed especially to be used as comparators. Some of the circuitry that is added to op-amps to make them behave more ideally as amplifiers will compromise the behavior when used as a comparator. That additional circuitry can simply be omitted. For many applications, particularly at lower speeds, an op-amp works just fine as a comparator. In what follows, no distinction is made between op-amps used as comparators and these devices which are especially designed as comparators.

If a small amount of positive feedback is added to a comparator, the result is a "comparator with hysteresis." The input/output relation is double-valued and depends on the previous history of the device, not just the instantaneous values of the inputs. Such a circuit is illustrated in Fig. 11.3a. Here, V_{in} is compared to the voltage at the non-inverting input, V_+, which is in turn given by

$$V_+ = V_{ref} \frac{R_p}{R_r + R_p} + V_{out} \frac{R_r}{R_r + R_p}. \tag{11.1}$$

Hence, V_+ depends on the present state of V_{out}. And in turn, V_{out} is near either the positive or negative power supply voltage (V^+ or V^-) depending on the previous state of the input. Tracing the behavior starting with $V_{in} = V^-$, sweeping up to V^+ and then back down to V^- will produce the double-valued behavior shown in Fig. 11.3b. The extra threshold to switch, created by the positive feedback, is useful for noise immunity. Comparators with hysteresis are available as a self-contained device for certain applications.[4]

---

[4] A comparator with hysteresis used for digital electronics is usually referred to as a "Schmitt trigger."

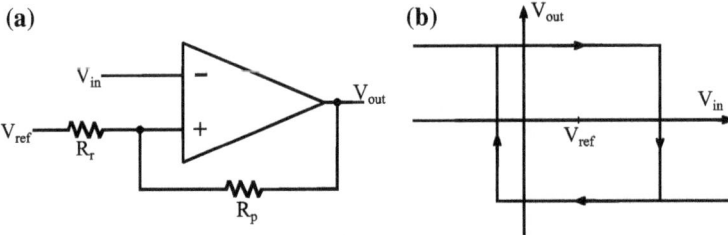

Fig. 11.3 A comparator with positive feedback (**a**), has a double valued output (**b**), with hysteresis—The output depends on previous values of the inputs

A number of applications using comparators follow. For many of them, hysteresis could also be included, but is not shown.

Example 1—Low-Level Warning

Comparators can be used to provide a simple indication of a signal level. In the simplest case, it might simply provide a warning when a signal, such as a battery voltage, is too high or too low.

A simple indicator that provides a warning if a supply voltage has dropped below a preset threshold is shown in Fig. 11.4. In this circuit, the diode on the right is an LED that glows when there is a forward current of a few mA. The forward voltage drop of the diode on the left, its turn-on voltage, is being used to provide a reference voltage (about 0.5 V) which is relatively independent of the power supply voltage. A Zener diode (with reverse bias) could be used instead. When the voltage at the inverting input drops below the diode voltage, the LED turns on. So, for example, if this circuit is used with a 9 V battery and a warning is desired when the battery voltage drops below about 8.8 V, choose $R_1 \approx 100$ k and R_2 so that

$$V_D = 8.8 \frac{R_2}{R_1 + R_2} V, \qquad (11.2)$$

where V_D is the (measured) diode reference voltage for your circuit.

A more complicated circuit may use several comparators to indicate a more continuous range of values, such as for the audio level on some audio devices. Figure 11.5 shows such a circuit that will indicate three levels. The same principle can be used for many more levels.[5] The resistors on the right are included to limit the current through the LEDs. They may not be necessary if the output current from the comparators (op-amps) is internally limited at a safe level.

---

[5] The LM3914 and LM3915 integrated circuits include 10 such levels built into a single integrated circuit device.

The Op-Amp Comparator

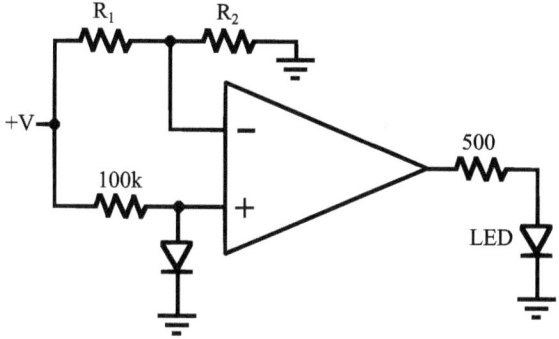

Fig. 11.4 Example 1—A simple low-voltage warning circuit using a comparator

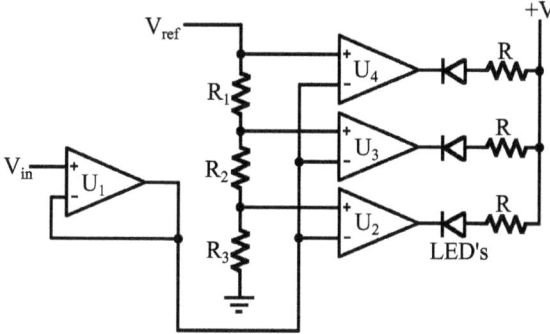

Fig. 11.5 A three-level voltage indicating circuit. Such a circuit can be expanded to many more levels

Example 2—Pulse Generator

The circuit of Fig. 11.6 can be used to make a voltage pulse. The pushbutton switch here is normally open (N.O.), meaning that there is no connection until the switch is pushed. When the switch is pressed (closed), the capacitor is rapidly charged to $+V$, which will exceed the value at the inverting input. Once the switch is released, the capacitor will gradually discharge through resistor R_0. The output of the comparator will remain high (i.e., near V^+) until the capacitor voltage falls below the voltage at the inverting input. Then the output goes low (i.e., near V^-).

A typical switch makes and breaks a mechanical connection between two conductors to perform the switching action. As it does so, the connection actually makes and breaks many times over a short time period (typically a few milliseconds), a phenomenon known as switch bounce. For many applications that is not a problem, but for high speed digital applications, each make and break is treated as a separate event, as if the switch were repeatedly pushed and released. This pulse generator can eliminate that effect, and when it is used for that purpose it is referred to as a switch de-bouncer.

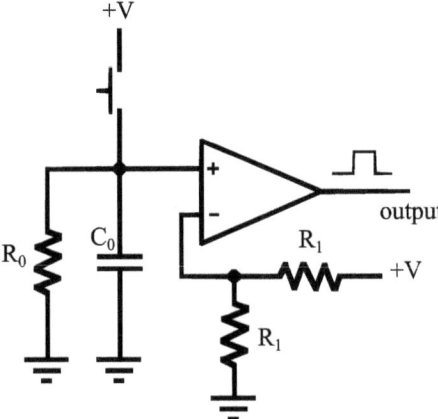

Fig. 11.6 Example 2—A single pulse generator. When the pushbutton is pressed and released, a square pulse is generated

Example 3—Simple Oscillator

Figure 11.7 shows a comparator with hysteresis with a capacitor and resistor added to supply the inverting input voltage. This creates a simple oscillator.

Start with the capacitor discharged and the op-amp output at either of its two possible values ("high" or "low"). Suppose it is high. Then the non-inverting input will see a positive voltage and the capacitor will begin to charge upwards. Once the capacitor reaches the value at the non-inverting input, the op-amp output will switch and go low. Then the non-inverting input will see a lower value and the capacitor will start to discharge towards that lower value. Once the capacitor reaches the lower value, the op-amp output once again goes high, and the process repeats. It is somewhat like a game of keep-away, where the capacitor is trying to get to the target value at the non-inverting input, but each time as it just gets there, the target changes.

There are two outputs indicated in Fig. 11.7. The output V_{out1} will be a square wave. The voltage across the capacitor, V_{out2}, will be that of the capacitor charging and discharging. If the hysteresis loop is not too large, V_{out2} will approximate a triangle wave.

Example 4—A Voting Circuit

The circuit in Fig. 11.8 will function to determine the outcome of a voting process. The first op-amp sums the voltages from all the inputs. In this case a closed switch is a "no" vote and an open switch is a "yes" vote. The output of the first op-amp varies from 0 V (all "no") to $-V_{ref}$ (all "yes"). The comparator checks to see if the

The Op-Amp Comparator

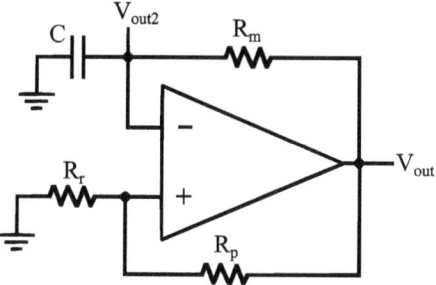

Fig. 11.7 Example 3—A simple oscillator made using a comparator with hysteresis that uses a delayed version of its own output as one of the inputs

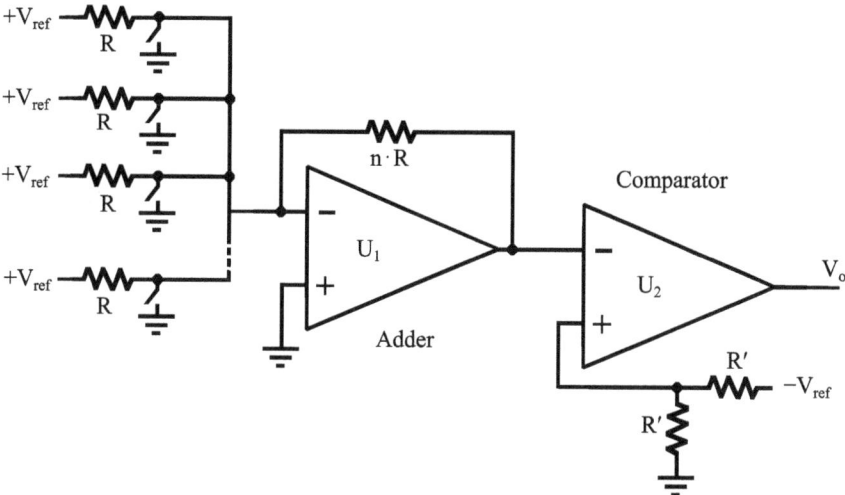

Fig. 11.8 Example 4—A voting circuit

result is more or less than $-V_{ref}/2$. The output, V_{out}, will be near the positive supply voltage if the majority say "yes" and near the negative supply voltage if the majority say "no."

Since real resistor values are not particularly accurate and may depend somewhat on temperature, such a circuit could only be used for a relatively small number of "voters." It would also be best to use it only for an odd number of voters since the results are unpredictable in the event of a tie. Such a "voting" procedure may not necessarily be part of a democratic process. For example, it might be desirable to have an alarm on an airlock that goes off if any 2 of 3 doors are open. In that case this circuit can be used with n = 3 and switches appropriately wired to the doors. For other uses, the resistors can be made unequal in value to make some "votes" count more than others, use a 2/3 majority, etc.

Example 5—Sine to Pulse Train Converter

Sometimes it is necessary to provide an on/off signal with a frequency determined by a sine wave. For example, a digital alarm clock may get its timing from a sinusoidal signal derived from the power outlet (i.e., 50 or 60 Hz, depending on location). Of course, the sine wave input must be small enough in amplitude so that it does not exceed the input ratings of the op-amp at any point in the cycle.

The circuit in Fig. 11.9 will produce a pulsed output between about 0 and 5 V from a sine wave input. The "duty cycle," which is the ratio of the "on" time to the total time, can be varied by adjusting the relative values of R_1 and R_2. If they are equal, a 50% duty cycle would be expected. The feedback resistor, R_f, may not be needed in all applications. It provides hysteresis to reduce false triggers, as described above, and it is large compared to R_1 and R_2. The lower diode shorts the output to ground if the op-amp output goes negative and the diode to the +5 V supply shorts the output to the 5 V power supply if the output of the op-amp exceeds 5 V. The resistor at the output, R, should be large enough to keep the op-amp output current within the range specified by the op-amp manufacturer.

For this application, an op-amp designed for 5 V, single-supply operation could be used directly without the diodes and resistor R, though the resistor and diode protection may be necessary on the *input* to avoid excessive input currents, particularly during the negative half of the cycle.

Example 6—Zero Crossing Detector

The circuit in Fig. 11.10 uses two comparators in order to provide a short pulse when a signal crosses zero with a positive slope. It is easily modified to work with a negative slope and/or to add an offset to detect when the signal crosses any desired voltage level. The function of this circuit is the same as the trigger circuitry used for an oscilloscope. The signals are illustrated in the figure are for a sinusoidal input though any time dependent input can be used.

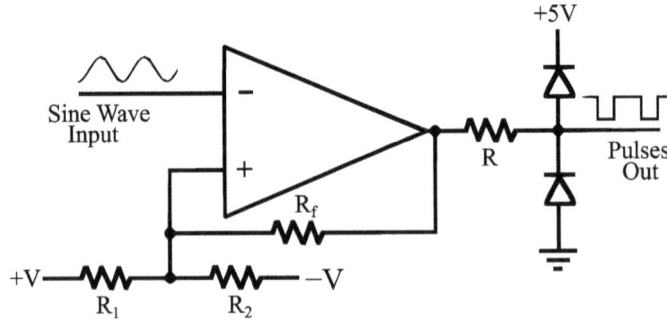

Fig. 11.9 Example 5—Sine wave to square pulse converter

The Op-Amp Comparator

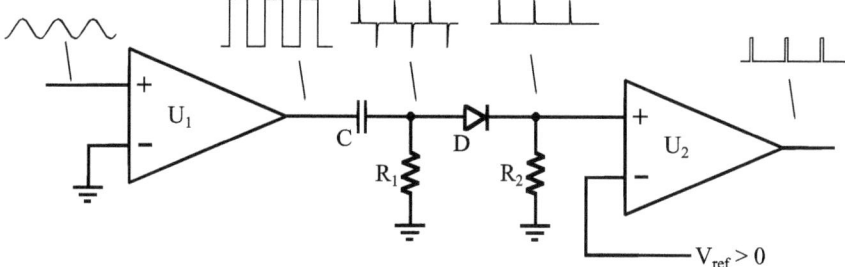

Fig. 11.10 Example 6—Zero crossing detector, such as what might appear in an oscilloscope triggering circuit

The first op-amp (U_1) converts the input to a square wave—high if the input is positive, low if it is negative. The capacitor and first resistor act as a high-pass filter, and let through much more signal during the rapid *changes* in the square wave than during the flat portions. The diode blocks the negative going pulses. The pulses at this point have a sharp rise and a (rapid) exponential decay. The second op-amp (U_2) compares the input to a reference voltage that is above zero, but possibly not above zero by very much. The output of U_2 then "squares up" the output, removing the exponential decay.

If the diode is reversed, the circuit is sensitive to the negative slope. If an offset is used, instead of ground, for the inverting input of U_1, then the trigger level will move away from zero. Positive feedback can be added to provide hysteresis for either or both comparators, as needed.

Example 7—Pulse Conditioner/Lengthener

Figure 11.11 illustrates a circuit useful for taking a short duration "messy" input signal, and converting it to a nice "clean" square pulse, possibly significantly longer in duration. The output signal might be used to signal that an event has occurred, such as a click from a Geiger counter or the sound of a clap picked up by a microphone. The circuit also has the feature that the strength of the signal can be reflected in the length of the output pulse—a stronger input signal yields a longer output pulse. A measurement of the pulse length is then a measure of the strength of the input pulse.

The first op-amp, U_1, is connected as an ideal diode circuit. Thus, when the input is positive, the output should follow the input. The capacitor, C, will then charge (or discharge) through R_1 based on the amplitude of the input. The $R_1 C$ time constant should be comparable to, or longer than, the expected duration of the longest input pulse. Once the short-duration signal is complete, the capacitor can only discharge through R_2. If R_2 is much larger than R_1, the discharge time can be much longer

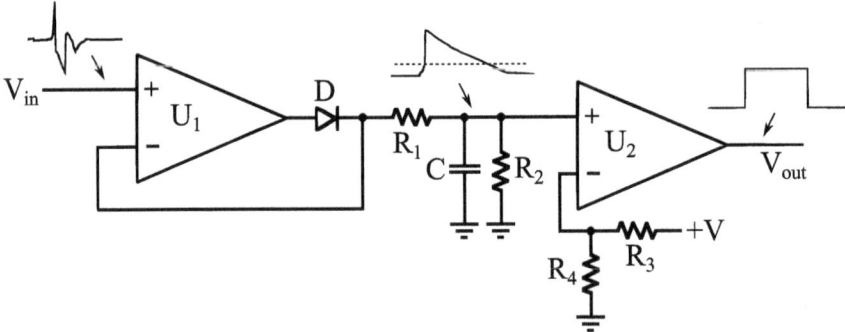

Fig. 11.11 Example 7—Pulse conditioner/extender

than the charging time. The second op-amp, U_2, is a simple comparator that outputs a positive pulse as long as the voltage on the capacitor exceeds the reference level set by R_3 and R_4.

Using the Comparator for Feedback

A comparator can be useful for feedback. A simple case would be a home heating system where the furnace is to be either on or off depending on whether the house temperature is above or below a preset value. A comparator of some sort checks the temperature. If it is too high, the furnace is turned off and if too low, the furnace is turned on. In such an application, some hysteresis would be used so that the furnace is not turning on and off too rapidly.

With some filtering, the comparator can be used for other feedback applications leading to behavior that has the appearance of being continuous, rather than on/off.

Automatic Gain Control Amplifier

Figure 11.12 shows a purely electronic example of a comparator used for feedback.[6] This circuit is designed to have a sinusoidal input and a sinusoidal output, but where the output amplitude is (almost) independent of the input amplitude. Such a circuit would be useful if there is a signal of unknown (or variable) amplitude and it is desirable to have a "reference signal" at the same frequency, but

---

[6]Specific values and components are given for reference and to serve as a starting point as there is some interdependence for some values and no simple formula that can be applied.

Using the Comparator for Feedback

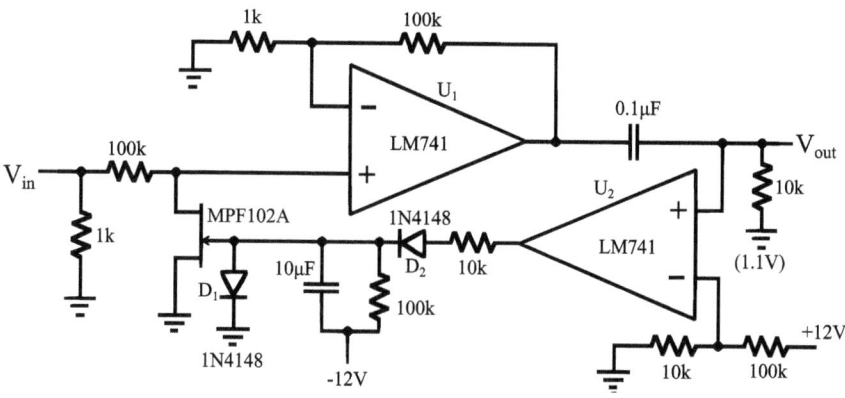

Fig. 11.12 An automatic gain control amplifier for sinusoidal signals. Feedback derived from the output of a comparator is used to keep the amplitude of the output signal constant for a large range of input signal amplitudes

with a fixed amplitude (e.g., for a lock-in amplifier). This idea might also be used as an automatic amplitude control for audio signals.[7]

Assume the op-amps are powered from ± 12 V and V_{in} is sinusoidal. The upper op-amp (U_1) functions as a linear amplifier with a nominal gain of ~ 100. The lower op-amp (U_2) compares the output from the first to a d.c. reference voltage, in this case approximately 1.1 V. Any time the output from the first op-amp exceeds 1.1 V, the output of the second op-amp goes from about -12 V to $+12$ V.

The diode D_2 blocks the comparator output when the comparator output is -12 V. In the absence of current through the diode, the capacitor connected to the gate of the FET will charge to supply -12 V to the gate of the FET, which is more than enough to pinch-off the FET. With the FET pinched-off, it is effectively removed from the circuit and so the input is amplified by ~ 100. However, if the comparator goes positive because some part of the signal has exceeded 1.1 V, D_2 conducts and the input voltage at the gate increases until the FET "turns on." When it is on, the FET is being used as a voltage variable resistor (since V_{ds} is kept small) and the input voltage applied to the op-amp is reduced—that is, if the output amplitude gets too large, the input is reduced to compensate. Diode D_1 is included to protect the FET gate from positive voltages and the capacitor on the output of U_1 blocks any d.c. offset voltages from getting to U_2.

---

[7]Note that loudness is a perceived quantity that depends on many factors in addition to peak amplitude, and so use as an automatic *volume* control may be problematic.

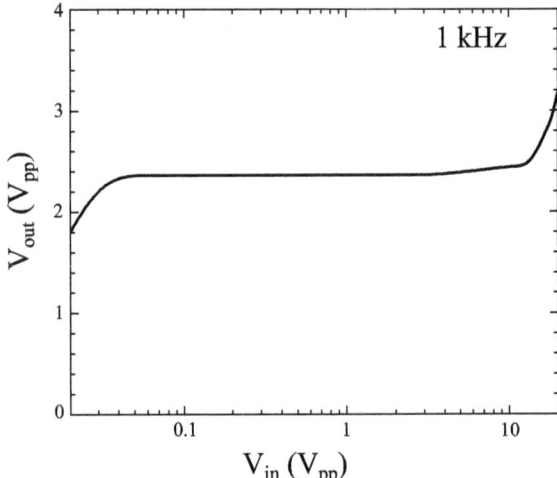

Fig. 11.13 The experimental input/output relationship measured for the circuit of Fig. 11.12. Note that the horizontal axis is logarithmic

Measured values obtained with this circuit and a sinusoidal input are shown in the graph in Fig. 11.13. The output amplitude is indeed quite constant as the input is changed over a range of a bit more than a factor of ten.

Putting Pieces Together

Many tasks in electronics are accomplished using a combination of circuits, each of which has a simpler, well-defined task to perform. The overall function requires they each do their job and work together. Here is one example.

Simple Phase Sensitive Detector

The circuit in Fig. 11.14 shows (schematically) how one might construct an inexpensive phase sensitive detector (a "lock-in amplifier") by combining op-amp circuits from this and the previous chapter.

If you trace through this circuit, the output (V_{out}) is a very low frequency signal (nominally d.c.) that is proportional to that part of the input signal (V_{in}) which has the same frequency as, and a certain phase relative to, a reference signal. Signals that are at a different frequency or are out of phase will exit the multiplier as positive as often as they do negative and will then average to zero in the low-pass filter.

Problems

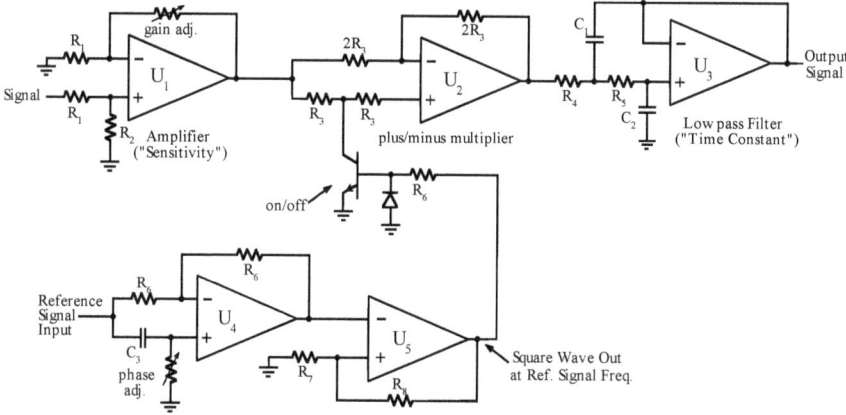

Fig. 11.14 Op-amp circuits with different functions can be combined to create more complicated instrumentation. Here, a design for a simple phase-sensitive lock-in amplifier is shown

Problems

1. Assuming the magnitude of the op-amp power connections are larger than V_{ref}, predict V_{out} as a function of V_{in} for the comparator circuit in Fig. 11.P1. All the resistors have the same value and ideal diodes may be assumed. Note that there is no connection between V_{in} and the resistors on the left even though the wires cross in the diagram.

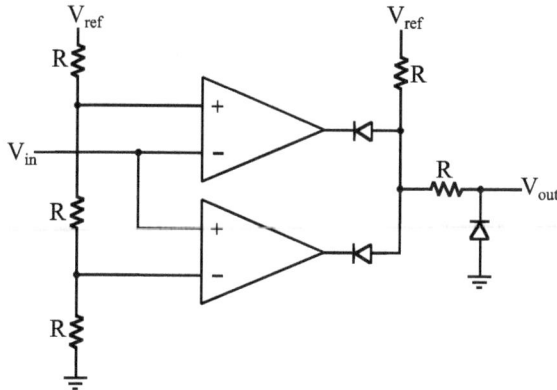

Fig. 11.P1 Problem 1

2. Describe the behavior of the output for the circuit in Fig. 11.P2 when the push-button switch is pressed (closed) and then after it is released (opened) at some later time. What happens if the button is pressed multiple times in rapid succession, say once every 0.1 s for 1 s?

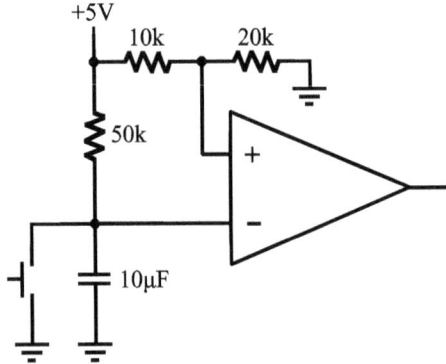

Fig. 11.P2 Problem 2

3. For the circuit in Fig. 11.P3, when do the LEDs turn on/off after the switch has been pressed (closed) for a while and then released (opened) at $t = 0$? The op-amps are powered with ± 12 V.

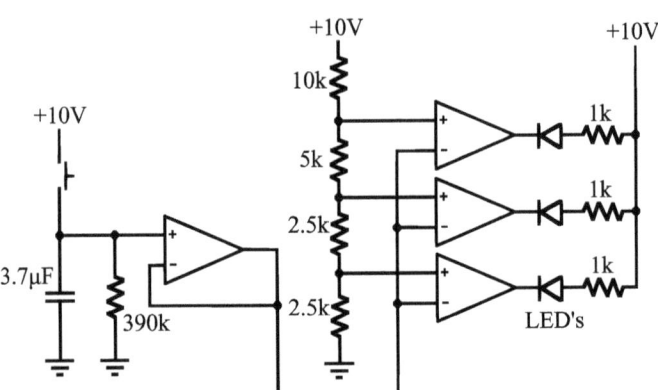

Fig. 11.P3 Problem 3

Problems

4. A comparator with hysteresis was constructed following the schematic shown in Fig. 11.3, using $V_{ref} = 0$ V and $R_r = 10$ k. The measured output is shown in Fig. 11.P4. What value of R_p was used?

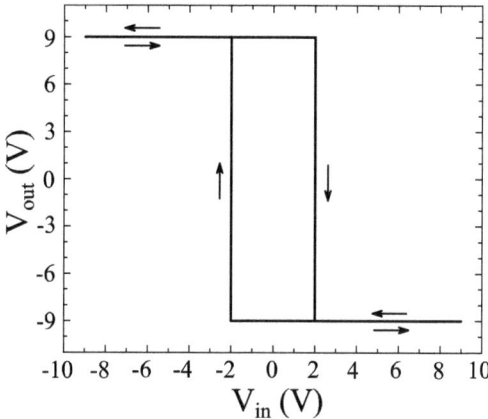

Fig. 11.P4 Problem 4

5. A triangle wave with a peak-to-peak amplitude of 2 V is used for the non-inverting input to a comparator. Sketch the output of that comparator for the three cases where the inverting input is −0.5 V, 0 V, and +0.5 V. This illustrates a voltage to pulse length conversion.

Chapter 12
Digital I

Digital electronics uses circuitry that has two possible states for inputs and outputs. You might call them On and Off, 0 and 1, or maybe True and False. The mathematics of logic, known as Boolean algebra, is based on the latter and is useful for digital circuit analysis, at least as a starting point. In this chapter, this logic-based understanding of digital circuits is introduced followed by the introduction of flip-flops, a first example in sequential logic, where the output depends on the inputs and the previous state of the system.

Boolean Algebra

In summary, Boolean algebra[1] includes the following:

- Elements that have one of two values—that is, they are "binary." Here True = 1 and False = 0 will be used as the two values, but other choices are possible.
- A bar over the top of a value means "the other value" (a "NOT" when using True and False). That is $\overline{1} = 0$ and $\overline{0} = 1$. The "NOT" operation acts on a single value.
- Two operations defined for *pairs* of values:

  "**OR**," designated with "+" and "**AND**" designated with multiplication dot[2] defined as follows:
  OR: Returns false if both the values are false, otherwise true.
  AND: Returns true if both the values are true, otherwise false.

---

[1]Boolean Algebra was named for the English Mathematician George Boole who first used it in the mid-1800s as a system for the analysis of logic, long before digital electronic circuits existed.
[2]As with multiplication, the dot is often omitted if the meaning is clear.

© Springer Nature Switzerland AG 2020
B. H. Suits, *Electronics for Physicists*, Undergraduate Lecture Notes in Physics,
https://doi.org/10.1007/978-3-030-39088-4_12

Note that since the result of the NOT, AND, and OR operations belong to the same set of values as the original arguments (i.e., true or false), the result of an operation can be used as an input for another operation, a subtle but important point.[3] Note also that the OR and AND operations are commutative, associative, and distributive.

By convention, when an expression contains multiple operations, the order of execution for the operations is similar to what is used for addition and multiplication. Quantities in parentheses are evaluated first, then ANDs followed by ORs. When a NOT bar covers an expression involving several values, the expression is to be evaluated first, before the "NOT" is applied. As examples, consider the following:

- $A + B \cdot C \rightarrow$ The AND is applied to B and C, and that result is OR'ed with A.
- $A \cdot (B + C) \rightarrow$ The OR is applied to B and C, and that result is AND'ed with A.
- $\overline{A \cdot B} + \overline{C} \rightarrow$ The AND is applied to A and B, then the NOT is applied to that result. The result of that NOT is OR'd with the result of the NOT applied to C.
- $\overline{\overline{A + B} \cdot C} \rightarrow$ The OR is applied to A and B, then the (lower) NOT is applied to that result. The result of that NOT is AND'ed with C, and then the (upper) NOT is applied to that result.

Notations that result in possible ambiguity should be avoided. For example, $A \cdot \overline{B + C}$ might be confusing. Adding parentheses often helps. For example, the order of operations for $A \cdot \overline{(B + C)}$ is unambiguous.

Since there are only four possible input combinations each for the AND and OR operations, it is not hard to write them out in a "truth table," such as Table 12.1. Truth tables are an alternate and simple way to show all the possible inputs and the corresponding outputs.

The double operations obtained by applying a NOT to the result of both AND and OR are seen in electronics often enough they are given their own names—NAND and NOR respectively.[4] For the NAND operation the output is false only if all the inputs are true, and for NOR the output is true only if all inputs are false.

Useful Rules and Theorems for Boolean Algebra

A number of useful relations for Boolean algebra are listed in Table 12.2. Many of these are relatively obvious. The last two entries are known as DeMorgan's theorems and are very important for the understanding and design of digital circuits. It is important to become proficient with the result of DeMorgan's theorems. All of the

---

[3] An example from physics where this does not occur is the vector dot product.
[4] Even though the NOT is applied last, the "N" representing the NOT operation is put at the beginning of the name. Possibly this is because ANDN and ORN are much less pleasant to pronounce.

Useful Rules and Theorems for Boolean Algebra

Table 12.1 Truth table for OR and AND

Input values		Result	
A	B	A + B	A · B
0	0	0	0
0	1	1	0
1	0	1	0
1	1	1	1

Table 12.2 Some basic relationships found in Boolean algebra

$\overline{\overline{A}} = A$	$A + B = B + A$
$A + 0 = 0 + A = A$	$A \cdot B = B \cdot A$
$A \cdot 0 = 0 \cdot A = 0$	$A + A \cdot B = A$
$A + 1 = 1 + A = 1$	$A + (B + C) = (A + B) + C$
$A \cdot 1 = 1 \cdot A = A$	$A \cdot (B \cdot C) = (A \cdot B) \cdot C$
$A + A = A$	$A \cdot (B + C) = A \cdot B + A \cdot C$
$A \cdot A = A$	$A + (B \cdot C) = (A + B) \cdot (A + C)$
$A + \overline{A} = 1$	$\overline{(A + B)} = \overline{A} \cdot \overline{B}$
$A \cdot \overline{A} = 0$	$\overline{A \cdot B} = \overline{A} + \overline{B}$

relations in Table 12.2 are easily proved using truth tables. A truth table shows all of the possibilities and if they all agree then the theorem is rigorously proved. For example, the proof for one of DeMorgan's theorems is shown in Table 12.3—the middle and last columns are the same for all possible input combinations, so they are equivalent. Such a proof may not seem elegant, but it is rigorous.

There are sixteen possible unique operations that can be defined that combine two binary values (A and B) to give a binary result. Some of these operations are trivial, but all of them can be expressed using the NOT, OR, and AND operations above. Hence, the Boolean operations form a complete set.[5] Table 12.4 shows all sixteen operations. Note that for the first six entries, the value of one or both of the arguments is immaterial—the result of the operation does not care.

The last two entries in Table 12.4 involve the so-called "exclusive or" (XOR). The exclusive OR more closely matches the use of the word "or" in everyday language. The result is true if *either* A or B is true, *but not both*. The usual Boolean OR, sometimes referred to as the "inclusive OR," will be true if either is true, *including* the case where both are true. For convenience, the special symbol "⊕" is sometimes used for the XOR function, though that operation is not really a part of the algebra. Be careful using XOR with Boolean algebra because XOR does not have all of the nice properties you might expect for a mathematical operation.

---

[5] In fact, one can write all of these operations using only the NAND or only the NOR operations. Hence, there is some redundancy to the operations. In terms of digital circuit design, this means there will often be many different ways to obtain the same result.

Table 12.3 Truth table proof for one of deMorgan's theorems—for all A and B, $\overline{A+B} = \overline{A} \cdot \overline{B}$

A	B	$A+B$	$\overline{A+B}$	$\overline{A}$	$\overline{B}$	$\overline{A} \cdot \overline{B}$
0	0	0	1	1	1	1
0	1	1	0	1	0	0
1	0	1	0	0	1	0
1	1	1	0	0	0	0

Table 12.4 The set of Boolean operations that are possible for two inputs

Name	Description of result
Null or False	0 (no matter what)
One or True	1 (no matter what)
A	A (i.e., B is ignored)
B	B (i.e., A is ignored)
NOT A	$\overline{A}$ (B is ignored)
NOT B	$\overline{B}$ (A is ignored)
A AND B	$A \cdot B$
A OR B	$A + B$
A AND (NOT B)	$A \cdot \overline{B}$
(NOT A) AND B	$\overline{A} \cdot B$
NAND of A and B	$\overline{A \cdot B}$
A OR (NOT B)	$A + \overline{B}$
(NOT A) OR B	$\overline{A} + B$
NOR of A and B	$\overline{A + B}$
XOR of A and B	$A \cdot \overline{B} + B \cdot \overline{A}$
XNOR of A and B	$A \cdot B + \overline{A} \cdot \overline{B}$

Where possible, the logic operations are usually pronounced as single words. XOR is pronounced with two syllables ("ex-or"). To be consistent, XNOR might be written NXOR, but that is too hard to pronounce—hence "ex-nor," with the X first, is used for the sake of pronounceability.

Digital Logic Circuits

Digital logic is implemented electronically using circuits designed to perform these Boolean operations. Hence, those electronic components are often referred to as "logic circuits" or "logic gates." Schematic symbols for the circuits that perform the basic operations are shown in the Table 12.5. As shown in this table, the inputs are supplied on the left and the output is on the right. Many of these circuits are also available with more than two inputs, in which case the description on the right applies. That is, an "eight-input NAND" circuit would produce F (false) only if all eight inputs were T (true).

Digital Logic Circuits

Table 12.5 Basic Digital Gate Schematics

Name	Schematic	Logic function
Buffer		Output equals input
Inverter		Output opposite of input (NOT)
OR		Output F only if all inputs F
NOR		Output T only if all inputs F
AND		Output T only if all inputs T
NAND		Output F only if all inputs T
XOR		Output T only if just one input T (Is F if both inputs same)
XNOR		Output T only if both inputs same

Notice the convention that in the schematic the "NOT" function is added using a small circle. A small circle on the input side would indicate that the "NOT" is to be performed on the input before the logical operation is performed. For example, consider the three schematics in Fig. 12.1 that have corresponding outputs $A \oplus B$, $\overline{A} \oplus \overline{B}$, and $\overline{A \oplus \overline{B}}$.

For digital logic circuits, the two logic states are associated with two voltages and/or currents. Most common is the use of 0 and 5 V or 0 and 3.3 V, with 0 V being "false" in both cases. More accurately, any voltage smaller than a certain maximum is considered "low" and any value higher than a certain minimum is considered "high." Thus, some truth tables will be written using H and L, rather than 1 and 0, corresponding to those voltage levels. Between high and low is a region where results are unpredictable. Digital logic circuits are produced in "logic families" that have matching electrical characteristics. For details on any particular circuit, refer to the manufacturer's datasheet for the device. When combining digital logic circuits, it is safest, though not always necessary, to stick to circuits from the same logic family.

Examples of one way to create logic functions using diodes and transistors are shown in Fig. 12.2. In 12.2a, a high input turns on the transistor, and so the output is low and a low input leaves the transistor off, so the output is high. That is the NOT function. In 12.2b if either or both of the inputs are low then the transistor is off, and the output will be high. This is the NAND function. And in 12.2c, if either input is high, the transistor is on, so the output is low. This is the NOR function. This diode-transistor logic (DTL) is easy to construct with discrete components, however its use within integrated circuits is very rare.

Fig. 12.1 Three examples of how the NOT can be shown in a schematic using a small circle. The NOT can be applied to the input or output. The three schematics correspond to the operations $A \oplus B$, $\overline{A} \oplus \overline{B}$, and $\overline{A \oplus \overline{B}}$, respectively

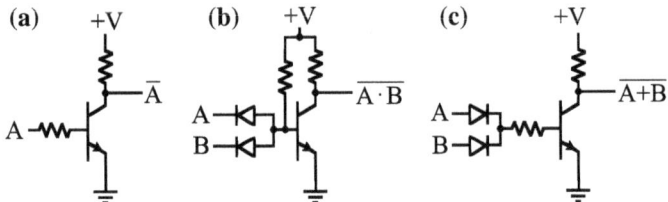

Fig. 12.2 Three examples of how simple digital circuits can be constructed using discrete diodes and transistors, (a) the NOT, (b) the NAND, and (c) the NOR

Note that power supply connections are rarely shown for digital logic circuits, but are necessary to make a circuit function. Logic that uses voltages levels will typically use those same levels for the power connections. That is, if logic levels are 0 and 5 V, then the power supply connections are also 0 and 5 V. When power supply connections are shown on a schematic, they are often shown completely separate from the logic circuitry.

Combinations of Digital Logic Gates

When logic gates are connected together, the resulting output can be determined using Boolean algebra or by writing out a truth table. Three examples follow demonstrating how this is accomplished.

Example 1—Solving with Boolean Algebra

Consider the circuit of Fig. 12.3. Boolean algebra can be used to figure out what it does. When solving, it often helps to include some of the intermediate states, as is shown in the figure. Also observe that this particular circuit is symmetric so that it should be the case that swapping the inputs A and B should not change the output value S.

Consider the inputs to the last NOR gate and (carefully) make use of DeMorgan's theorems:

Combinations of Digital Logic Gates

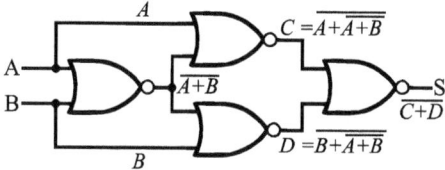

Fig. 12.3 Circuit for Example 1. Intermediate values have also been labeled

$$C = \overline{A + \overline{A+B}} = \overline{A} \cdot \overline{(A+B)} = \overline{A} \cdot (A+B) = \overline{A} \cdot A + \overline{A} \cdot B = \overline{A} \cdot B. \quad (12.1)$$

Similarly,

$$D = \overline{B} \cdot A, \quad (12.2)$$

and so, again using DeMorgan's Theorems:

$$\begin{aligned} S = \overline{C+D} = \overline{\overline{A} \cdot B + \overline{B} \cdot A} = \overline{\overline{A} \cdot B} \cdot \overline{\overline{B} \cdot A} \\ = (A+\overline{B}) \cdot (B+\overline{A}) = A \cdot B + \overline{A} \cdot \overline{B}. \end{aligned} \quad (12.3)$$

The final result is that of the XNOR operation—the output is true (i.e., 1) if both inputs are the same, and is false (i.e., 0) otherwise. Note that the symmetry predicted is indeed present in the answer.

Suppose there were three inputs. What circuit could be used to check to see if all three inputs are the same? How might such a circuit be constructed using only 2-input NOR logic gates? (Consider the statement "if A and B are the same, and B and C are the same, then A, B and C are all the same.")

Example 2—Solving with a Truth Table

Consider the digital circuit in Fig. 12.4. To figure out what it does, create a truth table. The intermediate value "D" is defined and is included in the table for

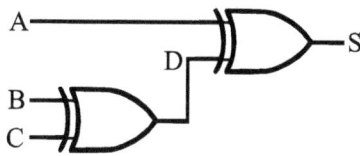

Fig. 12.4 Circuit for Example 2

Table 12.6 Truth table for Example 2

Inputs				Result
A	B	C	$D = B \oplus C$	$S = A \oplus D$
0	0	0	0	0
0	0	1	1	1
0	1	0	1	1
0	1	1	0	0
1	0	0	0	1
1	0	1	1	0
1	1	0	1	0
1	1	1	0	1

convenience. Recall that each exclusive OR gives a true when its inputs are different, and a false when they are the same. Table 12.6 shows resulting truth table, worked out one row (i.e., one set of inputs) at a time.

It is always desirable, though not always possible, to try to put into simple words what a circuit does. Such a description may not be unique. One description for this circuit would be that "the output is 1 if the number of non-zero inputs is odd, and 0 if even." Whether or not the number of 1's in a sequence of binary digits is even or odd is known as the "parity." Hence, this circuit tells you if the parity is even ($S = 0$) or odd ($S = 1$) for these three inputs. How can this circuit be generalized to determine the parity of N inputs?

Example 3—Solving Both Ways

It may not be obvious what function is performed by the digital circuit in Fig. 12.5, however it is straightforward to create a truth table. Since there are three inputs, there should be $2^3 = 8$ rows (plus the header). Intermediate values (D, E, and F) are included to help reduce errors. The resulting truth table is shown in Table 12.7. It can be seen that G, the result, is the opposite of C for all rows except the second, which is one of the two rows where both A and B are 0. Or to phrase it another way, the output is true if A and B are both false or if C is false. Thus, translating those words into the algebra, the solution should be

$$G = \overline{C} + \overline{A} \cdot \overline{B} = \overline{C} + \overline{A+B}. \tag{12.4}$$

The result can, of course, also be derived using Boolean algebra. In fact, one good way to ensure a solution is correct is to do it both ways. The result better be the same. The algebraic steps for this example are

$$D = \overline{A+B} = \overline{A} \cdot \overline{B}; \ E = C \cdot \overline{B} + \overline{C} \cdot B \tag{12.5a}$$

Combinations of Digital Logic Gates

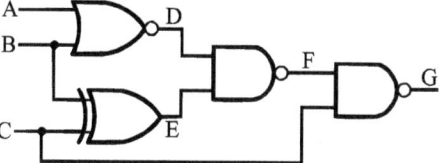

Fig. 12.5 Circuit for Example 3

Table 12.7 Truth table for Example 3

Inputs						Result
A	B	C	D	E	F	G
0	0	0	1	0	1	1
0	0	1	1	1	0	1
0	1	0	0	1	1	1
0	1	1	0	0	1	0
1	0	0	0	0	1	1
1	0	1	0	1	1	0
1	1	0	0	1	1	1
1	1	1	0	0	1	0

$$F = \overline{D \cdot E} = \overline{\overline{A \cdot B} \cdot \left(C \cdot \overline{B} + \overline{C} \cdot B \right)}$$
$$= \overline{\overline{A} \cdot \left(\overline{B} \cdot B \right) \cdot C + \overline{A} \cdot \left(\overline{B} \cdot B \right) \cdot \overline{C}} = \overline{\overline{A} \cdot \overline{B} \cdot C} \quad (12.5\text{b})$$

$$G = \overline{C \cdot F} = \overline{C \cdot \overline{\overline{A} \cdot \overline{B} \cdot C}}$$
$$= \overline{C \cdot ((\overline{A} \cdot \overline{B}) + \overline{C})} = \overline{C \cdot (A+B) + C \cdot \overline{C}} \quad (12.5\text{c})$$
$$= \overline{C + A + B} = \overline{C} + \overline{A} \cdot \overline{B}$$

DeMorgan's theorems were used numerous times, as well as some other identities from Table 12.2. A particular challenge for this example is keeping careful track of the operation order and the NOTs inside NOTs inside NOTs. The final result suggests that the same function could have been obtained using a much simpler circuit. Boolean algebra can be a useful tool to search for alternate and/or simpler designs that perform the same logical function.[6]

---

[6]Another method to simplify logic circuits uses a so-called Karnaugh map. It is easy to find many articles about how to use these maps.

Equivalent Circuits

With digital circuits, there is always more than one way to achieve a desired function. This can be seen by considering the examples in Fig. 12.6 constructed entirely using 2-input NAND gates, but that produce other (simple) functions (NOT, OR, AND). Since all of the basic operations can be constructed entirely from 2-input NANDs, any digital circuit can, in principle, be constructed solely from 2-input NANDs.

As another example consider the construction of the XOR function. The basic relationship for XOR is

$$A \oplus B = A \cdot \overline{B} + \overline{A} \cdot B. \tag{12.6}$$

The circuit of Fig. 12.7a follows immediately from the right-hand side. Other equivalent circuits can be discovered using Boolean algebra. One "trick" to use is that any expression that is always false can be ORed with another expression without changing its value, in the same way zero can always be added to a numerical equation. Starting with that idea, using the distributive property and one of DeMorgan's theorems, an alternate XOR circuit is found as follows:

$$\begin{aligned} A \oplus B &= A \cdot \overline{B} + B \cdot \overline{A} = A \cdot \overline{A} + A \cdot \overline{B} + B \cdot \overline{A} + B \cdot \overline{B} \\ &= A \cdot (\overline{A} + \overline{B}) + B \cdot (\overline{A} + \overline{B}) \\ &= (A + B) \cdot (\overline{A} + \overline{B}) = (A + B) \cdot \overline{(A \cdot B)}, \end{aligned} \tag{12.7}$$

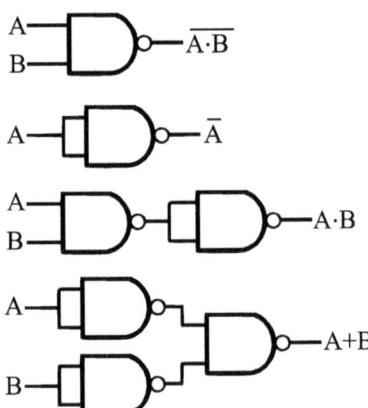

Fig. 12.6 Four examples showing how NAND circuits can be used to produce other basic functions. From top to bottom the results are NAND, NOT, AND, and OR

Equivalent Circuits

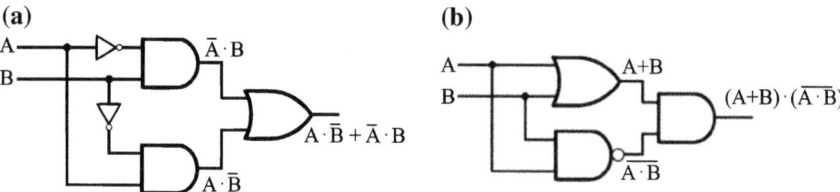

Fig. 12.7 Two examples showing the construction of the XOR function using basic Boolean operations

that is constructed as shown in Fig. 12.7b. It is also possible to make an XOR in many other ways. Suppose it is desired to make an XOR using only 2-input NORs. Start the algebra as before but work to turn all ANDs into ORs using DeMorgan's theorems. Then make sure there is at least one bar over them for the NOT. That is, starting the algebra as before,

$$A \oplus B = (A+B) \cdot (\overline{A} + \overline{B})$$
$$= \overline{\overline{(A+B)} \cdot \overline{(\overline{A} + \overline{B})}} = \overline{\overline{A+B} + \overline{\overline{A} + \overline{B}}}. \quad (12.8)$$

The far right-hand side can be implemented using only NORs. In fact, all the basic operations can be achieved using 2-input NORs, and if desired, any digital circuit can be constructed solely with NORs.

Gates Versus Logic Functions

With digital circuits, a single circuit can appear to have radically different functions when viewed in context. Consider the examples in Fig. 12.8. The first (12.8a) is a simple logical combination of two inputs that appear to be treated equally. The same circuit is shown in 12.8b, but labeled with a "control input" (cntrl) that either allows D to get through (when cntrl is true, then D' is equal to D) or forces the output to be 0 (when cntrl is false), thus blocking the data D from getting through. In this context, the inputs may not appear to be treated equally. A simple XOR circuit is shown in 12.7c where $D' = D$ if cntrl is 0, and $D' = \overline{D}$ if cntrl is 1. When used in this way, the XOR is called a "controlled NOT," or cNOT gate. The quantum cNOT gate is very important for quantum computing.

When presented as in Fig. 12.8b, c, the digital circuits are acting like gates that either let the data through, block it, or invert it. When used in such a manner, the "logic" does not seem as important as does the control function, though the electronics is identical. In practice, logic gates are most often, in fact, used to perform functions that are not based in logic, though they can be analyzed using logic. After all, digital logic gates are wired to perform some specific (electronic) function, not to ponder philosophy.

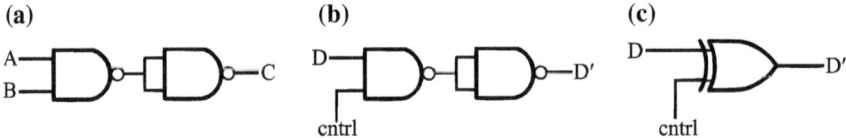

Fig. 12.8 The interpretation of the function of a circuit can depend on context. At (**a**) is a simple logic application. In (**b**) the same circuit appears to be used for control functions. In (**c**) the XOR is being used as a controlled NOT

Decoders and Encoders

Decoders and Encoders are often used to efficiently send data from one place to another, usually for control, input or display purposes. Decoders take a combination of inputs and produce a desired outcome based on that combination.

An example of an "active line low" decoder is shown in Fig. 12.9 and the function is shown in the truth table, Table 12.8. The indices have been chosen so that the two inputs can be combined mathematically to get a label (i.e., 2 × B + A) corresponding to the output that is forced low. Thus, the signals on two wires can be decoded to select one of four output lines.

While the circuit and wiring may get somewhat more complicated it is not difficult to increase the number of inputs. Consider such a decoder with n inputs. Since there are 2^n possible combinations of inputs, 2^n outputs can be used. The advantage for large n is obvious—with 10 wires you can control the output on $2^{10} = 1024$ wires.

Another type of decoder is used for displaying numbers using a 7-segment display. Refer to Fig. 12.10. Here there are four inputs[7] and 7 outputs. Six of the sixteen possible input combinations are not needed for this task and may or may not produce a meaningful display. The seven output lines are used to selectively turn on one of seven bars of the display. Should it be necessary, the first step to build such a decoder is to look at each segment in the display, one at a time, and write out a truth table for that segment. In some cases, it might be better to determine when the segment is off, rather than on. For example, segment a is on for all digits except 1 and 4, segment b is on for all except 5 and 6, and so on. Thus, it might be simpler to construct the logic to turn segments off, rather than on. The internal workings for such a decoder will not be reproduced here. Fortunately, inexpensive integrated circuit 7-segment decoders can be purchased that perform this function.

An encoder performs the inverse function from that of a decoder. The truth table will look similar to that for the decoder, but with inputs and outputs swapped. A simple 4-input encoder is shown in Fig. 12.11 with its truth table in Table 12.9. In this case a high (1) on *one* of the inputs causes a corresponding pair of binary values to appear on the outputs. An additional output, S, is included to indicate that

---

[7]Since there are 10 digits, and 3 inputs yields only 8 possibilities, there must be at least 4 inputs.

Decoders and Encoders

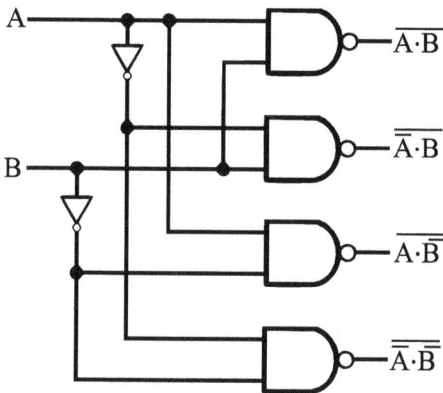

Fig. 12.9 Circuit for a simple decoder

Table 12.8 Truth table for decoder circuit

B	A	O₀	O₁	O₂	O₃	2B + A
0	0	0	1	1	1	0
0	1	1	0	1	1	1
1	0	1	1	0	1	2
1	1	1	1	1	0	3

Fig. 12.10 Numerical digits are often displayed using a 7-segment scheme. A decoder circuit will take a combination of four inputs and will activate the appropriate segments for each digit

one (or more) of the input lines is high—S provides an indication that an input has occurred.

To shorten the truth table, and to more easily understand the circuit, the symbol "X" is introduced to indicate a value that can be either 0 or 1 with no change in the output (given the other input values). That is, for the circumstances shown in that line of the truth table, that input does not matter. The "X" is sometimes referred to as "Don't Care." With that shorthand, the sixteen possible input combinations for the encoder can be written using many fewer rows, and the table is easier to read and understand.

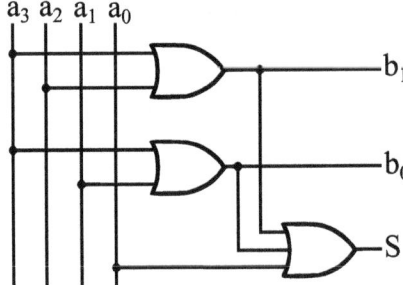

Fig. 12.11 A circuit for a four input encoder, where the two output lines will contain a code for which input line is high

Table 12.9 Truth table for 4-line encoder

a_3	a_2	a_1	a_0	b_1	b_0	S	$2b_1 + b_0$
1	X	X	X	1	1	1	3
0	1	0	X	1	0	1	2
0	0	1	X	0	1	1	1
0	0	0	1	0	0	1	0
0	0	0	0	0	0	0	0
0	1	1	X	1	1	1	3

One possible problem with this simple encoder occurs when two or more of the inputs are high at the same time. The corresponding decoder had only four possible states, but this encoder has sixteen. For this example, in all but one case the output with the larger index will appear at the outputs. That is, if a_3 and a_1 are high together, a_1 is ignored and the result is as if only a_3 were high. With a small amount of additional circuitry, the one remaining case (if a_1 and a_2 are simultaneously high, as seen on the last line of the truth table) can be "fixed" so that the output always represents that for the highest value index. Such an encoder that, by design, resolves all such conflicts in one way or another would be referred to as a "priority encoder."

One use of an encoder/decoder combination would be the efficient transmission of data. For example, 2^N separate inputs can be compressed into N data lines for transmission, then turned back into 2^N separate outputs at the other end. Decoders and/or encoders can also be used in pairs to create a "row/column" set-up used in "multiplexing." As will be seen in the next example, this is an efficient way to wire (for example) some displays. An example showing how this is used for a modern keyboard is shown later.

Multiplexing

Multiplexing allows many devices to use some common electronics and can result in simpler circuitry and significantly lower cost.

Consider the circuit in Fig. 12.12 used to illuminate 1 of 16 LED's. In this schematic, the convention used is that wires that cross are not connected unless there is a solid circle (a 'blob of solder') at the crossing point.

For this circuit, four values determine the row and column of the LED to illuminate. The two decoders are functionally the same as the one described in Table 12.8, though one is wired for selected line high rather than low. For any combination of the four inputs, only one LED will have a high level on the anode (the row) and a low level on the cathode (the column), and that one LED will light up. The resistors (which may not be necessary in some circumstances) limit the current (typically 5 to 15 mA for normal LED's). While only one LED can be illuminated at a time, if multiple LED's are illuminated in rapid sequence, they will appear to a human observer to be simultaneously illuminated.

The row and column design make it possible to "address" 2^N things (in this case LED's) using N inputs. While the savings going from 4 to 16 may not seem so impressive, imagine if $N = 16$, requiring 16 inputs (each with a wire) that allows $2^{16} = 65,536$ things to be addressed that might be, for example, "64k" memory locations—a trivial amount of memory but far from a trivial number of wires.[8]

Flip-Flop Circuits

The circuit of Fig. 12.13a seems to be showing a form of circular logic. There are two inputs, R and S, and two outputs Q and Q', however the outputs are also being used as inputs. This circuit is usually presented as shown in Fig. 12.13b, where the lower NOR gate has been twisted around so all the inputs are on the left and the outputs on the right.

A NOR will produce a false (0) if any input is high. Hence, if R is high, then Q is low, and if S is low, then Q' is high. Likewise, if S is high, Q' is low, and if R is low, Q is high. The remaining state, when both R and S are low, is where it becomes interesting. Consider what happens for that case if Q is high. In that case Q' must be low, which makes Q high, as was assumed. All is well. However, if instead it is assumed that Q is low, then Q' must be high, which forces Q to be low. All is well with the opposite assumption also. So, is Q high or low? Logically, there are two valid, though contradictory, solutions. The resulting unresolved truth table is shown in Table 12.10.

---

[8]In this context, the number 1k is $2^{10} = 1024$, that is close to, but slightly larger than, 1000. Hence, 64k is also slightly larger than 64,000.

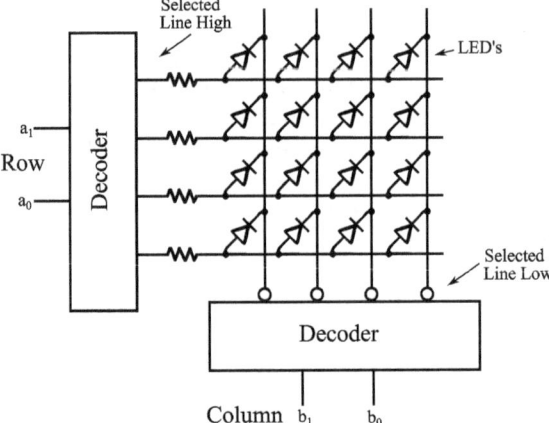

Fig. 12.12 Two decoders in a row-column arrangement can be used to choose one of sixteen LED's to illuminate based on four input lines

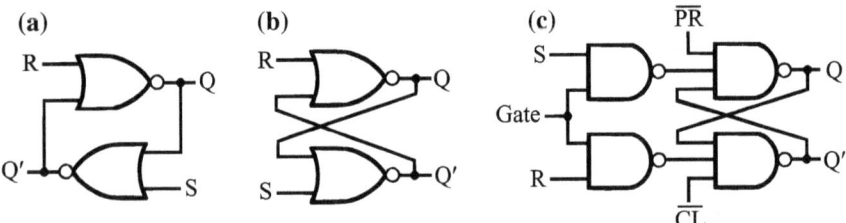

Fig. 12.13 A simple RS flip-flop is shown in (**a**) where circular logic is clearly shown. The circuit in (**b**) is identical to that in (**a**), except all the inputs have been moved to the left. In (**c**) an extra control input, the gate, is added to make a gated RS flip-flop

Table 12.10 Truth table for Fig. 12.12a

R	S	Q	Q'
1	0	0	1
0	1	1	0
1	1	0	0
0	0	?	?

A solution for a real electronic circuit is found by noting that before R and S were both low, they were something else. If those previous values were 0 and 1, or 1 and 0, the solution remains valid when they both R and S become 0. Hence, the outputs stay what they were before. The previous state is stored. The case where both R and S are 1 followed by both R and S 0 leads to unpredictable results. In any real circuit, both will not switch simultaneously—even a few nanoseconds of

Flip-Flop Circuits

difference is enough—and even then the output state with both Q and Q' zero is not a stable solution. The circuit will end up in one of the stable states. It is not obvious which one. Hence, going from both inputs true directly to both inputs false should be avoided.

Circuits such as in Fig. 12.13a, b, that have two stable states that can be stored, are known as flip-flops. This particular circuit is known as an RS flip-flop. If you think of Q as being the principle output, then S serves to set the value (to 1) and R serves to reset the value (to 0). Aside from the unstable situation when both R and S are 1, Q' is always the opposite of Q. While RS flip-flops are not all that common as separate devices in circuits, they serve as the building block for flip-flop circuits that are. Flip-flops are an example of sequential logic, in that the output does not only depend on what the inputs are now, but also on the sequence of events that occurred before "now."

With the addition of a control function out front, the RS flip-flop can be enabled and disabled (with a "gate" or "clock" signal). In that case, the set and reset functions will only occur if the gate signal allows them through. In practice, the gate will be used to synchronize actions to an external signal. The inputs (R and S) are set to their desired values before the gate is opened, then the gate opens and lets them through at a prescribed time. On the other hand, an overall set or reset might be desired that should override such a gate, so extra inputs can be brought out for those purposes.

Figure 12.13c shows a "gated RS flip-flop with preset and clear" constructed with NANDs. The preset and clear functions set the output to 1 or 0. Note that in this case the functions (or their abbreviations) have a bar over them in the diagram. The bar indicates that a low (0 or false) input preforms the desired action. For example, to preset the flip-flop, a low level is applied to the $\overline{PR}$ input—if $\overline{PR}$ is made false, then PR is true. It is a funny way to use negatives, to be sure, but that is a standard practice. The corresponding truth table for this circuit is shown in Table 12.11. The symbol Q_0 here is used to designate the previous value of Q—that is, the starting value of Q when the current state began. There are still some states that are "unstable" and should be avoided.

Two possible solutions to deal with the unstable states of the RS flip-flop are the so-called JK flip-flop and the D flip-flop. The JK flip-flop ultimately creates an alternative output to replace the unstable state, while the D flip-flop limits the inputs so the unstable state cannot be reached.

A JK flip-flop is shown in Fig. 12.14a. The circuit may also have preset and clear functions that are not shown. The JK flip-flop is essentially the gated RS flip-flop with the addition of the connections between the outputs and the inputs. Here J is the set and K is the reset input. The feedback from the outputs is such that the set function is disabled if the circuit is already set, and the clear function is disabled if it is already cleared. Hence, only one input will be effective at a time. An interesting thing happens if both inputs are true—since the only input that is active is the one that does not match the current output, the output will switch, or "toggle" to the other state. The truth table for this JK flip-flop is shown in Table 12.12. There

Table 12.11 Truth table for Fig. 12.12c

$\overline{PR}$	$\overline{CL}$	Gate	S	R	Q	Q'
0	1	X	X	X	1	0
1	0	X	X	X	0	1
1	1	0	X	X	Q_0	$\overline{Q_0}$
1	1	1	1	0	1	0
1	1	1	0	1	0	1
1	1	1	0	0	$\overline{Q_0}$	$\overline{Q_0}$
0	0	X	X	X	Unstable	
1	1	1	1	1		

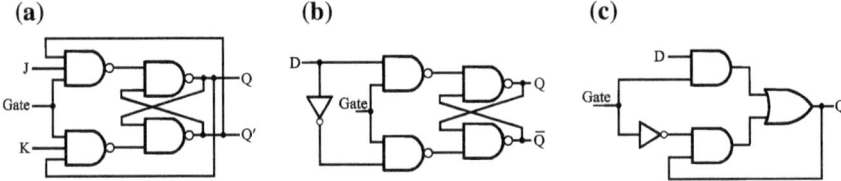

Fig. 12.14 Extensions of the RS flip-flop include the gated (a) JK flip-flop and (b) the D flip-flops. An alternative D flip-flop that uses fewer logic circuits is shown in (c)

Table 12.12 Truth table for Fig. 12.13a

Gate	J	K	Q	Q'
0	X	X	Q_0	$\overline{Q_0}$
1	0	0	Q_0	$\overline{Q_0}$
1	1	0	1	0
1	0	1	0	1
1	1	1	$\overline{Q_0}$	Q_0

Table 12.13 Truth table for Fig. 12.13b, c

Gate	D	Q
0	X	Q_0
1	0	0
1	1	1

is a problem in that the gate can only be kept high for a very short time, otherwise the output will continue to toggle back and forth as long as the gate is open. This issue will be resolved shortly.

A D flip-flop is shown in Fig. 12.14b. Here the two inputs, R and S, are replaced with one input, and opposite values are applied to the two inputs of the RS flip-flop, forcing $R = \overline{S}$. The unstable state cannot happen and it will always be the case that

$Q' = \overline{Q}$. The D flip-flop is a basic storage element for data (D). The truth table for the D flip-flop is shown in Table 12.13. An alternative, and somewhat simpler circuit that performs the same function is shown in Fig. 12.14c. Preset and clear functions can be added to either of these circuits, if desired.

Edge-Triggered Flip-Flops

A very useful device is constructed using two flip-flops in series. This will be illustrated using the D flip-flop, though a similar result can be obtained using other flip-flips. The details of the construction of the flip-flop are not important at this point, so the D flip-flop is represented by a simple box with the inputs and outputs labeled appropriately. The two flip-flops in series are connected as shown in Fig. 12.15. Due to the inverter, only one of the flip-flops can have its gate open at any given time.

Each of the D flip-flops have the property that while the gate is open, the output (Q) follows the input (D). If the gate is closed, the output is the stored value of D that was the last to occur before the gate closed. For the circuit of Fig. 12.15, when the clock is high, the first flip-flop has its gate open and its output follows the input. That output, Q_i, is presented to the second flip-flop, that ignores the value since its gate is closed. When the clock goes low, the first flip-flop stores and continues to output the last value it saw at the input. At the same time, the second flip-flop now opens its gate, and so its output follows its input. Since the second flip-flop's input is now the stored value from the first flip-flop, the output from the second flip-flop becomes the value of the input at the time the clock changed from high to low. If the clock goes back high, the output from the second flip-flop does not change. Only a transition from high to low can affect a change.

The truth table, that is, a description of what happens, for the circuit of Fig. 12.15 is shown in Table 12.14. Here the down arrow, ↓, indicates a change in value from high to low. Such a device is referred to as "edge-triggered." In contrast, the original flip-flop might be referred to as "level-triggered." It is certainly possible to modify the circuit to trigger on the rising edge (↑).[9] An edge triggered flip-flop appears to grab the input at a particular time, though internally it is really a hand-off between two flip-flops. An edge triggered JK flip-flop is constructed in a similar manner, eliminating the problem of having long gate signals when in toggle mode, mentioned above.

When the internal construction is not of concern, an edge-triggered D Flip-flop would appear in a schematic something like what is shown in Fig. 12.16. That it has an edge-triggered input is indicated with a small triangle. The circuit on the left (12.16a) is triggered on the rising edge (low to high or "going true") while the circuit on the right (12.16b) is triggered on the falling edge (high to low or "going

---

[9]In a truth table, an image representing a falling or rising edge may be used instead of an arrow.

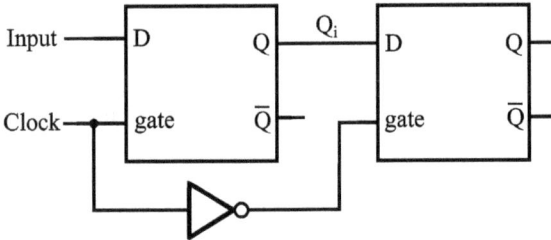

Fig. 12.15 Two gated D flip-flops in series, with an inverted gate input, can produce an edge-triggered D flip-flop, where the output is loaded only when the gate changes

Table 12.14 Truth table for Fig. 12.14

D	Clock	Q
X	1	Q_0
X	0	Q_0
1	↓	1
0	↓	0

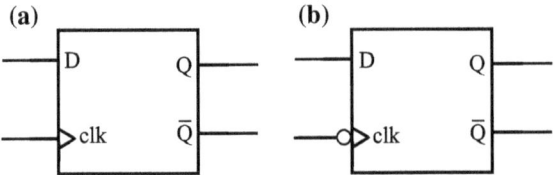

Fig. 12.16 Schematic representations of edge-triggered D flip-flops. At (a), the rising edge is used and in (b) the falling edge. Such flip-flops may have additional inputs for preset and clear functions

false"), as indicated by the "not" circle on the input. If the device also has preset, clear, or other functions, they are simply added to the box with appropriate labels. These edge-triggered flip-flops are available in various configurations as integrated circuit devices.

A Directional Electric Eye

A simple electric eye detects the presence of an object between a light source and a light detector. The circuit of Fig. 12.17a uses two such detectors and an edge-triggered flip-flop to also determine the direction of motion of the object. In this circuit, the light detector is a phototransistor that conducts when there is incident light. Hence, in the absence of an object the transistors are on and both

A Directional Electric Eye

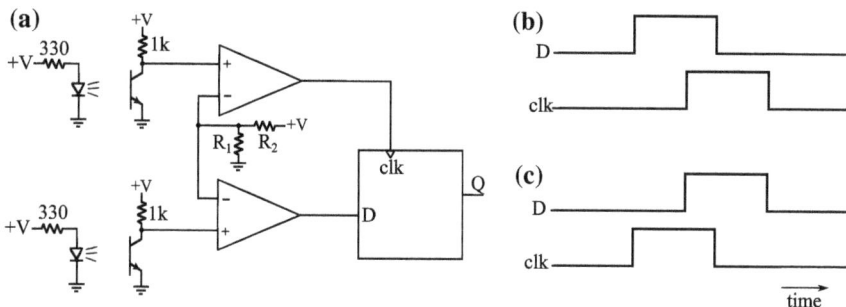

Fig. 12.17 Two optical detectors and an edge-triggered flip-flip can be used to detect the passage of an object and be able to determine the direction of motion

comparators will output a low. Assuming the object is larger than the spacing between the two sensors, the output of the two comparators as the object passes will be as shown in 12.17b or 12.17c, depending on direction. In 12.17b the object is traveling up the page, while in 12.17c it is going down the page. The flip-flop will store the value of the D input when the clock goes from low to high. In 12.17b, D will be high, and in 12.17c, D will be low at the time when the clock goes high.

Such circuitry is common in the design of early mechanical computer mice and trackballs. A wheel with periodic openings is placed between the sensors. The opaque region between the openings acts as the object. Similar designs are still used with some optical rotary (or shaft) encoders. Measuring the frequency of the pulses from either of the two detectors, plus the output of the flip-flop, provides a measure of the speed and direction of rotation. A much simpler application for this circuit is to measure the period of a pendulum. As the pendulum swings through the detectors, one pulse occurs (from the flip-flop) for each full swing.

Combinations of Flip-Flops

The D Flip-Flop is a very simple memory since it can store a single previous value, a "bit," indefinitely. An array of flip-flops can obviously be used to store more data. In addition, there are a number of interesting circuits based on interconnected edge-triggered flip-flops.

Shift Register

A particularly simple application, illustrated in Fig. 12.18, is the so-called shift register. The output of each flip-flop is connected as the input to the next. When the clock makes its transition, in this case low to high, the data moves one flip-flop to the right. That is, Q_4 changes to the previous value of Q_3, Q_3 changes to the previous value of Q_2, and so on. In addition to the applications discussed below, a shift has important applications in serial communications and multiplication, both discussed later. Shift registers are available as integrated circuit devices.

A shift register with feedback from the last stage forms a ring counter. If the feedback is inverted, as shown in Fig. 12.19, then it is referred to as a Johnson counter. If such a counter starts with all zeros (or all ones) and has n flip flops, it will divide the clock frequency by $2n$. Such a circuit has a potential for 2^n different output states, and is using only $2n$ of them. Provisions are usually included to ensure that the counter starts in a known state (e.g., all zeroes), or if it gets into a state it is not supposed to, it will be set to a state that is ok.

A shift register with "taps" to provide its own input from a combination of several of the later stages can generate useful sequences. These are examples of one type of "linear feedback shift register" (LFSR). For example, using the correct taps XOR'd (or XNOR'd) together from N flip-flops that are fed back to the beginning will lead to a "maximal length sequence" of $2^N - 1$ distinct values having statistics similar to those of random numbers. The sequence is completely determined and will repeat, however. Such a sequence is referred to as "pseudo-random." Correct tap positions for such a sequence can be found in published tables.[10] Figure 12.20 shows such a circuit with $N = 4$ that will produce 15 pseudo-random values before repeating. If a similar circuit with 31 flip-flops is constructed, with taps at the 24th and 31st flip-flops, a sequence of just over 2×10^9 pseudo-random 0's and 1's is produced before repeating.

The initial state of each flip-flop is set using a "seed" value using "preset" or "clear" inputs on the flip-flops (not shown in the schematic). For the circuit of Fig. 12.19 the only bad seed is all ones, that simply reproduces itself indefinitely. Pseudo-random number generators and maximal length sequences are an important component of spread-spectrum communications used by GPS (and other) satellites, some cell phones and other wireless technologies, as well as for many games. Random values can also be used as a tool to simulate various random processes in the physics lab.

Binary Counter

If an edge-triggered D flip-flop is connected so that the inverted output ($\overline{Q}$) is wired to its own data input, then with each clock pulse the output will toggle to the other state.[11] In Fig. 12.21 several such D flip-flops are connected so that the output (Q) of each flip-flop is connected to the clock input of the next. Thus, when the

---

[10]There are always an even number of taps, and, numbered starting from 0, the tap positions are "setwise prime," meaning no common factor, other than 1, divides all of them.

[11]Alternatively, a JK flip-flop in toggle mode can be used.

Combinations of Flip-Flops 269

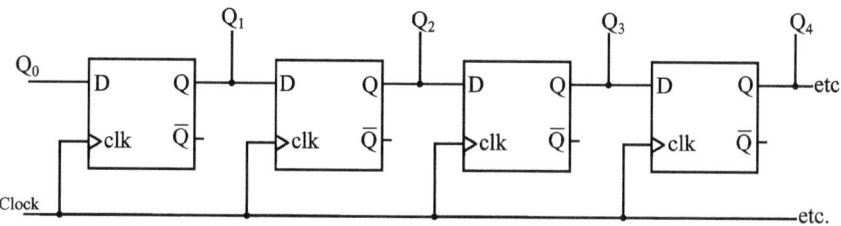

Fig. 12.18 Edge-triggered D flip-flops connected in series, where the output of each is connected to the input of the next, can form a shift register

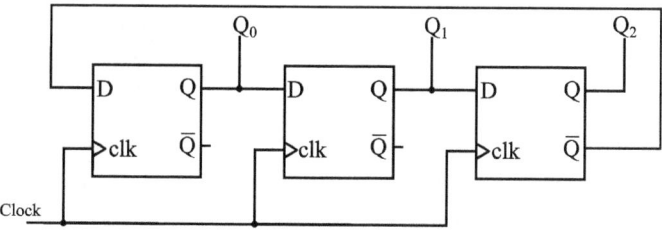

Fig. 12.19 A shift register where the output of the last stage is connected back to the first forms a ring counter. If the inverted output of the last stage is used, as shown, it is sometimes referred to as a Johnson ring counter

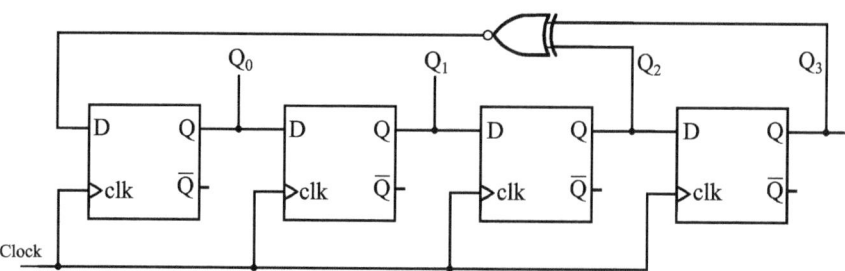

Fig. 12.20 A shift register where the output of several stages are combined with an XOR or XNOR can produce a pseudo-random number generator if the correct tap points are used. Such a circuit with N flip-flops will cycle through $2^N - 1$ states before repeating

output Q_0 goes from high to low, the output Q_1 will toggle, and when Q_1 goes from high to low, Q_2 will toggle, and so forth. Figure 12.22 illustrates the sequence of outputs starting from all low (0's). In the process, all $2^4 = 16$ output states are encountered for the four flip-flops, before the cycle repeats.

As a frequency divider, a circuit with N flip-flops can easily divide the clock frequency by a factor of 2^N. If clear inputs are also available, additional logic can reset the values back to all zeros at any point, allowing division by any integer value up to, and including 2^N.

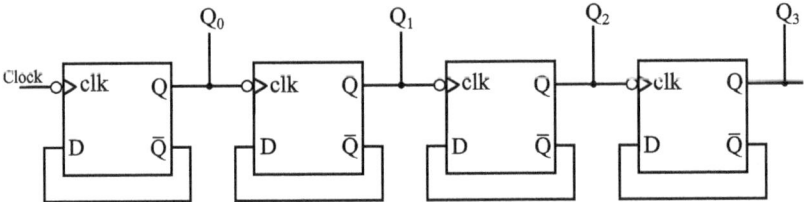

Fig. 12.21 A series of edge-triggered D flip-flops, where the output of each is connected to the clock input of the next, can be used to create a counter circuit that runs through all the possible output states

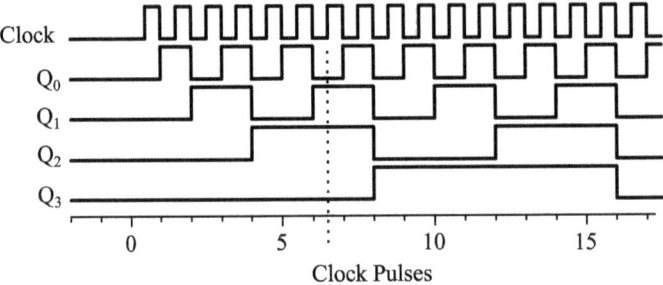

Fig. 12.22 The outputs for the circuit of Fig. 12.20 as they might be seen on an oscilloscope

If a high is taken to be 1 and a low 0, then the number of pulses, n, that occurred (since the outputs were last all zero) can be computed from the outputs using

$$n = Q_0 + 2 \times Q_1 + 4 \times Q_2 + 8 \times Q_3$$
$$= \sum_{k=0}^{N-1} 2^k Q_k. \tag{12.9}$$

For example, at the time indicated by the vertical dotted line in Fig. 12.22, $Q_0 = 0$, $Q_1 = 1$, $Q_2 = 1$, and $Q_3 = 0$, so $n = 6$. Note that in order for this simple scheme to work, when labeling the output lines, the first is "0," and not "1." The circuit is counting the pulses using base-2, also commonly referred to as "binary," and so the circuit is referred to as a binary counter. Various codes, including base-2 binary, will be discussed more in the next chapter. Binary counters are available as integrated circuit devices. Similar circuits designed to reset on the tenth pulse will count from 0 to 9 and are referred to as decade counters and are also available.

Combinations of Flip-Flops 271

Since the binary counter with N flip-flops runs through all 2^N possible combinations of N 0's and 1's, it can be used for the scanning keyboard circuit in Fig. 12.23.[12] This circuit is similar to the decoder of Fig. 12.12, except it is used for input. Clock pulses cause the counter to run through all of the possible outputs—in this case all of the possible row and column combinations. The decoders here are selected line low. If the (pushbutton) switch between the row and column wire is pressed (closed) then a low is sent to the two-input OR below. If the corresponding column line is also low, then that OR produces a false, causing the four-input NAND to output true. That stops the clock from getting to the counter, freezing the counter, and signals that a key has been pressed. At that point, the output of the counter can be used as a code that can be used to determine which key was pressed.

This type of scanning keyboard circuit, with a few extra embellishments, is what is commonly used for modern computer keyboards. It is a priority encoder in that whichever key is encountered first stops the process, and so pressing two keys at once does not cause an electrical conflict. In practice, the scanning process is done very quickly so that the keyboard response appears to be instantaneous to a human user.

Other Non-logical Applications

All logic circuits have some propagation delay—a time between when the inputs are set and when the output is valid. For modern digital electronics that may be just a few nanoseconds, though can be longer for some low-power devices. While all applications might need to consider this delay in the design, there are a few applications that use the delay to make them work.

Very Short Pulse Generator

The circuit in Fig. 12.24a exploits the natural propagation delay of a logic gate to create a small time delay between the inputs to an XOR gate. If the gates had no delay, the output of the final XOR would always be low since the inputs would always be equal. However, the delays cause the inputs to be unequal for a very short time, giving rise to a very short pulsed output, perhaps just a few nanoseconds long, whenever the input switches. The circuit of Fig. 12.24b uses external components to exaggerate the delays between input and output and can be used for somewhat longer pulses if one uses high input impedance (e.g., CMOS) digital gates.

---

[12] Any circuit that runs through all the possible outputs can be used, the binary counter often being a convenient choice.

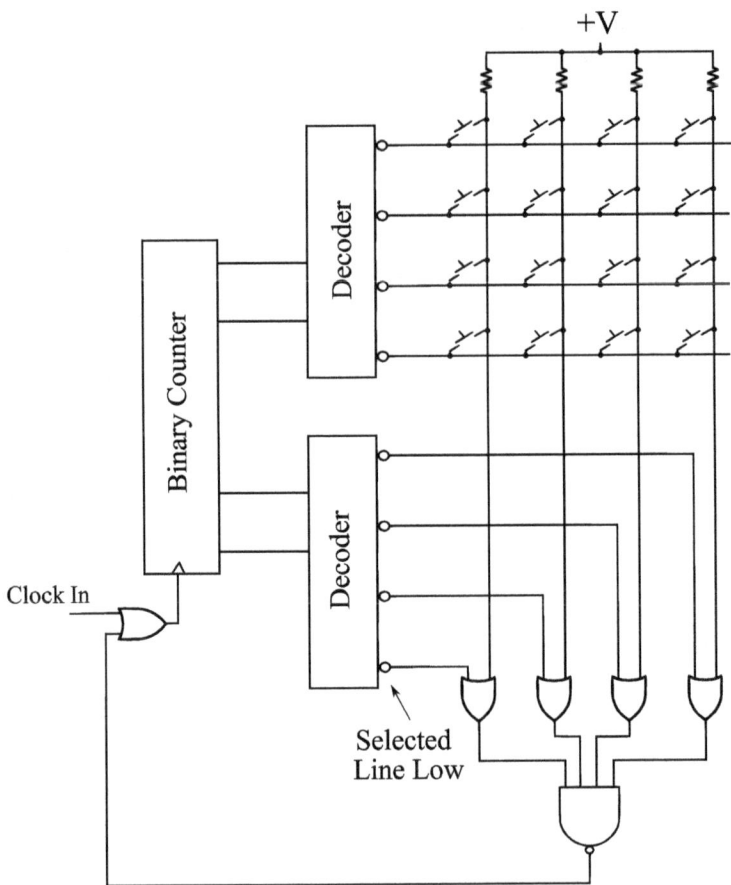

Fig. 12.23 A circuit that runs through all the possible outputs, such as a binary counter, can be used with decoders to form a scanning keyboard that checks for all the possible key presses one at a time

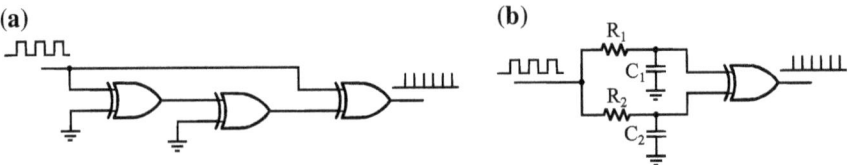

Fig. 12.24 The small delay between input and output, the propagation delay, can be used to create short pulse from longer pulses. In (**a**) the intrinsic delay (typically measured in ns) of an XOR is used whereas in (**b**) extra components have been added to enhance the delay

Other Non-logical Applications

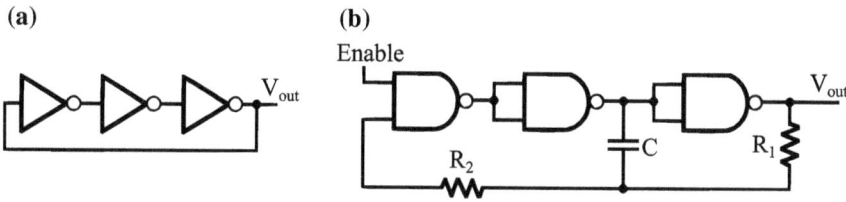

Fig. 12.25 The propagation delay can give rise to oscillations if an odd number of inverting circuits is used. In (**a**) the intrinsic delay is used to get a high frequency oscillator. In (**b**) external components are used to enhance the delay, and hence slow down the oscillations. In addition (**b**) includes an input that can turn the oscillator on and off. These oscillators work best with CMOS logic family devices

Oscillators

The delay between input and output can also lead to oscillations. Sometimes those oscillations are desirable, for example to produce a clock signal. The circuit of Fig. 12.25a uses three inverters, with feedback from the last supplying the first. Since there are an odd number of inverters, the signal that is returned is the opposite of what it was. Hence the circuit can oscillate due to the delay. This type of circuit is referred to as a ring oscillator. The circuit of Fig. 12.25b uses external components to exaggerate the delays and can make an oscillator with a frequency largely determined by the RC time constants. Oscillator circuits, such as these, work best using logic families that have symmetric electrical characteristics for the inputs and outputs—that is the CMOS or NMOS families, but not the TTL families.

Problems

1. For each of the digital logic circuits in Fig. 12.P1, use Boolean algebra to derive the outputs in terms of the inputs. Be sure to show intermediate steps and simplify the final result so the function of the circuit is clear. Also state the result in words (e.g., *"if all inputs are true, the output is false"*).

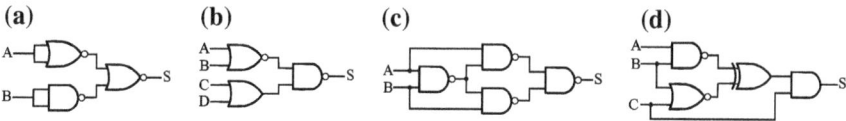

Fig. 12.P1 Problem 1

2. Boolean algebra has two operators, "and" and "or." A circuit that is the "exclusive or," XOR, was also discussed and is signified by "⊕." Use Boolean algebra or a truth table to prove or disprove the following. (Recall that $A \oplus B = A \cdot \bar{B} + \bar{A} \cdot B$, and then use known theorems for the "and" and "or" operations)

   (a) $A \oplus (B \oplus C) = (A \oplus B) \oplus C$ (associative rule).
   (b) $A \cdot (B \oplus C) = (A \cdot B) \oplus (A \cdot C)$ (distributive rule).
   (c) $A \oplus B = \bar{A} \oplus \bar{B} = \overline{A \oplus B}$ (see Fig. 12.1).
   (d) If $A \oplus B = C$ then $B \oplus C = A$.

3. Consider the 4-bit binary counter shown in Fig. 12.21. For that circuit, the flip-flops are triggered on the falling edge. Describe the sequence of base-2 values (e.g., from Eq. 12.9) that would occur if instead the flip-flops were triggered on the rising edge, as shown in Fig. 12.P3.

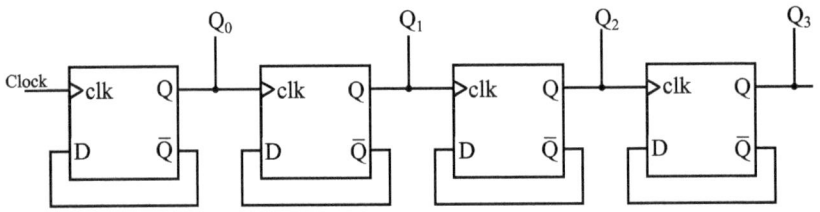

Fig. 12.P3 Problem 3

4. For the circuit in Fig. 12.P4, the flip-flops are reset (0) at $t = 0$. Assuming a 1 kHz clock pulse, that a logic low state is 0 V and a high state is 5 V, plot the output of the op-amp as a function of time for $t > 0$. At least 8 clock pulses should be considered (but note that 8 calculations should not be necessary).

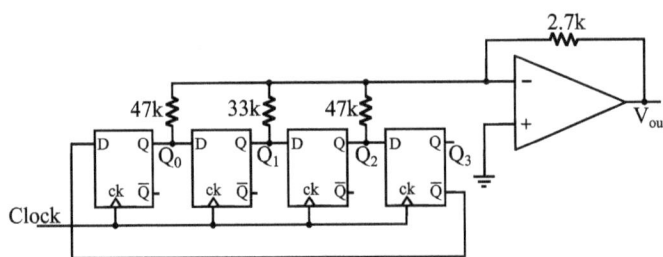

Fig. 12.P4 Problem 4

5. In Eq. 12.9, the Q outputs of the flip-flops are used. What happens if instead, for the same circuit, the values of the $\bar{Q}$ outputs are used to compute n?

Problems 275

6. Construct a digital XOR circuit using only 2-input NORs based on the right side of $A \oplus B = \overline{\overline{A+B} + \overline{A} + \overline{B}}$.
7. (Challenge Problem) The circuit of Fig. 12.P7 includes three D flip-flops. The diodes shown are LEDs arranged in a specific geometric pattern and that will illuminate when current flows through them. The resistors keep the current from becoming too large.

   (a) Assuming the initial state has $Q_0 = Q_1 = Q_2 = 0$, find the values for at least the next six (6) clock pulses. Draw out the pattern of lit LEDs for each of these six states.
   (b) Consider the same circuit except that the central LED ("d") is connected to Q_0 instead of $\overline{Q_1}$. How is it different?
   (c) If you have solved (a) correctly, you will probably see that $Q_0 = Q_2 = 1, Q_1 = 0$ might be considered an "undesired state." Such a state could arise during power up, for example. How can this circuit be improved, possibly using preset and/or clear functions (not shown), to automatically turn this state into a desirable state (one of the six from part a) without otherwise affecting the operation?

(Note: Wires that cross are not connected unless there is a solid circle present. Also assume that 0 V is the logic low, +V is the logic high, and that +V is greater than twice the LED turn-on voltage.)

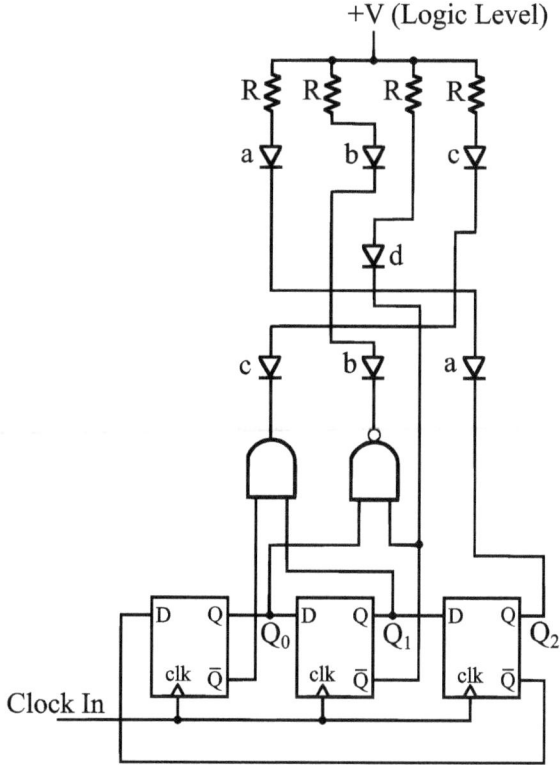

Fig. 12.P7 Problem 7

Chapter 13
Digital II

In this chapter, additional uses for logic gates are presented—in particular how the 0's and 1's can be used in combination as digital codes. In addition to codes useful for numerical computation, several codes that are used for other purposes are also presented. The process of converting between codes, including analog signals and digital codes, is also briefly presented.

Many binary codes are shown in this chapter, most of which are presented to help provide a broad understanding of the different ways that 0's and 1's can be used in combination. Most of these schemes are not to be memorized. It is good, however, to have some idea about what these codes are about and why they might be useful. The codes to know in detail are base-2 binary and the related BCD code. It is also useful to be able to read and use base-2 binary numbers written in different formats, especially including base-2 and hexadecimal notations.

Binary and BCD Numbers

Binary Numbers

All codes based on a combination of 0's and 1's are binary. However, when referring to numerical values, "binary" is generally used to refer to the specific case of base-2 numerical values. Base-2 values were briefly touched upon in the previous chapter and are included here in more detail. Base-2 values are represented using a series of 0's and 1's, and hence are a natural numeric format for digital circuits. Rather than the 1's, 10's, 100's, etc., places used in base-10, base-2 numbers use 1's, 2's, 4's, 8's, etc., places. Base-2 numbers are written so that the 1's place is to the right, just as is done with base 10. To convert from base-2 to base-10, add the appropriate powers of 2.

© Springer Nature Switzerland AG 2020
B. H. Suits, *Electronics for Physicists*, Undergraduate Lecture Notes in Physics,
https://doi.org/10.1007/978-3-030-39088-4_13

Example Convert the base-2 value 10110 to base 10:

$$10110 = 0 \times 2^0 + 1 \times 2^1 + 1 \times 2^2 + 0 \times 2^3 + 1 \times 2^4 = 2 + 4 + 16 = 22$$

For a binary number, each 0 or 1, that is, each digit, is referred to as a bit. The bit on the far right is the least significant bit (LSB or lsb), and has the smallest weight; the digit on the far left is the most significant bit (MSB or msb), as it has the largest weight. For the value 10110, the LSB is 0 and the MSB is 1. Note, however, that for the value 010110 the MSB is 0 and not 1. For electronics and data storage, a fixed number of bits will be used. To determine the MSB, it is necessary to know how many bits (i.e., digits) are available.

BCD Numbers

In some cases, it is more convenient to use the base-10 digits we are all used to. In order to represent those digits using 0's and 1's with digital circuits, some sort of alternate binary code is used for the digits. Any such code that uses 0's and 1's to represent base-10 digits is a form of "binary coded decimal" or BCD. A minimum of four-bits is necessary to represent the digits 0–9. In that case six of the possible four-bit codes are not used.

Normally, when referring to BCD, the scheme is to represent each digit with its base-2 representation, and then those four-bit digits are used as the digits of a base-10 value. That is, instead of using the normal base-10 symbols, each digit is represented by a sequence of four bits, which is also the base-2 representation of that digit.

Example Convert 375 decimal (base 10) to BCD

$$3 = 0011; \quad 7 = 0111; \quad 5 = 0101$$
$$375 \rightarrow (0011)(0111)(0101) = 001101110101$$

In the same way that, when referring to numerical values, "binary" is generally used to mean base-2, BCD without any other qualifiers usually refers to this use of the base-2 representation of the decimal digits. There are many other ways to use 0's and 1's to represent the digits 0 to 9 that are also forms of BCD. Some examples are illustrated below.

Hexadecimal and Octal Notation

While electronics works well using base-2 binary, people like to use decimal (base-10). As a compromise, base-8 ("octal") or base-16 ("hexadecimal") notation is sometimes used to write base-2 binary numbers in a more compact way. Each of these is, in some sense, "closer" to base 10 than is base-2.

In octal, the digits 0–7 are used and each octal digit represents three base-2 digits. The digits 8 and 9 are illegal. In hexadecimal, there are 16 digits: the digits 0–9 are used plus A, B, C, D, E, and F for the digits corresponding to the base-10 values 10, 11, 12, 13, 14, and 15 respectively. Each hexadecimal digit corresponds to four base-2 digits. For octal and hexadecimal there are no unused binary codes.

Example Convert the base 2 value 10110101 to octal and hexadecimal.

- Octal: group bits in threes starting on the *right* (add zeros on the left as necessary), and write the corresponding digit.

$$(010)(110)(101) = 265.$$

- Hexadecimal: group bits in fours starting on the *right* (add zeros on the left as necessary), and write the corresponding digits, using A through F for the digits corresponding to 10–15.

$$(1011)(0101) = B5.$$

It is much more common to see hexadecimal than it is to see octal. To distinguish hexadecimal values from decimal (base-10) values, an additional character is sometimes included. For example, 123h, $123, #123, and 0x123 are some of the many notations that are used to indicate that the value is to be interpreted as hexadecimal.

To convert to decimal, multiply each digit by the appropriate power of the base, starting on the right. Remember to start the powers with 0 and not 1.

Example Convert 1B3h to base-10

$$3 \times 16^0 + 11 \times 16^1 + 1 \times 16^2 = 3 + 176 + 256 = 435.$$

Other Weighted Binary Codes

A weighted binary code consists of N "bit" positions and each position is assigned a numerical weight. The base-2 binary code is an obvious example where the weights are (from right to left) increasing powers of two. Binary Coded Decimal (BCD) is a mixed weighted code. For each group of four bits, BCD is identical to base-2

Table 13.1 4221 codes for the digits 0 to 9

Digit	Code	Digit	Code
0	0000	5	0111
1	0001	6	1100
2	0010	7	1101
3	0011	8	1110
4	1000	9	1111

binary. If the bit position contains a 1, that weight is added to the result, if the position holds a 0 that weight does not contribute. Then each group of four bits has a weight which is a power of ten.

There are other weighted codes that are sometimes useful.

The 4221 Code

Consider a four-bit code with weights 4, 2, 2, and 1. This code is to be used to represent the decimal digits 0–9. It is clear that no matter what selection of bits is chosen, the result will never be outside the range from 0 to 9. Some digits can be represented in more than one way. In the 4221 code, the ambiguity is resolved using the following rules.

- For the digit 5, use 2 + 2 + 1 instead of 4 + 1.
- If the digit is less than 5, use the rightmost 2 first, if needed.
- If the digit is greater than 5, use the leftmost 2 first, if needed.

The resulting codes for the decimal digits are shown in Table 13.1.

Such a code could also be considered a form of BCD, though in this case "binary" does not refer to base-2, but simply that there are two values (0 and 1). Again, some possible codes are illegal and are not used. This code has the property that for any digit N, the code for 9-N (known as the "9's complement") is also the one's complement of N. That is, to get the 9's complement, change all zeros to ones and all ones to zeros, a task that is easily accomplished in the electronics with NOT gates (inverters). For a multi-digit number, do this for each digit. This makes the 4221 code convenient for implementing some decimal arithmetic using electronics (via the 1's complement, or NOT gate).[1] Another code that has this same property is the "excess 3" code, where each digit is represented by its binary value plus three

---

[1]To subtract Y from X, add the 9's complement of X to Y, then take the 9's complement of the result. If X = 492, then its 9's complement is 507. Hence 492-237 = 9's complement of (507 + 237 = 744) = 255. The computation can also be accomplished using the 10's complement of Y, which is its 9's complement plus 1. Add the 10's complement of Y to X, then discard any extra digits generated on the left. So 492-237 → 492 + (762 + 1) = 1255. Discard the extra "1" on the left to get 255. In either case, the subtraction has been changed to addition.

(e.g., the digits 4 = 0111, 5 = 1000, etc.). The excess 3 code is not a weighted code due to the offset by 3, but is close.

2 of 5 Codes

Consider a five-bit code to be used to represent the digits 0 to 9 only. A code can be constructed where exactly two of the five bits are 1's, the remaining being zero. There are exactly ten unique ways to select two things out of five. This gives 10 unique codes that can represent the ten digits used for numerical values. The assignment of which code corresponds to which digit can be done in an arbitrary way, however it is usually done in a way that can be interpreted as a weighted code. For example, the weights 7, 4, 2, 1, and 0 can be used. In this scheme each code corresponds to a digit obtained by adding the appropriate weights for the two locations where the bits are 1's. The exception is the digit "0," that is assigned the one remaining code corresponding to weights 7 + 4 = 11. The weights 6, 3, 2, 1, and 0 could be used instead. The weights may be used in any order—largest to smallest and smallest to largest being the most common.

For multiple digit numbers, simply run the codes together. It is hoped that the code is unique enough that any stray signals can easily be discarded as "illegal codes."

Example For 7, 4, 2, 1, 0 weighting, decode "000111100010010"
Put in groups of 5 − (00011)(11000)(10010), and apply the weights to get "108."

For this code, a sequence of five 1's in a row will never occur. That fact can be used to create a special code of five 1's in a row that can be used to indicate the beginning and/or the end of the data transmission. While not always utilized, it also provides a means for the receiver of the code to be synchronized to the transmitted data before the data starts.

Note that, in principle, a 5-bit code can represent 32 different items. Like other BCD codes, the 2 of 5 codes do not use all the possibilities. This can be useful to detect errors. For the 2 of 5 code most errors result in an illegal code which is easily detected. Other similar codes exist, though may not be weighted. For example, a 3 of 6 code has 20 possible unique codes making it suitable for sending hexadecimal digits (with a few leftover), and is "balanced" in that there are the same number of zeros as ones. Balanced codes have an advantage for some transmission methods.

Some more advanced coding schemes will encode several digits at once, rather than one at a time. For example, "5 of 13" and "8 of 13" codes are used as one part of some current postal bar code schemes. There are over 1000 possible unique codes for each.

Table 13.2 Decimal to gray code representation

Value Decimal	Gray code	Value Decimal	Gray code
0	0000	8	1100
1	0001	9	1101
2	0011	10	1111
3	0010	11	1110
4	0110	12	1010
5	0111	13	1011
6	0101	14	1001
7	0100	15	1000

Non-weighted Codes

Gray Code

The Gray code[2] is a simple example of a non-weighted code where the conversion to and from base-2 binary (a weighted code) is straightforward. The Gray Code is designed so that only one bit changes from one number to the next. One way to express the basic rule for counting up from zero is that for each new value, change the lowest (rightmost) bit that results in a new code. Counting from 0 using a (four bit) gray code would generate the values in Table 13.2.

Such a code is useful for some analog to digital conversions so that noisy data near a boundary between two values will result in one or the other of two adjacent values. An example of such a system is to encode a rotating mechanical position using a Gray code wheel. One possible 4-bit wheel is illustrated in Fig. 13.1a. The wheel might be installed on a shaft with sensors such as shown in a side view in Fig. 13.1b. Here the white areas in the disk above would be clear and the dark areas opaque (or vice versa). If a base-2 binary code is used, accurate results would require that multiple sensors switch simultaneously at the boundaries. For example, to go from the number 7 (0111) to the number 8 (1000) in base-2 requires all four bits to switch simultaneously. If they do not switch at exactly the same time, some other value, possibly very far away from both 7 and 8, will be present during the transition. If Gray code is used, then only one bit changes and the result at that boundary will be either 7 or 8, and will never venture far away.

---

[2] Named after Frank Gray (1887–1969) who patented a coding method using the code in 1947. He referred to the code as a "reflected binary code." Gray codes are also an important part of Karnaugh maps, that are sometimes used to simplify digital circuitry. There are now a number of variants of the original code that is described here.

Non-weighted Codes

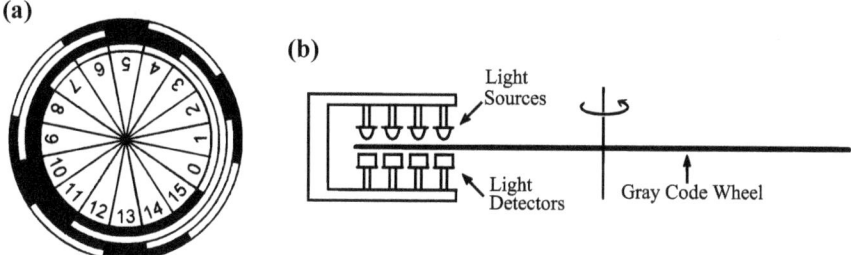

Fig. 13.1 (a) An example of one type of Gray code wheel and (b) how such a wheel might be used

The ASCII Code

Other non-weighted codes are simply assignments between certain codes and certain "values" that may not even be numeric. One such example is the ASCII code, which is by far the most common way to represent letters and characters using 1's and 0's. The original standard US ASCII code uses 7 bits. Modern computers extend most character sets to 8 bits or more. Tables of ASCII codes for different alphabets (or symbols) are easy to find, and can be used to look up the code for each character.

The original ASCII code included many control functions. Many of these control functions are obsolete and only a few will actually appear in practice. In many applications, the control codes are now replaced with printable characters. Initially these codes were to be produced by holding down the control (Ctrl) key on the keyboard while typing the corresponding letter.[3] Some of these control codes have evolved to behave differently depending upon context. For example, on many systems the Ctrl-G code ("bell") will ring a bell only when it is being "displayed" and not when it is being "entered" and while Ctrl-I used to be the same as "Tab," in some systems it is now used to toggle italic fonts on and off.

Bar Codes

For binary data transmission it is common to send and receive serial data—data sent one bit at a time—sometimes at an arbitrary rate and sometimes in "noisy" environments. Since the receiver does not necessarily know the speed of transmission, or even if a transmission has occurred, the receiver must be able to figure that out from what is received. The following are illustrations of codes designed to help do

---

[3]This is where the "Ctrl" key comes from. To generate the control codes, one would hold down the Ctrl key and press the appropriate letter. For example, Ctrl-H is a backspace, Ctrl-M is a carriage return, Ctrl-[is the Esc key, etc.

this. What are shown are examples of bar codes, which are easily examined visually. The same practices can be used for binary data sent between devices over cables or via radio signals.

Interleaved 2 of 5

A bar code can be implemented using fat bars and thin bars to represent 1's and 0's. Noting that the white space between the bars can also be fat and thin, this gives rise to the "interleaved 2 of 5 code" where each group of 5 bars represents a digit, where two of the bars are fat and three thin, and the 5 spaces represent a second digit where two of the spaces are fat and three are thin. A start code (commonly a thin bar-thin space-thin bar-thin space) and a stop code (commonly a fat bar-thin space-thin bar) are also defined, as is a minimum "quiet zone" near the bar code, where no bars or other markings are allowed. The differing start and stop codes allow the code to be read from either end and still be decoded correctly. An example of an interleaved 2 of 5 bar code is shown in Fig. 13.2.

Codes such as, and similar to, the interleaved 2 of 5 code can observed on many cardboard shipping cartons.

UPC Codes

The UPC codes used for identifying products in stores are somewhat more complicated than the 2 of 5 code, though are still forms of BCD. Only the 12 digit "UPC - A" is discussed here. It has the following properties for coding of digits:

- The code for each decimal digit is a sequence of 7 bits.
- As you traverse the 7 bits, the bits *change* exactly 4 times—the first change being the first bit, that by definition changes from whatever preceded it. There are 20 such codes—only 10 if one also considers codes in pairs.
- Whether you start with a "1" or "0" is irrelevant, only the changes matter.

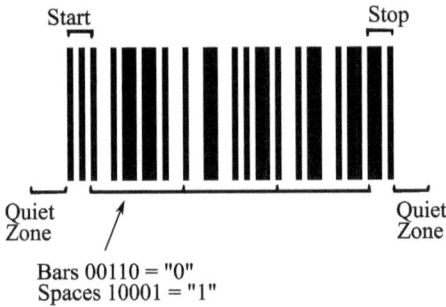

Fig. 13.2 An example of an interleaved 2 of 5 bar code

Bar Codes

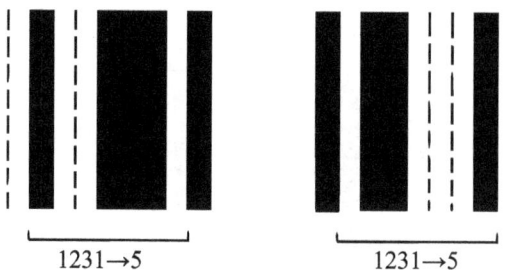

Fig. 13.3 Examples of the two ways to represent the digit 5 in a UPC code

The codes for the digits in UPC written in terms of the number of like bits in a row are shown in Table 13.3. There is no simple relationship between the code and the digit. The codes were chosen in a manner that was thought to reduce the possibility of read errors during normal use.

Note that the codes in reverse are also unique and correspond to the same digit – digits will not be confused when the code is read backwards. When this code is turned into a bar code, the values above for the condensed code correspond to the width of bars and spaces. Each digit is 7 units across, with two bars and two spaces. Figure 13.3 illustrates an example, showing the two codes for the digit 5.

The UPC code also has special codes at the beginning, middle and end, referred to as "guard bars." These bars also make it so that the six digits on the left will always start with a space while the six on the right start with a bar.

The first digit of the UPC code is used to describe the type of product. The next 5 digits are a manufacturer's code, the five after that are a product code and the last digit is a check digit. To compute the check digit, add the digits in the 2nd, 4th, etc., positions, plus three times the digits in the 1st, 3rd, 5th, etc., positions. Then the smallest integer that can be added to get to a multiple of ten is the check digit. When scanning, the check digit helps to ensure an accurate scan. An example of a UPC code is shown in Fig. 13.4. Similar ideas are used to send "packets" of information, which can contain much more information, over cables or using radio signals. The decimal digits across the bottom of the UPC code are for reading by a human, in the event that automatic scanning does not work.

Table 13.3 7-bit codes used to represent decimal digits for UPC symbols

Digit	Code (condensed)	7 bit codes
0	3211	1110010 or 0001101
1	2221	1100110 or 0011001
2	2122	1101100 or 0010011
3	1411	1000010 or 0111101
4	1132	1011100 or 0100011
5	1231	1001110 or 0110001
6	1114	1010000 or 0101111
7	1312	1000100 or 0111011
8	1213	1001000 or 0110111
9	3112	1110100 or 0001011

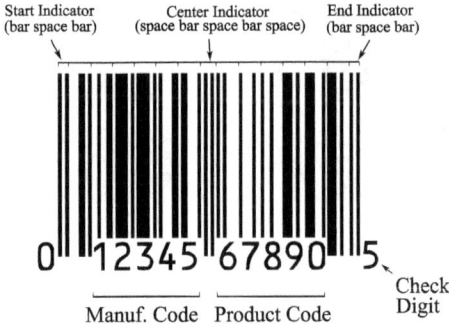

Fig. 13.4 An example UPC code

Many binary code transmission protocols are also based on the use of changes to indicate bit values. That is, a voltage level of 0 V or 5 V is not important, but when the voltage level either changes, or does not change, from one to the other, that is where the information is contained.

Many fancier "bar codes" are the 2-dimensional bar codes, such as "QR codes," which are not discussed here. Those codes are much more sophisticated, can represent both numbers and characters, and include error correcting in that even a damaged code can be correctly decoded.

Some Numeric Code Conversions

When codes are developed systematically, rather than by random assignment, it is usually possible to convert from one to another relatively simply. Of particular interest here are conversions that are easy to do within digital electronics, and not necessarily those that are done using a human with a hand calculator. In what follows, to "complement" a value is to change all of its 0's to 1 and 1's to 0, and recall that LSB and MSB refer to the least significant (right-most) and most significant (left-most) bits of a stored value with a fixed number of bits.

Binary to Gray Code

A procedure to convert from base-2 binary to Gray code is as follows:
Initially, set X = binary number (integer) to convert and $Y = 0$.

(1) Shift X and Y left by one bit; MSB of $X \rightarrow$ LSB of Y
(2) If LSB of $Y = 1$, then complement X
(3) Repeat 1 and 2 for all bits of X
     Y will now contain the result.

Gray Code to Base-2 Binary

One way to convert from Gray code to base-2 binary is as follows:
Initially set Y = Gray Code Number (integer) to convert and $X = 0$.

(i) Shift Right both X and Y by one bit; LSB of $Y \to$ MSB of X
(ii) if LSB of $Y = 1$, complement X.
(iii) Repeat steps i and ii for all bits of Y.
X will now contain the result.

The two conversions between base-2 binary and Gray code can be implemented with electronics as shown in Fig. 13.5 (shown for 4 bit values, but extends to any length).

Decimal to Gray Code

While a conversion from decimal to Gray code could be accomplished by first converting to base-2 binary, and is probably faster when done that way, the conversion can be done directly. Since Gray code is not a weighted code, it is not obvious that such a conversion should be possible.

The procedure is as follows:

Set Y equal to the positive decimal integer to be converted into an N-bit Gray code value ($2^N > Y$).

Fill in the N bit positions of the result from the *left* as follows:

(1) Add $1/2$ to Y (this takes care of rounding issues).
(2) If $Y \geq 2^{N-1}$ then the MSB is 1, otherwise it is 0.
(3) Set $m = N - 1$.
(4) Set $X = |Y - 2^m|$
(5) if $X \leq 2^{m-1}$ the m-th bit is a 1, otherwise it is 0.

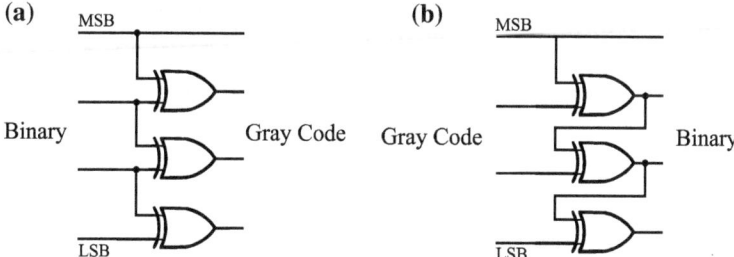

Fig. 13.5 Circuits to (a) convert base-2 binary to Gray code and (b) convert Gray code to base-2 binary

(6) Set $Y = X$, and decrement m by 1.
(7) Go back to step 4 and repeat until $m = 1$.

Note that the first comparison, in step 2, a greater than or equal sign is used, whereas all later comparisons, e.g., in step 5, it is a less than or equal to sign. Because of the added ½ in step 1, the "equal" part of the those signs is somewhat superfluous.

Example Convert 37 base-10 to 6-bit Gray Code.

- $37 + 0.5 = 37.5$
- $37.5 \geq 2^5$, so 6th bit is 1.
- $|37.5 - 2^5| = 5.5$
- $5.5 \leq 2^4$, so 5th bit is 1.
- $|5.5 - 2^4| = 10.5$
- $10.5 \geq 2^3$, so 4th bit is 0.
- $|10.5 - 2^3| = 2.5$
- $2.5 \leq 2^2$, so 3rd bit is 1.
- $|2.5 - 2^2| = 1.5$
- $1.5 \leq 2^1$, so 2nd bit is 1.
- $|1.5 - 2^1| = 0.5$
- $0.5 \leq 2^0$, so 1st bit is 1.

The result is that 37 base-10 is 110111 in Gray code.

A circuit to implement this is shown later as part of the discussion of analog to digital conversion.

BCD to Binary Conversion

Here BCD is used in its usual sense to mean "base-2 binary coded decimal." Of interest here is a direct conversion from BCD to binary, without an intermediate conversion to decimal. That is, how it would be done with electronic circuits.

Initially set X = integer value to convert (written as a string of BCD digits) and set $Y = 0$.

(1) Shift X right by one bit, with LSB of X = first (next) bit of binary number, Y
(2) Examine each new 4-bit digit of X. If a digit is greater than 7, subtract 3 from it.
(3) Repeat steps 1 and 2 until X contains only zeros.

   Y now contains the result.

Example Convert 49 to base-2 binary.

$X = 49$ written as a two-digit BCD number = (0100)(1001), (Parentheses added to show groups of 4) and $Y = 00000000$ initially.

```
Shift X right by 1, LSB → Y.
     X = (0010) (0100)      Y = 00000001 (1st digit of Y = 1)
            2      4
neither digit is greater than 7, so proceed to do again.
     X = (0001) (0010)      Y = 00000001 (2nd digit of Y = 0)
            1      2
     X = (0000) (1001)      Y = 00000001 (3rd digit of Y = 0)
            0      9  → 9 > 7
     subtract 3 from this digit so now
     X = (0000) (0110)
            0      6
Continue with new value
     X = (0000) (0011)      Y = 00000001 (4th digit of Y = 0)
            0      3
     X = (0000) (0001)      Y = 00010001 (5th digit of Y = 1)
            0      1
     X = (0000) (0000)      Y = 00110001 (6th digit of Y = 1).
     X = 0 so done.
```

Thus, the BCD value 01001001 equals the base-2 value 00110001.

Binary to BCD Conversion

To go from binary to BCD, run the above algorithm in reverse. Let Y = binary number to convert, $X = 0$ initially.

(1) Examine each group of four bits of X, add 3 to all that are greater than 4 (this is trivial the first time through, since they are all 0).
(2) Shift X and Y left one bit, $0 \to$ LSB of Y, MSB of $Y \to$ LSB of X.
(3) Repeat steps 1 and 2 until $Y = 0$.
    X now contains the result.

Example Y = 110001 in (base-2) binary, convert to BCD

```
    X = 0000 0000    Y = 110001
         0    0
Shift left by 1, 0 into LSB of Y, MSB of Y to LSB of X
    X = 0000 0001    Y = 100010
         0    1
None of the groups of 4 are greater than 4, so continue.
    X = 0000 0011    Y = 000100
         0    3
    X = 0000 0110    Y = 001000
         0    6
Right group of 4 is greater than 4, so add 3 to it.
    X = 0000 1001
         0    9
Now continue
    X = 0001 0010    Y = 010000
         1    2
    X = 0010 0100    Y = 100000
         2    4
No groups larger than 4, so keep going
    X = 0100 1001    Y = 000000
         4    9
Y = 0 so done.
```

X is now the BCD representation of 49 (0100 = 4, 1001 = 9).

Digital to Analog Conversion

It is often desirable to convert stored digital information into a continuous analog signal (i.e., a non-digital signal). For example, music stored on a digital device (a CD, an MP3 player, a downloaded data stream, etc.) is stored as a series of zeros and ones, however the sound those numbers represent, and the sound you want to listen to, is a continuous analog signal. The conversion of a digital value to an analog value is referred to as digital to analog, or D/A ("D to A"), conversion.

The 1-Bit D/A

A very simple and inexpensive digital to analog D/A scheme uses just 1-bit for output. This scheme works as long as the maximum frequency components within the signal are quite low in frequency compared to the maximum frequency of the digital electronics. Music and voice recordings are prime examples where this can be true.

The simplest D/A of this sort is based on the duty cycle. During an interval, T, the output is a logic 1 for a fraction of the time proportional to the desired output, and off the rest of the time. The average during the time T represents the analog value. This simple scheme is called "pulse length encoding" or "pulse width encoding" and is illustrated in Fig. 13.6a. The output is then sent through a low-pass filter resulting in the time averaged signal. If the highest desired frequency is f_{max}, then it must be the case that $1/T \gg f_{max}$ and the filter would be designed to have a low-pass cut-off to block frequencies higher than f_{max}.

A second similar scheme uses pulses of constant duration, but the spacing between the pulses is variable. Once again, the analog value is represented by the duty cycle and is extracted by sending the pulses through a low-pass filter. Such a scheme would be referred to as "pulse density encoding" and is illustrated in Fig. 13.6b.

The so-called delta-sigma 1-bit D/A (and also A/D) is a variant of this simple scheme that switches the output on and off many times during the time interval T. In this case, the values are related to the time rate of change (the slope) of the signal, and to convert to analog, the sum of the changes (a numerical integration) is used.

Consider two people traveling along a sidewalk. One of them can walk forward and/or backward continuously at any speed. The other, perhaps due to a childhood curse, can only hop by 1 m each second, though the hop can be either forward or backward. To travel together, the second will hop forward if behind the first, and hop backward if ahead. Clearly, if the first walks faster than 1 m/s, then the second will not be able to keep up. However, as long as the first is not traveling too fast, the

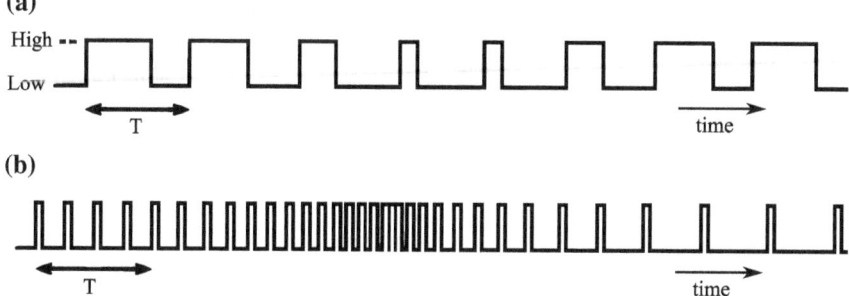

Fig. 13.6 Examples of (**a**) pulse width encoding and (**b**) pulse density encoding. In each case the signal is sent through a low-pass filter to extract the average value

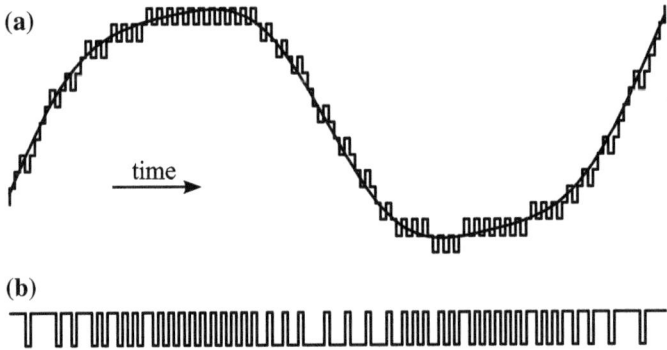

Fig. 13.7 At (**a**) a signal (solid line) is approximated with a "hopping" digital value. The direction of each hop is stored, as shown in (**b**)

second will always be within 1 m. If a record is kept of the direction of each hop, the approximate location, within 1 m, of the first person can be reproduced from the recorded directional values. To reconstruct the values, that is to convert the digital data back to analog, add the number of forward hops and subtract the number of reverse hops. This scheme is illustrated in Fig. 13.7. In Fig. 13.7a, the signal and the digital approximation are shown. Figure 13.7b, shows the stored data, that indicates the direction for each successive hop (a "1" is forward and a "0" is reverse).

The real advantage of all of these 1-bit schemes is that only digital logic gates are required—no linear circuits other than a simple low-pass filter are necessary. In fact, for audio signals the mechanical properties of a speaker or earphones are often used to provide much of the low pass filtering required.

A Summing D/A

A simple D/A that can respond at higher speeds uses a summing op-amp circuit. The resistors are chosen to provide the appropriate binary weights to the signals. Using appropriate resistors, such a circuit can be used with any binary weighted code. Figure 13.8 shows an example using a binary base-2 code with four bits.

Due to the wide range of resistance values required, which results in a requirement for a high level of precision for the resistors, these circuits are practical only when there are a relatively small number of bits.

An improved circuit for binary base-2, that expands nicely to as many as 12 or even 14 bits, uses a series of identical resistors in a "ladder" circuit such as is shown in Fig. 13.9 (also see the R-$2R$ ladder example in Chap. 2). To get a good match, two resistors with resistance R in series can be used to get an equivalent with resistance $2R$, or two "$2R$" resistors in parallel to get R. Using this trick, all the

Digital to Analog Conversion

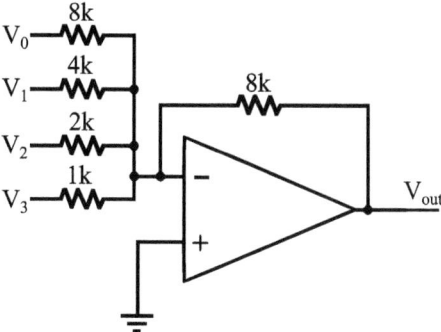

Fig. 13.8 A simple summing D/A converter for a base-2 binary weighted code

resistors used can have the same value. With current technology, such resistors can be matched to each other very precisely. In addition, rather than using the logic levels as voltage sources, the logic levels are used to control (transistor) switches connected to a highly regulated reference voltage in order to obtain more controlled and precise values. Such D/A devices are available as a single integrated circuit.

Analog to Digital Conversion

When collecting data from a measurement, digital codes generated and stored should represent the corresponding measured values (i.e., voltages or currents). For digital storage, the first step is to convert from the non-digital (analog) signal to a digital signal. This is called analog to digital conversion or A/D ("A to D"). Digital data can only have discrete values and what one really wants to do is to obtain the best digital approximation to the continuous analog data. The Gray code wheel already discussed above is a form of A/D that can provide, for example, a digital representation of the continuous angle of the rotation of the wheel.

A few schemes are discussed below related primarily to conversion of analog signals (e.g., voltages) to binary base-2 values. Some of these schemes can be adapted to other codes as well.

Voltage to Frequency Conversion

A particularly simple, though slow, method to convert a voltage to a digital value is to use a voltage-controlled oscillator (VCO). A VCO produces a periodic signal with a frequency that depends on an input voltage. VCO's are available as integrated circuit devices. The frequency from the VCO is measured to obtain a digital value. That is, output pulses from the VCO are sent to a counter circuit for a fixed time interval. For measurements of very slowly changing values, such measurements can be inexpensive and may have very high precision.

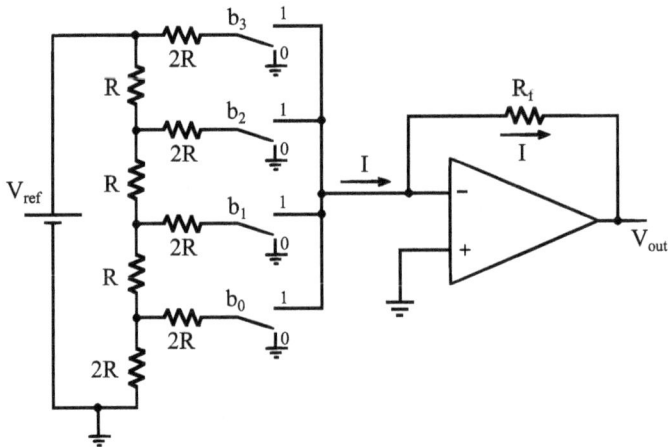

Fig. 13.9 An "R-2R ladder" approach to convert base-2 binary values to analog values

Timing Schemes

Another simple and inexpensive, though relatively slow scheme, is to generate a sequence of voltages using a linear ramp, and then count how long it takes for the generated voltage to pass the input voltage. To obtain a linear ramp, a constant current source is used to charge a capacitor, as shown in Fig. 13.10. While the capacitor is charging, clock signals are sent to a counter to measure the elapsed time. A comparator detects when the ramp passes the input voltage and then stops the counter. The digitized result is taken from the outputs of the counter.

Note that for this conversion scheme it is also a simple matter to obtain results in BCD, such as is appropriate for a simple digital voltmeter. Readily available BCD counters are used instead of binary counters. Also note that the switch to ground that discharges the capacitor, shown here as a manual switch, is normally implemented with a transistor and some control circuitry to start and stop the measurement.

Search Schemes

In a search scheme, a digital guess is made, converted to analog using a D/A, and then compared to the input. Based on the result, a new guess is formed until the desired accuracy is achieved. This is illustrated in Fig. 13.11.

The simplest search might start with zero and count up, which is almost identical to the timing scheme mentioned above. It is much more efficient to start with the most significant bit and work backwards. To start with zero and count up is analogous to finding an entry in the dictionary by starting at the beginning and reading each entry until a match is found. When starting with the most significant

Analog to Digital Conversion

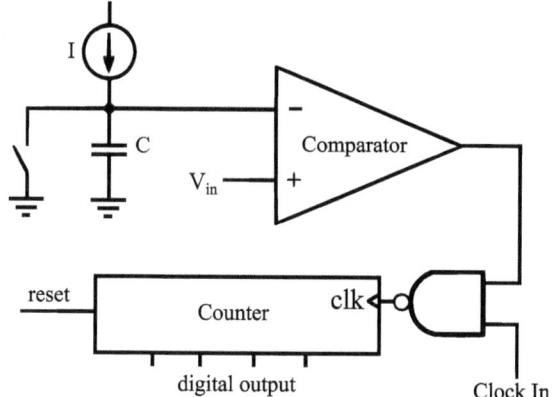

Fig. 13.10 Timing how long it takes a voltage ramp to match an input signal is an inexpensive A/D method. The ramp is constructed here with a constant current source and a capacitor

digit, the dictionary is opened to the middle and it is determined whether that was too far or not far enough. Based on that result, proceed forward or backwards half of the remaining distance to the beginning or end. Then based on the result there, go half of what is left toward the beginning or end. If there are 2^N entries, this latter scheme will take N searches, whereas the former would take, on average 2^{N-1} searches. If N is large, the difference can be quite significant.[4]

Circuits that do this type of conversion with good accuracy are available as integrated circuit devices.

Analog to Gray Code Conversion

Converting directly to Gray code, a non-weighted code, is more difficult to set up, but can result in a very fast conversion. Like the simple summing D/A, the precision of components will limit the number of bits to a half dozen or fewer. This scheme shown here was developed by Yuen (1978). The method is a literal electronic implementation of the decimal to Gray code conversion scheme seen earlier and the circuit is shown in Fig. 13.12.

In Fig. 13.12, the differential amplifiers have a gain of 1. The absolute value circuits can be any of those described previously (see Chap. 10 and the beginning of 11). The reference voltages are given by

$$V_{N-1} = \frac{V_0}{2^N}\left(2^{N-1} - \frac{1}{2}\right); \quad V_{N-m} = \frac{V_0}{2^N} 2^{(N-m)}, \qquad (131)$$

---

[4]There are about 200,000 words in a comprehensive English dictionary. Thus, $N = 18$ and any word can be found with no more than 18 of these operations.

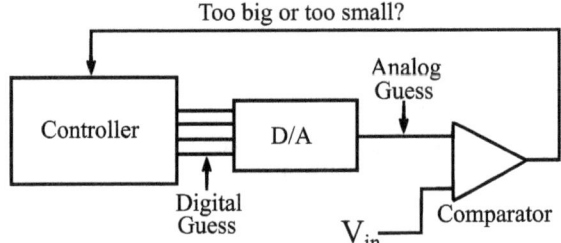

Fig. 13.11 A searching A/D provides a digital guess and then refines that guess based on the results

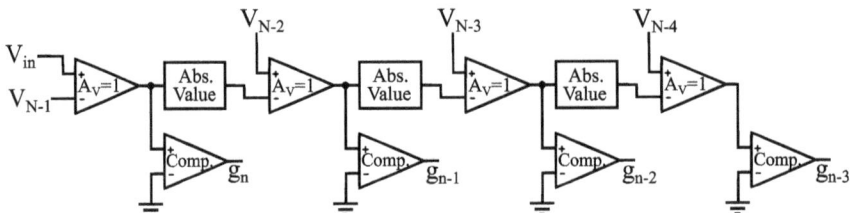

Fig. 13.12 A circuit to convert analog levels directly to Gray code

where the input range is 0 to V_0, and $1 < m \leq N$. Subtracting ½ for the first value is equivalent to adding ½ to corresponding input value, which is done to take care of rounding errors. The Gray code output is taken from the comparators.

Quantization Noise

With N digital bits of information, only 2^N discrete values can be represented precisely. Thus, the best any analog to digital converter can do will be to find which of the 2^N values is closest to the analog value. The possible digital values are discrete and quantized. The difference between the analog value and the closest digital value is referred to as quantization error or quantization noise. For example, with 7 bits, 128 distinct values can be produced. If these values are spread evenly through the measurement range, then the accuracy that any digital result will likely match an analog signal is no better than 1 part in 128, or about 1%, of the full measurement range.

For many types of measurements, the measurement procedure can be repeated multiple times with the expectation that the average of many results will more precisely represent the "true" value. With a low-noise signal and quantization error, that may not happen. If the variations in the signal are smaller than the spacing between the quantized levels, the same error can occur with every measurement—it is not a random error and will not average to zero. In fact, in such situations, one strategy for improving the accuracy is to add random noise to the signal. If the added noise averages to zero and is larger than the quantization spacing, then the average of many measurements can, in fact, be more accurate than measurements made without the additional noise.

Problems

1. Convert the decimal number 1234 into base 2, octal, and hexadecimal.
2. Convert the base-2 binary value 001001001001 into decimal.
3. Convert the BCD value 001001001001 into decimal.
4. Convert the decimal value 1234 into BCD.
5. The nine's complement of a digit N is $9 - N$, where $0 \leq N \leq 9$. For a multi-digit value, the nine's complement is formed by taking the nine's complement of each digit. The ten's complement of a decimal value is its nine's complement plus 1.

    (a) Compute 573 minus 250 using your usual method.
    (b) Compute 573 plus the (three-digit) ten's complement of 250, and then keep only the lowest 3 digits in the result.
    (c) Compute the sum of the nine's complement of 573 with 250, then take the nine's complement of the result.

6. Web pages often use hexadecimal values to define colors. The values represent 24 bits. The 24 bits are three groups of 8 bits to specify, in order, the amount of red, green, and blue that contribute to the color. Thus, in decimal, each of the three contributions can range from a minimum of 0 to a maximum of 255. Suppose a color is specified as "ffc033." What are the weights, expressed in decimal, for the three contributing colors?
7. If an A/D produces a 6-bit result, providing evenly spaced values representing values for the range from 0 V to 5 V, what is the spacing, in volts, between the evenly spaced values?
8. If analog values in the range between 0 V and 10 V are to be recorded and stored using an N-bit digital value, what is the minimum value of N that can be used if an accuracy no worse than 0.1% is desired?

Reference

Y.K. Yuen, Analog-to-gray code conversion. IEEE Trans. Comput. **27**, 971–973 (1978)

Chapter 14
Calculators and Computers

Calculators, computers, microprocessors, and the like are all based on digital circuitry. This chapter discusses how digital circuits can be combined to perform numerical calculations and the basics of how computer processing is accomplished. Along the way, tri-state devices are introduced as well as some unique additional applications that they enable.

Adding Base-2 Numbers

The truth table for the addition of two one-bit values, A and B, is shown in Table 14.1, corresponding to $0 + 0 = 0$, $0 + 1 = 1 + 0 = 1$, and in base-2, $1 + 1 = 10$. A simple circuit to add two bits is shown in Fig. 14.1a. Such a circuit is referred to as a "half adder."

For multiple-bit values, however, provisions for a possible carry from a previous operation need to be included—there will be three bits to add. That is accomplished using a "full adder." A full adder can be constructed from two half adders as shown in Fig. 14.1b. To add two N-bit values requires N full adders, as illustrated in Fig. 14.2 for $N = 4$. The result of the addition, S, also has N bits. If the final carry is 1, then the final carry is lost from S and S is said to have "wrapped around" zero. This behavior can be a problem but can also be used to advantage for subtraction using the two's complement method.

Two's Complement Arithmetic

Subtraction using binary numbers can be accomplished using a subtraction circuit, similar to the "full adder" shown previously. More commonly, however, subtraction is done as an addition using two's complement values—a process somewhat similar

Table 14.1 Truth table for 1 bit addition

A, B	Sum
0, 0	0
0, 1	1
1, 0	1
1, 1	0, carry a 1

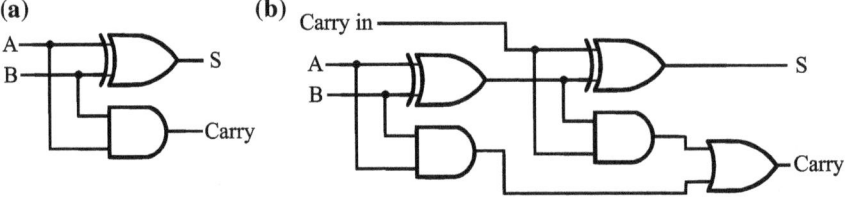

Fig. 14.1 Circuits for (a) half adder and (b) full adder

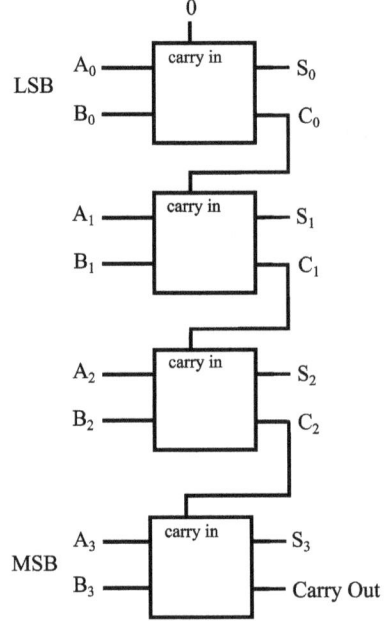

Fig. 14.2 An adder for four bit values, constructed from four full adders

to, but not the same as, performing subtraction by the addition of a negative value. With the two's complement method there is no need for a separate subtraction circuit.

When a digital circuit is used to add two numbers, the circuitry will always add a fixed number of binary digits (0's and 1's)—that is a fixed number of bits. This will be true no matter how many bits are actually required to represent the values being added. Often, binary values are used in the electronics (the "hardware") in groups of 8 bits, known as "bytes," 16 bits, which are called "words," 32 bits, which are "double words" or 64 bits, sometimes referred to as a "quadword."[1] That there is a fixed and finite number of bits is a limitation, but also an opportunity.

If there are N bits, each of which can be a 0 or a 1, they can represent 2^N binary numbers ranging from 0 to $(2^N - 1)$. Alternatively, one of the bits can be used to be a minus sign and then there are 2^N values from $-(2^{N-1})$ to $+(2^{N-1} - 1)$. Rather than the use of a simple minus sign, there is another way to represent negative values for simple integers that is called the method of two's complements. The two's complement method is only effective when there are a fixed number of bits.

To form a two's complement of a base-2 number, do the following two steps:

- Make the "one's complement" of each binary digit by changing all 0's to 1's and all 1's to 0's.
- Add 1 using normal binary addition, throwing away any carry from the MSB.

Example Find the two's complement of 0101100.

- Make the one's complement: 1010011.
- Add 1: 1010011 + 1 = 1010100.

Subtraction is then accomplished by adding the two's complement.

Example Compute 67−13 for 8-bit binary integers.

- *Convert decimal values to 8-bit binary*:

$$67 = 64 + 2 + 1 = 2^6 + 2^1 + 2^0 = 01000011$$
$$13 = 8 + 4 + 1 = 2^3 + 2^2 + 2^0 = 00001101$$

- *Form the 2's Complement for 13, the "subtrahend"*:

$$1's \text{ Complement} = 11110010$$
$$\text{Add } 1 = 11110010 + 1 = 11110011$$

- *Add the two values, keeping only the lowest 8 bits (the "1" in parentheses is thrown away, there is no place for it)*.

---

[1]Though rarely seen in hardware any more, a group of 4 bits, being half of a byte, is referred to as a nibble.

$$\begin{array}{r} 1 \quad 11 \\ 01000011 \\ +\ 11110011 \\ \hline (1)00110110 \end{array}$$

- *Convert back to decimal to check*:

$$2^1 + 2^2 + 2^4 + 2^5 = 2 + 4 + 16 + 32 = 54$$

Two's complement arithmetic is a "clock" arithmetic. For example, on a twelve-hour clock 1 o'clock minus 2 h is 11 o'clock. Hence, one can also interpret 11 o'clock as "−1 o'clock" since $1 - 2 = -1$. A negative value expressed as a two's complement will always have a "1" in the MSB, however that is not simply a minus sign. For example, for 8-bit two's complements, consider the values, $-1 = 11111111$, $-2 = 11111110$, etc. The initial 1 signals that the value is negative, but the remaining binary number is not the magnitude of that value.

A Simple Arithmetic Logic Unit (ALU)

Rather than create separate circuitry to perform each operation, many potential operations can be built into one circuit. That circuit has additional inputs that tell it which operation to apply this time. One example of such a circuit is an Arithmetic Logic Unit (ALU). An ALU performs the basic functions for computations and logic operations.

Figure 14.3 shows a simple example of such a circuit that includes 4 "command" input bits that determine what operation is to be applied to the bits of the two input values (A and B). The outputs include a value S along with some "flag" bits. The flag bits indicate general features of the result. Here A is an *n*-bit binary (base-2) value represented by the bits A_i, $i = 0$ to $(n - 1)$, and similarly for B, S, a, and b. The single-bit flag values "c" and "z" will depend on the values of the input arguments and commands. In the diagram, a "register" is (anything equivalent to) a set of D flip-flops that store values after a clock pulse. A description of what the ALU circuit does can be presented as an abbreviated truth table, such as Table 14.2. Notice how easy it is to add the two's complement of B to A, a subtraction, through the use of the command bits and the full adder.

Inside a (simple) microprocessor, it is common to find the ALU output register directly wired back to one of the inputs, such as the A input. That A input will then be referred to as the accumulator. That is, the output from any operation is wired to (potentially) become one of the inputs for the next operation. If it is assumed that the output bits (S_0 to S_{n-1}) are indeed connected directly to the A input bits (A_0 to A_{n-1} respectively), and the B inputs (B_0 to B_{n-1}) are coming from a register called "B,"

A Simple Arithmetic Logic Unit (ALU)

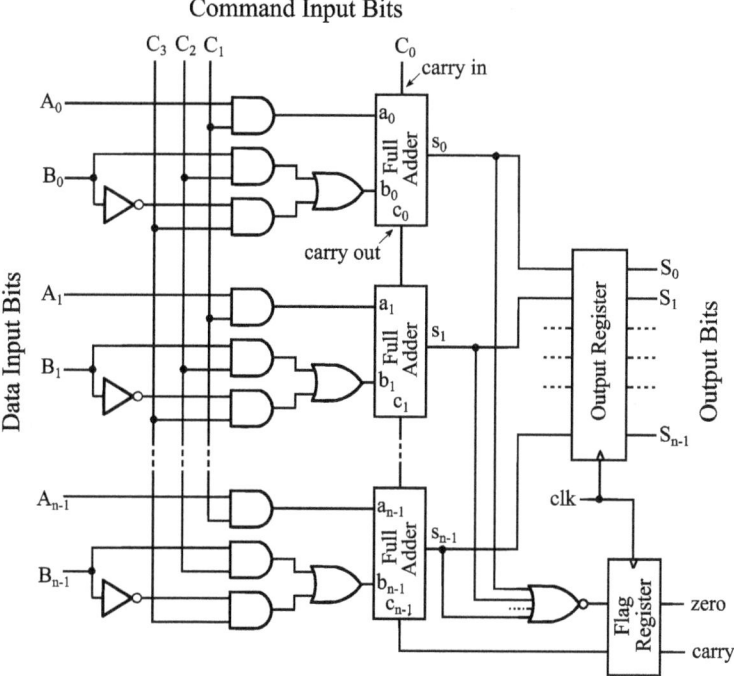

Fig. 14.3 A simplified arithmetic logic unit (ALU) that operates on two input values, A and B, and performs an operation determined by the command values C

then the sixteen commands for this simple ALU can be described as shown in Table 14.3.

Of course, in a real microprocessor (or similar device) many additional commands would be expected. Additional operations would include shifts, various logic operations (AND, OR, XOR, etc.), and others. This ALU is only a simplified example to illustrate the principal that is used.

In addition to the command for the ALU, a microprocessor will have commands that move values between other various locations (e.g., from a memory location to the register B). Each command is specified using 0's and 1's. The 0's and 1's within the processor are referred to as the machine code. The command mnemonics are used for human consumption and are used as part of assembly language programming.[2] Assembly language has a direct connection to the 0's and 1's, but is somewhat more readable. Most programming is done using a higher-level language, such as C/C++, Python, etc., which serves as a more convenient, and consistent, interface to the user. That higher-level language will ultimately be converted to the 0's and 1's appropriate for the processor being used.

---

[2]The mnemonics here are for this simplified ALU for example only, and may not appear in any established programming language.

Table 14.2 Summary of the functions of the circuit in Fig. 14.3

Command input bits				Full adder inputs		Outputs			Description
C_3	C_2	C_1	C_0	A	b	S	Carry	Zero	
0	0	0	0	0	0	0	0	1	Clears to zero
0	0	0	1	0	0	1	0	0	Sets to one
0	0	1	0	A	0	A	0	z	Copies A
0	0	1	1	A	0	A + 1	c	z	Increments A
0	1	0	0	0	B	B	0	z	Copies B
0	1	0	1	0	B	B + 1	c	z	Increments B
0	1	1	0	A	B	A + B	c	z	Adds A and B
0	1	1	1	A	B	A + B + 1	c	z	Adds A and B and increments
1	0	0	0	0	$\overline{B}$	$\overline{B}$	0	z	One's complement of B
1	0	0	1	0	$\overline{B}$	$\overline{B}+1$	c	z	Two's complement of B
1	0	1	0	A	$\overline{B}$	$A+\overline{B}$	c	z	A plus one's complement of B
1	0	1	1	A	$\overline{B}$	A − B	c	z	Subtract B from A
1	1	0	0	0	1	−1	0	0	All bits set to 1
1	1	0	1	0	1	0	1	1	Sets carry and zero
1	1	1	0	A	1	A − 1	c	z	Decrements A
1	1	1	1	A	1	A	1	z	Copies A with carry set

Table 14.3 A summary of the available commands for the ALU of Fig. 14.3

Command				Hex	Command mnemonic	Description
C_3	C_2	C_1	C_0			
0	0	0	0	00	CLRA	Clear Accumulator to Zero
0	0	0	1	01	ONEA	Set Accumulator to 1
0	0	1	0	02	CLRC	Clear Carry (set carry to 0)
0	0	1	1	03	INCA	Increment Accumulator
0	1	0	0	04	MOVB	Move B to Accumulator
0	1	0	1	05	INCB	Move B + 1 to Accumulator
0	1	1	0	06	ADDB	Add B to Accumulator
0	1	1	1	07	ADPB	Add B + 1 to Accumulator
1	0	0	0	08	CMPB	Move $\overline{B}$ to Accumulator
1	0	0	1	09	NEGB	Move -B to Accumulator
1	0	1	0	0A	ADCB	Add $\overline{B}$ to Accumulator
1	0	1	1	0B	SUBB	Subtract B from Accumulator
1	1	0	0	0C	SETA	Set all Bits of Accumulator to 1
1	1	0	1	0D	CLCA	Clear Accumulator, Carry Set
1	1	1	0	0E	DECA	Decrement Accumulator
1	1	1	1	0F	SETC	Set Carry to 1

Base-2 Multiplication

Base-2 binary multiplication can be accomplished using the same basic procedure as for base ten, however the multiplication table is greatly simplified to include only 0 and 1.

Example: Multiply 5 times 3.
- 5 base 10 = 101 base 2; 3 base 10 = 011 base 2
- Multiply them, one term at a time, shifting by one place, then add.

```
   101
 ×  11
   101
 +101
  1111
```

To perform this operation using digital logic, it is possible to make a single circuit that does multiplication as long as the number of bits is not too large, however it is usually more expedient to perform the operation serially—i.e., as a series of steps rather than all at once. Consider the following description of multiplication that is more easily translated into circuitry:

1. Set A and B to the numbers to be multiplied
2. Initialize C to be zero ("Reset" or "Clear" C)
3. Examine the LSB of B, If LSB is a 1, then add A to C
4. Now shift A left by one bit, B right by one bit (the previous LSB of B is lost).
5. If B is not zero, loop back to step 3 and do it again. Otherwise, go on.
6. C now has the result.

Example Multiply 19 by 9 (19 base 10 = 10011 base 2; 9 base 10 = 1001 base 2)

Step	A	B	C	
1,2	10011	1001	0	
3	10011	1001	10011	LSB of B is 1 → Add A
4,5	100110	100	10011	$B \neq 0$
3	100110	100	10011	LSB of $B = 0$
4,5	1001100	10	10011	$B \neq 0$
3	1001100	10	10011	LSB of $B = 0$
4,5	10011000	1	10011	$B \neq 0$
3	10011000	1	10101011	LSB of $B = 1$ → Add A
4,5,6	100110000	0	10101011	$B = 0$, C has result

Figure 14.4 is a partial schematic for a circuit that does this for four bits. Note that four bits are shown for A, B, and for the Result. It is more typical to allow for twice as many bits in the result as are used for the inputs—the product of two four-bit values may require up to eight bits for the result.

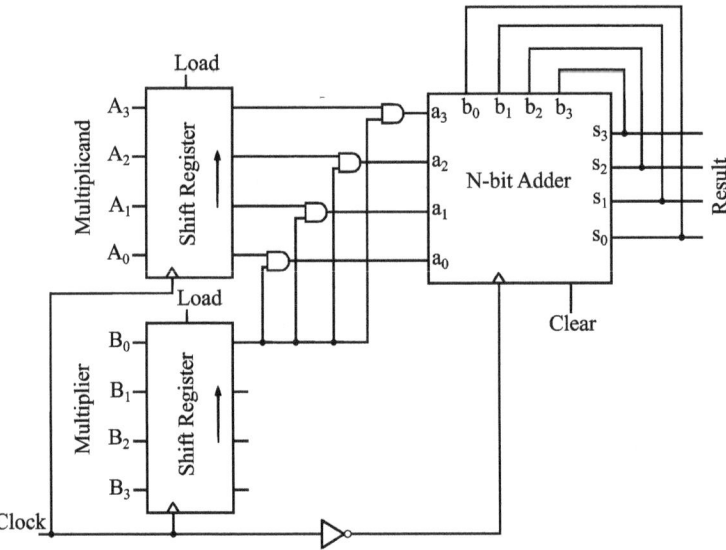

Fig. 14.4 An iterative multiplying circuit. The adder is initially cleared and the results of each step are added to the previous result

Earlier computers implemented the multiply function using software rather than hardware. Some modern microcontrollers sometimes also rely in whole or in part on software multipliers.

So-called "floating point" real number calculations are done using base-2 scientific notation and integer operations. The "point" referred to is the decimal point (which is commonly still referred to as a "decimal point" even though the number is not in base ten). For floating point values in base two the number is written as a mantissa (a value that is between 0.5 and 1.0) and an exponent for the power of two. The computer memory stores the mantissa and the power together in memory. Floating point values typically use a more traditional minus sign (here a 0 or 1 to indicate positive or negative) rather than the two's complement method.

Example Base-2 Scientific notation

Write the square root of two using base-2 scientific notation.

To do this, subtract the largest power of two that gives a positive (or zero) remainder and put a 1 in the appropriate place. Then decrease the power to the next largest power, etc. Recall that digits to the right of the decimal place correspond to negative powers. In base 10, $\sqrt{2} = 1.414213562$.

Base-2 Multiplication

Largest power of 2 less than 1.414... is 2^0	1.414213562 − 1 = 0.414213562	First digit is in the "2^0's" place = 1
Next largest power of 2 is 2^{-1} = 0.5, which is too big		The next digit is 0 1.0
2^{-2} = 0.25 is less	0.414213562 − 0.25 = 0.164213562	The next digit is 1 1.01
2^{-3} = 0.125 is less	0.164213562 − 0.125 = 0.039213562	The next digit is 1 1.011
2^{-4} = 0.0625 is too big		The next digit is 0 1.0110
2^{-5} = 0.03125 is less	0.039213562 − 0.03125 = 0.007963562	The next digit is 1 1.01101
2^{-6} = 0.015625 is too big		The next digit is 0 1.011010
2^{-7} = 0.0078125 is less	0.007963562 − 0.0078125 = 0.000151062	The next digit is 1 1.0110101
2^{-8} = 0.00390625 is too big		The next digit is 0 1.01101010
etc...		1.01101010...

Hence, $\sqrt{2} = 1.01101010\ldots$ in base 2. To put this into base-2 scientific notation with a mantissa between 0.5 and 1 the value must be divided by 2, that is it is shifted right by 1 digit, so the mantissa becomes 0.101101010... and the exponent, the number of multiples of 2 needed to restore the original decimal point, is 1. Hence, in base-2 scientific notation, write $\sqrt{2} = 0.101101010 \times 2^1$.

These base-2 values stored in scientific notation are referred to as "floating point values." To add two floating point values the decimal points need to be lined up first. To multiply two floating point values, multiply the mantissas as if they were integers, put the decimal point to the far left of the result, then add the exponents as integers.

When stored in computer memory, the first 1 to the right of the decimal point does not need to be stored since it will always be present. If this shortcut is used to gain one extra binary digit, a special code needs to be used to refer to the value zero. Often another special code is used to represent "not a number" or NAN. Such a code is generated, for example, if the computer is asked to divide a value by zero.

Some Recursive Computations

Wiring a single digital circuit to perform some operations is not cost effective in many cases. Computers (and calculators) will often use recursive numerical computations to provide what is needed. These calculations can be implemented using a combination of software and hardware.

The example algorithms below are for computation using digital logic capable of at least NOT, addition, and shift operations, from which one can also obtain subtraction and multiplication. The methods shown here illustrate how these operations might be accomplished and may not be the most effective to use in practice. Modern computers may include built-in specialized circuitry to perform many of the more common functions. That these methods work can be verified using a hand-held calculator. There is certainly no need to memorize the details of these examples.

Compute 1/x

Division is often accomplished by multiplying by the inverse. To compute the inverse of a value x, do the following:

(a) Write x in base-2 scientific notation: $x = d \times 2^n$, where $0.5 \leq d < 1$.
(b) Make a reasonable initial guess for $1/d$, called c. For example, $c = 1/0.75 = 1.333$. (Or one can use a look-up table and use a more accurate guess depending on the value of d).
(c) Set $b_0 = c$, $a_0 = b_0 \times d$.
(d) Compute

$$b_{i+1} = b_i \times (2 - a_i) \text{ and}$$
$$a_{i+1} = a_i \times (2 - a_i)$$

repeatedly (i.e., i = 0, 1, 2, 3, ...) until the result, b_i, does not change within some specified accuracy (a_i will converge to 1.000).
(e) Then $\lim_{i \to \infty} b_i = 1/d$ and so $1/x = (1/d) \times 2^{-n}$.

Compute $(1/x)^{½}$

One way to compute square roots, is to compute the inverse of the square root, then use the inverse function above.

(a) Write x as $x = d \times 2^n$, where $4 \leq d < 16$ and n is an even integer.
(This particular range for d is only necessary because of the specific initial guess shown below).
(b) Make a reasonable initial guess for $1/\sqrt{d}$, called c. For example, $c = 2.5/d$.
(c) Set $b_0 = c$, $a_0 = b \times b \times d/2$

(d) Compute

$$b_{i+1} = b_i \times (1.5 - a_i)$$
$$a_{i+1} = a_i \times (1.5 - a_i)^2$$

repeatedly until the result does not change within some specified accuracy.

(e) Then $\lim_{i \to \infty} b_i = 1/\sqrt{d}$ and $1/\sqrt{x} = (1/\sqrt{d}) \times 2^{-n/2}$.

Compute $x^{1/2}$

Method I:

(a) Compute inverse of $1/\sqrt{x}$ using the above routines, or
(b) Make an initial guess, for example, $b_0 = 1$ or $b_0 = x$. (This guess does not need to be particularly good).
(c) Then compute

$$b_{i+1} = (b_i + x/b_i)/2$$

repeatedly until the changes in the result become smaller than some specified accuracy. Then b_i has the result. This latter method takes a relatively long time since the inverse is used many times.[3]

Method II:

This is basically a search method that works in base-2 binary with a fixed number of bits. If the number of bits is not too large, this method can be surprisingly fast since no inverse operations are required. This uses a search method starting with the MSB.

(a) Start with $y = 0$.
(b) Change the most significant bit of y to 1. Compute y^2.
(c) If $y^2 > x$, then change the bit back to a 0, otherwise leave it a 1.
(d) Change the next most significant bit to a 1. Compute y^2.
(e) Loop to step c until all the bits of y have been tried. The result will be y.

---

[3]Some interesting integer approximations can be obtained by following this method using pencil and paper. For example, start with $b_0 = 1$ and iterate by hand to get $\sqrt{2} \approx 577/408$ accurate to about 6 decimal places. Do just one more iteration to show $\sqrt{2} \approx 665{,}857/470{,}832$ which is accurate to about 12 decimal places.

Compute tan(x)

The value of x should be expressed in radians and in the first octant (0 to 45°). (All other angles can be put in this range and the result converted using trig identities. Values of x outside this range may work, but possibly not as efficiently.)

(a) (i) set a variable *tiny* to a very small non-zero number, say $tiny = 1 \times 10^{-30}$.
    (ii) store the value $a = -x^2$ for later use.
(b) set $f_0 = x$, $c_0 = x/tiny$, $d_0 = 1$, $b_0 = 3$.
(c) compute as follows, repeatedly until the result, f_i, doesn't change significantly (or equivalently, the product $c_i d_i$ is "close enough" to 1.000).

$d_{i+1} = 1/(b_i + a \times d_i)$, if the denominator is zero, use $d_{i+1} = 1/tiny$
$c_{i+1} = b_i + a/c_i$, and if the result is zero, use $c_{i+1} = tiny$
$f_{i+1} = f_i \times (c_{i+1} \times d_{i+1})$, and if the result is zero, use $f_{i+1} = tiny$
$b_{i+1} = b_i + 2$

Note: One can compute $\tan(x/2)$ and use trigonometric relationships to get $\tan(x)$, $\sin(x)$, etc. for the entire first quadrant. Using $a = +x^2$ instead of $-x^2$, yields the hyperbolic tangent of x.

There are similar methods available to compute the inverse tangent, and hence all inverse trig functions, as well as for exponentials (e^x). For more information see Press (1992).

Compute K(k)

(The complete elliptic integral of the first kind).

Set $x_0 = 1 - k$, $y_0 = 1 + k$ and then iterate to compute the so-called "arithmetic geometric mean,"

$$x_{n+1} = (x_n + y_n)/2$$

$$y_{n+1} = \sqrt{x_n y_n}.$$

The difference between x and y becomes very small after surprisingly few iterations. Then $K(k) = \pi/(2x_n)$. The complete elliptic integral of the first kind shows up in many physics problems. For example, when computing the period, T, for a pendulum of length l when the amplitude, θ_0, is not small, $T = (2/\pi) K(\sin(\theta_0/2))T_0$ where T_0 is the period for very small angles.

Communications

In addition to numerical computations, a computer is a device that sends data from one place to another. Some of the locations are memory, some are arithmetic logic units or displays, etc. That is, some locations will leave the data untouched while others will act on it in some way. Thus, data communication is a major factor governing how fast a computer can go.

Serial communications—data is spread out in time. Send 1 bit at a time (includes schemes for indicating "start" and "stop" of data). Examples: Serial (com) port on computer, USB, most network communications including digital cell phones. As digital electronics has gotten faster and faster, serial communications has become quite common even between devices mounted on the same circuit board.

Parallel communications—data is sent simultaneously over parallel wires (or parallel optical fibers, etc.). Send n bits at a time (plus control signals). Examples: Printer port, IEEE-488. Parallel communication can be much faster, but as the number of bits grows, the hardware must grow along with it and can become cumbersome.

Mixed—a mixture of parallel and serial. Example: dual tone numbers used for phone dialing.

Tri-state Outputs

In order to facilitate the rapid movement of data from one place to another, a third possible output is added to the digital logic circuits that is a "high impedance state," often labeled "Z." Such devices are referred to as being "tri-state" or "3-state." Basically, the output is simply turned off or effectively disconnected in this third state. In that way, millions of outputs can be wired together and as long as only one of them is on at any given time, there will be no conflicts. The group of common connections between the (potentially) millions of outputs is referred to as a "bus."

A simple example of the use of tri-state outputs is shown in Fig. 14.5. Each circuit connected to the bus has an "output enable" input. When that input is false, the device cannot send data out to other devices—the output is disconnected. In Fig. 14.5, it is assumed that when the output is disabled, the connections act as inputs. For some devices, a separate "input enable," or equivalent, may also need to be supplied.

Here, the "command" 011000 (reading left to right) corresponds to "copy the contents of A to B." Similarly, 100001 means "copy the contents of C to A." Since A, B, and C could be simple memory locations, inputs or outputs to an arithmetic logic unit, a display device, etc., the ability to move data around using commands is very powerful. Decoders and encoders (discussed in Chap. 12) can be used to make the "command" more efficient and useful.

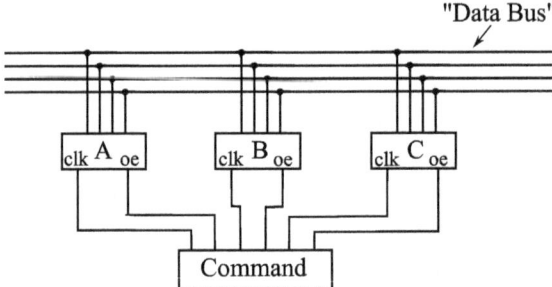

Fig. 14.5 With tri-state outputs, many circuits can be connected to a common bus. The output enable (oe) input along with the clock input, can be used to transfer data from one register (set of flip-flops) to another

Simplified CPU

The basic features of a computer's central processing unit (CPU) are illustrated in the diagram of a very simple (and hypothetical) CPU shown in Fig. 14.6. There are a number of registers, to store intermediate values, an ALU to do some numerical work, and control circuitry to run it all. Data and commands are retrieved (i.e., moved) from external memory locations and data may be sent to memory, other processors, displays, etc.

Program execution is accomplished by the CPU by, quite literally, simply moving data from one place to another in a prescribed sequence. Some of the locations are simply for storage, others cause some action to take place.

Other Uses for Tri-state Devices

The output of a typical tri-state device will be 0 V, a fixed voltage, V_0 (such as 3.3 V or 5 V), or a very high-impedance state, "Z". A very high impedance is equivalent to an open circuit. While perhaps not originally developed for such uses, that third output can be used in some clever ways.

Measuring a Small Capacitance

A simple circuit that can be used for measuring a small capacitance is shown in Fig. 14.7, where C_x is the small valued capacitor to be measured and C_1 is a much larger capacitor of known value.[4] To make a measurement, first S_2 and S_3 are closed to discharge both capacitors. Then, with S_3 left open, switches S_1 and S_2 are

---

[4] See, for example, Philipp (1999), where this technique is used as part of a proximity detector.

Other Uses for Tri-state Devices

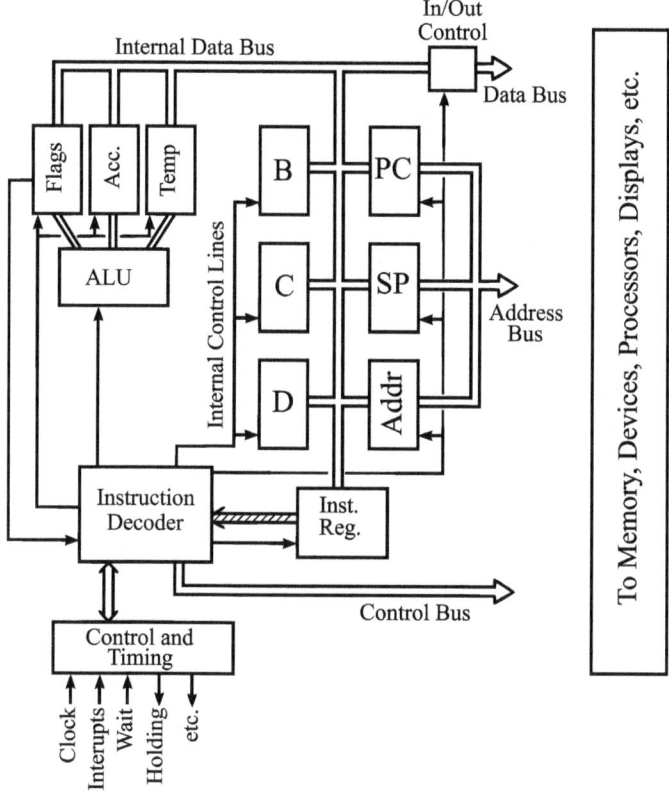

Acc = Accumulator or "A Register"
B, C, D = Internal storage registers
Temp = temporary storage used during computation
ALU = Arithmetic Logic Unit
PC = Program Counter (points to next instruction in memory)
SP = Stack Pointer (for "stack" operations, points to top of the stack.)
Addr = Address Register (points to a spot in memory to store or retrieve "data" sent via the data bus).

Typical Flag Bits
Zero
Carry
Parity
Sign
Etc...
Used for conditional instructions
(IF...THEN...ELSE...ENDIF).

Fig. 14.6 A simplified CPU

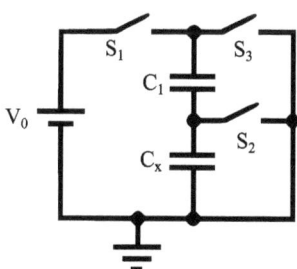

Fig. 14.7 A simple switching circuit that can be used to measure a small capacitance

alternately closed and opened, taking care not to have them both closed at the same time. After a number of cycles, the voltage across C_1 is measured with switch S_2 closed. Alternatively, the number of cycles for the voltage across C_1 to reach some preset value, as determined by a comparator, can be used as a measure of C_x. The basic operation is the same as determining the size of a small bucket by measuring how many times the small bucket must be filled in order to fill a larger bucket. This method is also related to the switched capacitor circuit mentioned in Chap. 4.

The capacitors are in series so the effective capacitance when S_2 is open is $C_{eff} = C_1 \| C_x \approx C_x$. Now if C_1 is initially charged at a potential V_i with charge $Q_i = C_1 V_i$ and C_x has been previously discharged through S_2, then when S_2 is reopened and S_1 closed, the extra potential difference applied across the series capacitors is $V_0 - V_i$, placing an additional charge of $Q_{i+1} = C_{eff}(V_0 - V_i)$ on C_1. If this occurs over a time Δt, then the average current into C_1 is $I = \Delta Q/\Delta t = C_{eff}(V_0 - V_i)/\Delta t$ which looks like a simple RC circuit with $R = \Delta t / C_{eff}$. So, starting with the capacitors discharged, the number of switching cycles necessary to get the potential on C_1 up to a preset threshold can be used to determine C_x.

Since the switches are either open or are connected to either a fixed voltage or to ground, such a circuit might be implemented using tri-state digital logic rather than switches. A simple tri-state buffer is shown schematically as a buffer with an extra control input (Fig. 14.8). The behavior is summarized in Table 14.4. The output "Z" corresponds to the high impedance state and for this buffer, $\bar{a}_n$, would correspond to "output enable." When the output is enabled, the output is either a fixed voltage (e.g., 3.3 V or 5 V) or ground (0 V), depending on the input b_n. The switched capacitor circuit can thus be implemented by replacing the switches with tri-state logic and the control signals, a_n, are used to switch the circuits on and off. In Fig. 14.9, a "high" on the control signal opens the switch and a "low" closes it. If necessary, a small resistor can be placed in series with the logic output in order to keep the current from becoming too large. Such capacitance measurements are easily implemented using a programmed microcontroller.

Fig. 14.8 Schematic for a tri-state buffer

Table 14.4 Truth table for tri-state buffer, Fig. 14.8

a_n	b_n	o_n
L	L	L
L	H	H
H	X	Z

Other Uses for Tri-state Devices

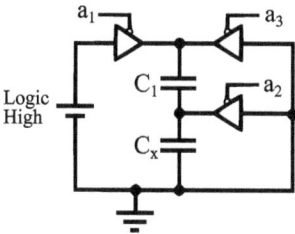

Fig. 14.9 The circuit of Fig. 14.7 implemented using tri-state buffers in place of switches

Charlieplexing

Another clever application for tri-state outputs is called "Charlieplexing." Charlieplexing, named for its inventor, Charles Allen, is a way of illuminating LED's one at a time. It is an efficient method, in terms of resource usage, to light LED's directly from microcontroller tri-state outputs. With n outputs from the microcontroller, you can individually address as many as $n^2 - n$ LED's without the use of any external decoder circuitry. For example, 8 outputs could individually illuminate up to 56 LED's.

The basic idea relies on the tri-state high impedance output and on the fact that LED's have a significant turn-on voltage. Between every pair of outputs, there are two LED's wired in opposite directions. In use, one output will be H (1), one L (0), and the rest will be high impedance (Z).

Figure 14.10a illustrates the use of 3 outputs to address 6 LED's. In order to illuminate LED D_2, for example, set O_1 to high impedance, O_2 high and O_3 low which gives the equivalent circuit (where NC means "No Connection') shown in Fig. 14.10b. Diodes D_1, D_5, and D_6 are clearly in reverse bias and will not illuminate. There is a forward path from O_2 to O_3 through diodes D_4 and D_3. However, the voltage across that pair is limited by the turn-on voltage of D_2, a single LED, and is thus insufficient to turn-on both D_4 and D_3 to any significant degree. Hence,

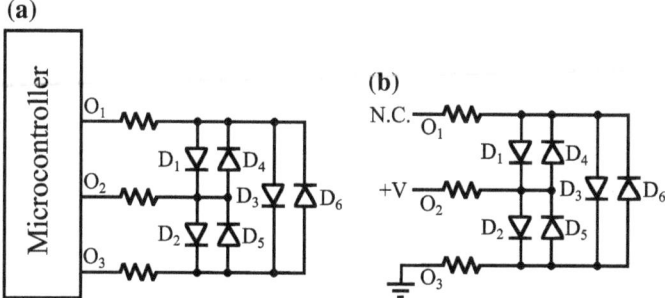

Fig. 14.10 (a) The connections of LED's to a microcontroller's output lines for Charlieplexing. To illuminate diode D_2, the outputs are set as shown in (b)

they remain dark. For this use, the LED's should be of a similar type, so they have a similar turn-on voltage.

There may also be a set of outputs that can illuminate two LEDs at a time, though clearly not all possible pairs can be simultaneously illuminated. For example, D_1 and D_4 could never be illuminated at the same time. The illusion that multiple LEDs are lit simultaneously is created by rapidly switching through the appropriate output states, illuminating the LEDs one at a time in such quick succession that they appear to the eye to be on simultaneously.

Problems

1. Starting with $a_0 = 1$, $b_0 = 2^{-1/2} = 0.70710678...$, and $s_0 = 0$, iterate a few times using the following computations,

$$a_n = \frac{a_{n-1} + b_{n-1}}{2}; \quad b_n = \sqrt{a_{n-1} \cdot b_{n-1}}; \quad s_n = s_{n-1} + 2^{n+1}\left(a_n^2 - b_n^2\right),$$

and when a_n and b_n are close, compute

$$z = \frac{4a_n^2}{1 - s_n}.$$

The result should be recognizable.

2. For the six-LED Charlieplexing example discussed at the end of this chapter, fill in the missing values in the following table. Here "0" is ground, "1" is a positive voltage and "Z" refers to the high-impedance state. (There is more than one way to have no LEDs lit.)

O_1	O_2	O_3	LED illuminated
			None
1	0	Z	D_1
Z	1	0	D_2
			D_3
0	1	Z	D_4
		1	D_5
0			D_6

3. Write π in base-2 scientific notation to at least 8 places to the right of the base-2 decimal point.

4. What is the two's complement of 0110100?
5. Compute -13×9 using 8-bit base-2 binary numbers and the 8-bit 2's complement method to represent -13. Show that the 2's complement of the 8 lowest bits of the result is the base-2 binary representation of $13 \times 9 = 117$.

References

H. Philipp, *Charge Transfer Capacitance Measurement Circuit*, US Patent 6,466,036 (1999)
W.H. Press, S.A. Teukolsky, W.T. Vetterling, and B.P. Flannery, *Numerical Recipes*, 2nd edn. (Cambridge University Press, 1992)

Appendix

SI Units

Many of the standard international (SI) units that are encountered in electronics are summarized in Table A.1.

Table A.1 SI Units often encountered in electronics

SI Units (MKSA)
- 1 amp = 1 A = 1 coulomb/sec
- 1 volt = 1 V = 1 m^2 kg s^{-3} A^{-1}
- 1 ohm = 1 Ω = 1 volt/amp = 1 m^2 kg s^{-3} A^{-2}
- 1 siemens = 1 S = (1 Ω)$^{-1}$ = 1 mho = 1 amp/volt
- 1 farad = 1 F = 1 amp sec/volt = 1 s/Ω = 1 m^{-2} kg^{-1} s^4 A^2
- 1 henry = 1 H = 1 V s/amp = 1 Ω·s = 1 m^2 kg s^{-2} A^{-2}
- 1 hertz = 1 Hz = 1 repeat/s = 1 cycle/s = 1 cps = 2π rad/s
- 1 watt = 1 W = 1 V A = 1 N m/s = 1 m^2 kg s^{-3}

Common Unit Prefixes

Prefixes are used with units to convey a power of ten multiplier. Many are defined and are commonly used with SI units. Some of the more common prefixes encountered in electronics are listed in Table A.2.

Table A.2 Some of the more common unit prefixes and their meanings

Symbol	Prefix	Multiplier
p	pico-	10^{-12}
n	nano-	10^{-9}
µ	micro-	10^{-6}
m	milli-	10^{-3}
c	centi-	10^{-2}
d	deci-	10^{-1}
k	kilo-	10^{+3}
M	mega-	10^{+6}
G	giga-	10^{+9}
T	tera-	10^{+12}

Fourier Series

The Basics

Near the turn of the 19th century, Joseph Fourier (1768–1830), developed many important theorems, including several related to vibrations. Fourier showed that for any periodic signal, call it $S(t)$, there is a choice for constants A_0, A_1, A_2, etc., called amplitudes, and φ_1, φ_2, etc., called phases, such that

$$S(t) = A_0 + \sum_{n=1}^{\infty} A_n \sin(n\omega_0 t + \varphi_n), \tag{A.1}$$

where the periodic function has a repeat frequency $f_0 = 2\pi\omega_0$. The series of terms on the right would be referred to as the "Fourier series representation of $S(t)$." With minor adjustments, trigonometric identities can be used to find equivalent series representations using cosine functions, a combination of sine and cosine functions, or any equivalent, instead of sine functions. Aside from the use of these trigonometric identities, the series representation of $S(t)$ is unique.

In some cases, mathematical equality in (A.1) is only achieved in the limit that an infinite number of constants and an infinite number of sinusoidal (e.g., sine) functions are used. Despite that, when representing any signal, the result of the sum is finite. A finite number of terms will usually suffice to achieve any desired level of accuracy for any electronic signals measured in the real world.

It should be obvious that if enough of the constants are known, $S(t)$ can be computed for any time, t, using the equation above. All that is required is a calculator with a "Sin" function. What is less obvious is that if $S(t)$ is known for at least one repeat cycle, it is possible to determine each of the constants for the series. In practice, if N distinct values of $S(t)$ are known, equally spaced in time over one repeat cycle, that is sufficient data to compute N constants. To determine those N constants, a Fourier transform (FT) is used.

Appendix

Fourier's theorem also shows that there is no difference between a periodic function produced directly, whatever its origin, from the result obtained by adding a large number of sine functions, with appropriately chosen amplitudes and phase constants, created by separate signal generators. Hence any periodic signal can be considered to be made up of, or composed from, a summation of a number of sinusoidal signals, no matter how that signal was actually generated in the first place. Using the superposition principle, when it applies, each of those sinusoidal signals can be considered separately and the results added together—the sum of the solutions is the solution to the sum. Analyzing a circuit one sine function at a time is sometimes referred to as analysis "in the frequency domain."

Four well-known mathematical examples are shown in Fig. A.1: the sine, square, triangle, and sawtooth functions. The functions shown each have amplitude 1 and repeat with frequency f_0. The Fourier series representations of these functions can be written:

- Sine function:

$$S(t) = \sin(\omega_0 t) \qquad (A.2)$$

- Square function:

$$S(t) = \frac{4}{\pi} \sum_{n \, odd}^{\infty} \frac{1}{n} \sin(n\omega_0 t)$$
$$= \frac{4}{\pi} \left(\sin(\omega_0 t) + \frac{1}{3}\sin(3\omega_0 t) + \frac{1}{5}\sin(5\omega_0 t) + \cdots \right) \qquad (A.3)$$

- Triangle function:

$$S(t) = \frac{8}{\pi^2} \sum_{n \, odd}^{\infty} \frac{(-1)^{(n-1)/2}}{n^2} \sin(n\omega_0 t)$$
$$= \frac{8}{\pi^2} \left(\sin(\omega_0 t) - \frac{1}{9}\sin(3\omega_0 t) + \frac{1}{25}\sin(5\omega_0 t) + \cdots \right) \qquad (A.4)$$

- Sawtooth function:

$$S(t) = -\frac{2}{\pi} \sum_{n=1}^{\infty} \frac{(-1)^n}{n} \sin(n\omega_0 t)$$
$$= \frac{2}{\pi} \left(\sin(\omega_0 t) - \frac{1}{2}\sin(2\omega_0 t) + \frac{1}{3}\sin(3\omega_0 t) + \cdots \right) \qquad (A.5)$$

where $\omega_0 = 2\pi f_0$, as before.

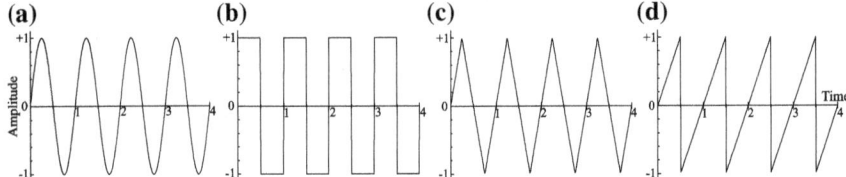

Fig. A.1 Sine, square, triangle and sawtooth functions shown as functions of time, t, with t in units of $1/f_0$

There are a several cases where care must be exercised when using Fourier series representations and analysis in the frequency domain. In particular where there are non-linearities so that superposition no longer applies. Also, for a signal that is changing in time—for example it is turned on and/or off—the addition of non-periodic time dependence can also make matters much more complicated.

How a Fourier Transform Works

There is considerable mathematical theory behind the Fourier transform that goes beyond this discussion. The principle that makes the transform work can be understood by considering the trigonometric identities

$$\sin(\omega_1 t)\sin(\omega_2 t) = \tfrac{1}{2}(\cos((\omega_1 - \omega_2)t) - \cos((\omega_1 + \omega_2)t)) \tag{A.6a}$$

$$\cos(\omega_1 t)\cos(\omega_2 t) = \tfrac{1}{2}(\cos((\omega_1 + \omega_2)t) + \cos((\omega_1 - \omega_2)t)) \tag{A.6b}$$

$$\sin(\omega_1 t)\cos(\omega_2 t) = \tfrac{1}{2}(\sin((\omega_1 + \omega_2)t) + \sin((\omega_1 - \omega_2)t)). \tag{A.6c}$$

When such terms are averaged over time, the result will be zero unless $\omega_1 = \omega_2$. If $\omega_1 = \omega_2$, then the first two (but not the third) will each average to one-half (because $\cos(0) = 1$).

Now start with any signal, $S(t)$, and multiply it by a sine (or cosine) function that has any frequency, $2\pi\omega$. Mathematically, that is equivalent to multiplying the Fourier series representation of $S(t)$ by that same sine, even if you do not know what that representation is. That is[1]:

$$\begin{aligned}\sin(\omega t)S(t) = &A_0\sin(\omega t) + A_1\sin(\omega_0 t)\sin(\omega t)\\ &+ A_2\sin(2\omega_0 t)\sin(\omega t) + \cdots.\end{aligned} \tag{A.7}$$

---

[1] The phase factors have been left off for clarity. More generally they are included for the signal, in one way or another, and multiplication by a cosine, averaged over time, will also be necessary.

Appendix

Now average that result over time. On the right it can be seen that the expression will average to zero unless ω is equal to one of the frequencies ω_0, $2\omega_0$, $3\omega_0$, ... and so on. That must be true on the left as well. Now sweep ω starting from zero, taking a new time average for each new value of ω. When $\omega = \omega_0$, the result is proportional to A_1. Increase ω some more and the whole expression averages to zero until $\omega = 2\omega_0$, at which point the result is proportional to A_2, and so on. The time-averaged response as a function of the variable ω is the Fourier transform (aside from an overall constant) and it yields the coefficients, A_n.

If a signal is not periodic, but finite in time, then it can be treated as if the signal were periodic over a longer time. That is, if $S(t)$ is known for $0 < t < T$, then it can be taken to repeat for $T < t < 2T$, and so on, for the sake of the calculation. That idea can be formally extended to the limit that $T \to \infty$, and thus all time-dependent signals, $S(t)$, can be represented by a sum of sinusoids.

With digital sampling of a (finite length) signal, and the use of a computer, the transform can be obtained relatively quickly. If the data is equally spaced in time and the total number of data values is chosen appropriately, a common choice being 2^N where N is a positive integer, an algorithm known as a fast Fourier transform (FFT) can be used. A FFT uses a particularly efficient way to organize and perform the numerical calculation.

Complex Numbers—A Review

The Basics

A complex number, Z, can be written using two real numbers, A and B,

$$Z = A + iB \tag{A.8}$$

where i is the square root of minus 1 ($i^2 = -1$). Some authors will use j instead of i. A is called the "Real Part" and B the "Imaginary Part" of the number. The "Re" and "Im" functions take a complex number as an argument and return the real and imaginary parts respectively. That is

$$\text{Re}(Z) = A; \quad \text{Im}(Z) = B. \tag{A.9}$$

Using the identity[2]

$$e^{ix} = \cos(x) + i\sin(x) \tag{A.10}$$

---

[2] Here "e" is Euler's number, 2.7182818... .

it is possible to write

$$Z = Ce^{i\varphi}, \tag{A.11}$$

where C and φ are both real numbers. The magnitude, C, is related to the real and imaginary parts through the Pythagorean theorem,

$$C = |Z| = \sqrt{A^2 + B^2}, \tag{A.12}$$

and

$$A = C\cos(\varphi); \; B = C\sin(\varphi), \tag{A.13}$$

so

$$\tan(\varphi) = B/A, \text{ or } \varphi = \tan^{-1}(B/A). \tag{A.14}$$

Note that A and B look like the x and y components of a 2-dimensional vector of length $|Z|$. Even though Z is not a vector, it is often convenient to think of it as a vector with the real part corresponding to the x-component and the imaginary part corresponding to the y-component. The phase angle, φ, would then be the angle counter-clockwise from the x-axis. See Fig. A.2.

Multiplication and addition of complex values proceed as for real numbers, though with the inclusion of the definition of i. Hence if $Z = A + iB$ and $Y = C + iD$ are two complex numbers, then

$$\begin{aligned} Z + Y &= (A + C) + i(B + D) \\ Z \cdot Y &= (AC - BD) + i(AD + BC). \end{aligned} \tag{A.15}$$

Note that multiplication is *not* the same as a dot product for vectors, though it may be tempting to do so. Multiplication and division are often easier using the magnitude and phase. For example, if

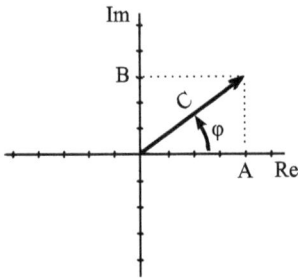

Fig. A.2 The relationship between real and imaginary parts, A and B, and the magnitude and phase, C and φ, for a complex number visualized using an x-y graph

Appendix

$$Z = |Z|e^{i\varphi}; Y = |Y|e^{i\vartheta}, \tag{A.16}$$

then the product is simply

$$Z \cdot Y = |Z||Y|e^{i(\varphi+\vartheta)} \tag{A.17}$$

Note that the amplitudes multiply and the phases add.

Multiplication and addition of complex values have the usual commutative, distributive, and associative properties that are present for real numbers.

The complex conjugate of a complex number Z is designated with an asterisk, Z^*. Compute the complex conjugate by changing all "i's" to "$-i$'s" or change the sign of the phase angle, φ. For example, if $Z = A + iB = Ce^{i\varphi}$ then $Z^* = A - iB = Ce^{-i\varphi}$. One consequence of this definition is that

$$|Z|^2 = Z \cdot Z^* = A^2 + B^2. \tag{A.18}$$

This last relation is often used as a convenient way to find $|Z|$.

Some Complex Identities

Some identities that are relatively easy to derive are:

$$i^2 = -1, i^3 = -i, i^4 = 1, \ldots, i^{100} = (i^4)^{25} = 1, i^{101} = i, \text{etc.} \tag{A.19}$$

$$e^{i\pi} = -1, e^{2i\pi} = 1, \text{etc.} \tag{A.20}$$

$$1/i = -i \tag{A.21}$$

$$\sin(x) = \frac{1}{2i}(e^{ix} - e^{-ix}); \quad \cos(x) = \frac{1}{2}(e^{ix} + e^{-ix}) \tag{A.22}$$

$$e^{Z+Y} = e^Z \cdot e^Y \tag{A.23}$$

where Z and Y are any complex values and x is any real value.

Index

A

Active region
    BJT, 164
    FET, 145
Alternating Current (AC), 54
Ammeter, ideal, 9
Amplifier
    automatic gain control (AGC), 240
    complementary symmetry, 191
    differential, 192
    op-amp, 199
    transistor, 153, 174
Analog to digital conversion, 293
    analog to Gray code, 295
    quantization noise, 296
    search scheme, 294
    timing schemes, 294
    voltage to frequency, 293
Angular frequency, 55
Arithmetic Logic Unit (ALU), 302
ASCII, 283
Audio taper, 24

B

Balanced code, 281
Bar codes, 284
Barkhausen criterion, 224
Bias, forward and reverse, 124
Biasing, 171
Biasing resistors, 171
Bias line
    BJT, 170
Binary, base-2, 277
    addition, 299
    multiplication, 305
    scientific notation, 307
Binary code
    non-weighted, 282
    weighted, 279
Binary Coded Decimal (BCD), 278
Binary counter, 270
Bipolar Junction Transistor (BJT), 163
    DC current gain, 166
    model, 166
    saturation region, 181
BJT amplifiers
    common base, 174
    common collector, 174
    common emitter, 174
Boolean algebra, 247
    basic results, 248
    DeMorgan's theorems, 248
Boolean operations, 247
Branch method, 14

C

Cables, *see* transmission lines
Capacitors
    parallel, 66
    series, 66
Cascode pair, 190
Central Processing Unit (CPU), 312
Charlieplexing, 315
Circuit
    definition, 3
    open circuit, 3
Circuit reduction, 10, 11
Code conversions
    base-2 binary to Gray code, 286
    BCD to binary, 288
    binary to BCD, 289
    decimal to Gray code, 287

Gray code to base-2 binary, 287
Common Mode Rejection Ratio (CMRR), 192
Communications
　parallel, 311
　serial, 311
Comparator, 232
　with hysteresis, 233
Comparator circuits
　level indicator, 234
　oscillator, 236
　pulse conditioner, 239
　pulse generator, 235
　zero crossing detector, 238
Complementary symmetry amplifier, 191
Complex number, 62, 323
Condenser, *see* capacitor
Conduction band, 122
Current, 2
Current divider equation, 7
Current mirror, 193
Current source
　ideal, 19
　schematic symbol, 19

D
Darlington pair, 190
Decade counter, 270
Decoders and encoders, 258
Delta-sigma 1-bit D/A, 291
Delta-Y conversion, 42
DeMorgan's theorems, 248
Depletion region, 124
Differential amplifier, 192
Differential amplifier, non-ideal, 220
Differential voltage gain, 192
Digital codes, 277
Digital logic circuit elements, 250
Digital to Analog (D/A), 290
Diode, 121
　analytic model, 126
　ideal model, 127
　light emitting (LED), 138
　models, 125
　p-n junction, 123
　ratings, 136
　semiconductor, 121
　specialty, 136
　turn-on voltage, 125
Diode clamp, 130
Diode limiter, 129
Direct Current (DC), 54
Dissipation angle, 79
Dissipation factor, 79

Duality, 100
Duty cycle, 238, 291
Dynamic memory, 159

E
Earth, *see* ground
Effective resistance, 9
　parallel resistors, 9
　series resistors, 9
Electric eye, 184, 266
Electromotive force (emf), *see* voltage
Energy gap, 122
Equivalent digital circuits, 256
Equivalent resistance, *see* effective resistance

F
Feedback, amplifier, 200
FET amplifiers, 153
　common drain, 153
　common gate, 153
　common source, 153
Field Effect Transistor (FET), 143
　AC model, 150
　JFET, 143
　ohmic region, 156
Filter
　LRC, 92
　op-amp, 207
　state variable, 207
Flip-flop circuits, 261
　binary counter, 270
　D flip-flop, 264
　edge-triggered, 265
　gated RS flip-flop, 263
　JK flip-flop, 263
　Johnson counter, 268
　ring counter, 268
　RS flip-flop, 263
　shift register, 267
Floating point binary numbers, 306
Fourier series, 320
Fourier transform, 322
Frequency, 55
　angular, 55
Frequency, complex, 105
Frequency divider, 268, 269
Full adder, 299
Full-wave rectifier, 132

G
Gain-bandwidth product, 189
Gray code, 282
Ground, 21

Index

H
Half adder, 299
Half-wave rectifier, 128
Hexadecimal, 279
Hole, 122
Howland configuration, 213
Hybrid parameters, 169

I
Ideal diode model, 127
Ideal transformer, 88
IGFET, 157
Impedance, complex, 65
Impedance match, 99, 196
Inductors
    parallel, 66
    series, 66

J
Johnson counter, 268

K
Kelvin bridge, 44
Kirchhoff's laws
    current law (KCL), 5
    voltage law (KVL), 5

L
Laplace transform, 105
    table, 106
LC circuits
    parallel, 90
    series, 90
Light Emitting Diode (LED), 138
Linear device, 53
Linear feedback shift register, 268
Linear taper, 24
Load line
    BJT, 170
    FET, 148
Loop method, *see* mesh method

M
Maximum forward current, 136
Mesh method, 14
Micro-Electromechanical Systems (MEMS), 81
Microphone
    condensor, 73
    electret, 74
Miller's theorem, 187
MOSFET, 156
    depletion, 157
    enhancement, 157

Multiplexing, 261
Mutual inductance, 86, 87

N
Nodal analysis, 17
Node, 4
Norton equivalent, 31
Null measurement, 38

O
Octal, 279
Ohm's law, 4
Ohms per square, 48
Op-amp, 199
    ideal, 199
    rail-to-rail, 231
Op-amp circuits
    absolute value, 219, 232
    buffer, 201
    capacitive sensor, 210
    constant current source, 213
    difference amplifier, 204
    ideal diode, 215
    instrumentation amplifier, 209
    integrator, 205
    inverting amplifier, 202
    log amplifier, 217
    low-pass filter, 207
    more output power, 220
    negative resistance, 212
    non-inverting amplifier, 203
    peak follower, 216
    summing amplifier, 204
    twin-T oscillator, 225
Operating point, 146
Operational amplifier, *see* op-amp
Oscillations, amplifiers, 224
Oscillators, 273

P
Parity, 254
Partial fractions, 109, 116
Passive devices
    capacitor, 55
    inductor, 55
Peak Inverse Voltage (PIV), 136
Period, 55
Phase sensitive detector, 242
Phase shift, 60
Photodiode, 139
Phototransistor, 266
Poles, 117
Position sensor
    capacitive, 81

inductive, 89
Potentiometer, 24
  Kelvin-Varley, 25
Power
  AC, 71
  electrical, 23
  RMS, 72
Power factor, 73
Power sources, idealized, 56
Power Supply Rejection Ratio (PSRR), 192
Priority encoder, 260
Propagation delay, 271
Pseudo-random sequence, 268
Pulse density encoding, 291
Pulse length encoding, 291
Pulse width encoding, 291
Push-pull amplifier, 191

Q
Quality factor (Q), 79
Quantization noise, 296

R
Reciprocity theorem, 39
Recursive numerical computations, 307
Resistance, 3
Resistance measurement
  four wire, 44
  van der Pauw technique, 47
Resistors, 5
  in parallel, 7
  in series, 5
Reverse saturation current, 126
Rheostat, 24
Ring counter, 268
Ring oscillator, 273
RLC circuits, 90
Root Mean Square (RMS), 55
R-2R ladder, 40, 292

S
Saturation region
  BJT, 164
Scanning keyboard, 271
Schematic
  diagram, 4
  symbol, 4
Schematic symbol
  amplifier, op-amp, 199
  battery, 4
  bipolar junction transistor, 163
  capacitor, 55
  common, 21
  digital gates, 250

diode, 124
edge triggered flip-flop, 265
ground, 21
ideal sources, 56
inductor, 55
JFET, 143
MOSFETs, 158
resistor, 4
SCR, 194
switch, 4
triac, 194
Self inductance, 86
Semiconductor, 122
  doping, n- and p-, 123
  electrons and holes, 123
  intrinsic, 123
Seven-segment display, 258
Shift register, 267
Silicon Controlled Rectifier (SCR), 193
  threshold voltage, 194
Smith chart, 98
Standard International (SI) units, 319
State variable filter, 207
Summing D/A, 292
Superposition theorem, 22
Switch bounce, 235
Switch debouncer, 235
Switched capacitor methods, 84
  using tri-state, 312

T
Thevenin equivalent, 31
Thyristors, see silicon controlled rectifier
Time constant, 58
  L/R, 58
  RC, 58
Transconductance
  FET model, 150
Transconductance amplifier, 225
Transfer characteristic, 145, 165
Transfer function, 118
Transformers, 86
  dot convention, 89
Transmission lines, 93
  equivalent impedance, 95
  impedance of finite, 97
  wave speed, 97
Triac, 194
Tri-state, 311
Truth table, 248
  solving circuits with, 253
Turn-on voltage, diode, 125
Two's complement, 299

Index

U
Unit prefixes, 319
Unity gain bandwidth, 189

V
Valence band, 122
Voltage, 2
Voltage divider equation, 6
Voltage drop, 4

Voltmeter, ideal, 7

W
Wheatstone bridge, 36

Z
Zener diodes, 136
Zeros, 117